Residential Hot Water and Steam Heating: Gas, Oil, and Solid Fuels

Residential Hot Water and Steam Heating: Gas, Oil, and Solid Fuels

Martin L. Greenwald

Montclair State College
Upper Montclair, New Jersey

Prentice-Hall, Inc., Englewood Cliffs, New Jersey 07632

Library of Congress Cataloging-in-Publication Data

Greenwald, Martin L., (date)
Residential hot water and steam heating.

Includes index.
1. Dwellings—Heating and ventilation.
2. Hot-water heating. 3. Steam-heating. I. Title.
TH7467.G74 1987 697'.4 86-4928
ISBN 0-13-774712-8

Editorial/production supervision
and interior design: *Kathleen M. Lafferty*
Cover design: *20/20 Services, Inc.*
Manufacturing buyer: *Rhett Conklin*
Cover illustration: *Domestic hot water produced by a residential boiler coupled to an external heat exchanger. (Courtesy of Weil-McLain, a Marley Company.)*

Printed in the United States of America

10 9 8 7 6 5 4 3 2 1

ISBN 0-13-774712-8 025

Prentice-Hall International (UK) Limited, *London*
Prentice-Hall of Australia Pty. Limited, *Sydney*
Prentice-Hall of Canada Inc., *Toronto*
Prentice-Hall Hispanoamericana, S.A., *Mexico*
Prentice-Hall of India Private Limited, *New Delhi*
Prentice-Hall of Japan, Inc., *Tokyo*
Prentice-Hall of Southeast Asia Pte. Ltd., *Singapore*
Editora Prentice-Hall do Brasil, Ltda., *Rio de Janeiro*

Contents

3 Selecting the Peripheral Components 55

4 Oil-Fired Heating Systems 104

5 Gas Heating Systems 165

Preface

The residential plumbing and heating trades have always been rather loosely defined. This broad, technical area encompasses a wide variety of interrelated cognitive and practical skills. For example, a technician may be asked to wire something as simple as a replacement element on an electric hot water heater or as complex as a series of interconnected switching relays on a multifuel boiler installation. A plumbing installation may require only a cold water line for an additional basement wash sink or a complex configuration of a multizone add-on wood or coal gasification boiler that must be interconnected to an existing heating system.

Hot water and steam heating systems represent a major component of the residential plumbing and heating industry. While never static, recent upheavals brought about by the energy crisis have resulted in significant design changes in almost every aspect of the industry. Hot water and steam boilers fired with conventional fuels are far more efficient than their earlier counterparts. Both fuel oil and gas-fired appliances now feature ignition and combustion systems that can offer efficiencies thought almost impossible just a short time ago. The technology of these modern heating systems requires all those involved in their installation and maintenance to be thoroughly familiar with the intricacies of their operation in order to maximize safe and efficient system operation.

Technological improvements have not been limited to conventionally fueled appliances. The resurgence of wood heating has lead to considerable sophistication applied to the design of solid- and multifuel boilers. The use of these appliances in conjunction with conventionally fueled heating systems requires a thorough knowledge of the best interface to use in a particular heating situation.

The information in this text, *Residential Hot Water and Steam Heating: Gas, Oil, and Solid Fuels,* is both broad in scope and comprehensive in nature. Its features include an introduction to hot water and steam heating systems including piping layouts and appliance certification procedures; sizing techniques for heating systems based on heat-loss analysis procedures designed for quick, accurate results; an examination of a wide variety of system peripheral components and accessories used in most hydronic, steam, and solid-fuel installations; gas heating system fundamentals covering an examination of induced draft, standard atmospheric combustion, and electronic ignition systems; oil heating technology, focusing on the operation of modern high-speed flame retention oil burners; the design and installation of a variety of add-on solid-fuel and multifuel boilers; steam heating system layout, design, operation, and maintenance; and upgrading techniques for improving the operation of existing heating systems.

It is hoped that this text will be of value to all those who are involved, or hope to become involved, in the plumbing and heating trades. While designed primarily for vocational/technical courses within the hydronics industry, this book can be easily understood by both the novice and layperson with little technical background; included is a foundation of heating system design logic and system-sizing techniques applicable to the majority of hot water and steam heating systems. All information relating to system operation and troubleshooting procedures is succinctly written and enhanced with numerous system drawings and schematics resulting in a comprehensive and easily understood presentation of both concept and application.

ACKNOWLEDGMENTS

The author is greatly indebted to many individuals whose efforts and suggestions have had a major influence on the quality of the manuscript and the accuracy of its contents.

First and foremost, the author wishes to acknowledge the efforts of Joseph Doak, field representative for the Hydronics Division of Burnham Corporation. Joe's ongoing revisions, suggestions, and comments helped to refine this text and make it the most comprehensive coverage of material available within the plumbing and heating trades.

My friend and colleague, Tom McHugh, vice-president of Heliotherm, who coauthored our text, *Practical Solar Energy Technology,* tirelessly reviewed the manuscript and helped refine the format as well as the contents.

Jack Turner, president of Chimney Sweep Energy Corporation, reviewed the manuscript and specialized on all information relating to solid-fuel and multifuel boilers. His comments and suggestions are also embedded throughout this text, and they helped make the material comprehensive and up-to-date.

The products of many corporations and companies involved within the plumbing and heating trades are reflected in this text. In many instances, these companies were called upon to offer reviews of relevant manuscript material for accuracy and timeliness. Their efforts are also greatly appreciated and reflect the high quality and longevity of products manufactured for use within the plumbing and heating trades.

Last, but not least, I wish to express my gratitude and love to my wife. Writing a book is never an easy task; however, it is often harder on the family than the author with constant rearrangement of schedules and so on. Her flexibility and belief in the fact that what I did in this text is worthwhile and that it can be a significant contribution to education in general aided in bringing this project to a successful conclusion.

Martin L. Greenwald

1

Introduction

Modern hot water space heating systems are the result of years of manufacturing and engineering expertise devoted to *hydronics,* the science of heating with water. As conventional sources of fossil fuels become more and more scarce, the necessity of maintaining a heating system at optimum performance levels assumes greater importance.

To achieve efficient performance levels within the heating system requires an understanding of the full breadth of hydronic technology: the physics of heat transfer; plumbing circuitry; operation of controls and relays; and combustion characteristics of the burner mechanisms involved. Once these basics have been mastered, a decision as to which components will operate most efficiently with one another can be made.

Basic to an understanding of hydronic heating systems is a knowledge of the physics of heat transfer. All heating systems rely on a number of different methods to transfer heat from the burner mechanism to the interior of a house.

PRINCIPLES OF HEAT TRANSFER

Heat is transferred from one area, or object, to another in three fundamentally different ways, known as radiation, conduction, and convection. Each of these principles will now be examined in greater detail.

Radiation

Figure 1.1 illustrates the transfer of heat by radiation. Radiant energy travels through space in a manner similar to waves moving over the surface of a pond after a stone has been dropped into the water. The amount of radiant heat transfer from a surface depends on both the temperature and nature of the surface of the object. Generally, shiny surfaces are poor emitters of radiant energy because the heat tends to be reflected back into the object. Black, dull, nonreflective surfaces are good absorbers and emitters of radiant energy.

Conduction

The transfer of heat by conduction requires that two objects be in direct contact with one another (Figure 1.2). The nature of conduction can be illustrated by the action that takes place when one plays a game of pool. When the cue ball collides with other balls on the table, the energy in the cue ball is transferred to the other balls through random collisions. The amount of energy that is transferred in this process depends on the force and the angle of the striking cue ball, as well as the number of other balls encountered. Similarly, the speed at which heat is transferred through any material is dependent on the *thermal conductivity* of that substance. The thermal conductivity, *K*, can be expressed in Btu per inch per square foot per degree Fahrenheit (Btu/in./ft^2/hr/°F). *K* is different for each substance. It has become common to speak in terms of a material's *R* value instead of its *K* value. *R*

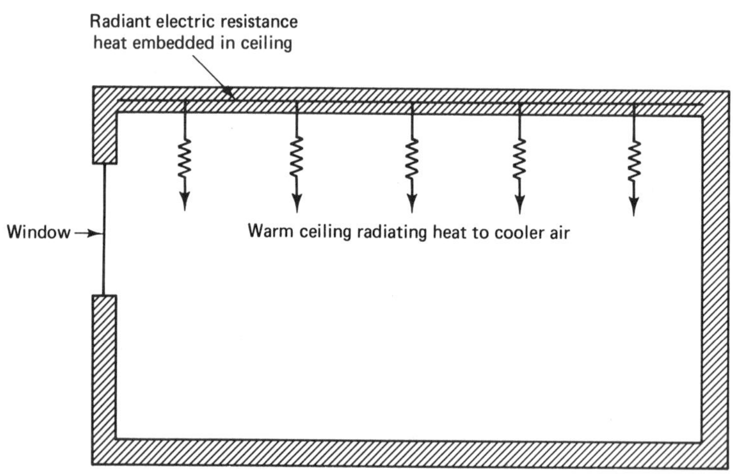

Figure 1.1 Transfer of heat by radiation. (Courtesy of Martin L. Greenwald and Thomas K. McHugh, *Practical Solar Energy Technology,* © 1985, p. 12. Reprinted by permission of Prentice-Hall, Inc., Englewood Cliffs, NJ.)

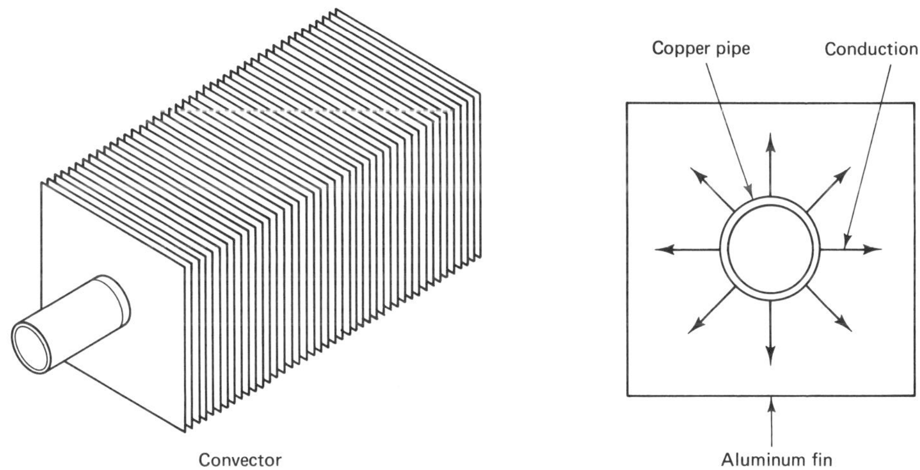

Figure 1.2 Transfer of heat by conduction. (Courtesy of Martin L. Greenwald and Thomas K. McHugh, *Practical Solar Energy Technology*, © 1985, p. 12. Reprinted by permission of Prentice-Hall, Inc., Englewood Cliffs, NJ.)

values are the mathematical reciprocals of K values and describe the ability of a material to resist the flow of heat. A material with an R value of 10 will resist the flow of heat twice as effectively as will a material with an R value of only 5. K values, on the other hand, describe the ability of a material to conduct heat rather than retard it.

In conductive heat transfer, heat flows from the hotter object to the cooler object at the point where the two surfaces touch. Objects that are good conductors of electricity, such as aluminum and copper, are also good conductors of heat, because many excess electrons are available within the atomic structure of these materials for heat conduction. Note in Figure 1.2 that a typical cross section of finned baseboard radiation contains a copper trunk line with aluminum fins in contact with the copper pipe. The heat conduction from within the copper pipe to the aluminum fins is highly efficient, due to the excellent conduction properties of these two materials.

The quantity of heat transfer from the circulating fluid to the baseboard is a function of the following factors:

1. The fluid temperature entering the baseboard
2. The rate of flow of the fluid within the baseboard
3. The linear footage and type of baseboard used in the system
4. The rate of heat transfer from the baseboard to the surrounding air

Convection

Convection is the method of transferring heat through the movement of a fluid. The fluid can either be a liquid or a gas. One of two types of convection is used in most hot water heating systems: natural convection or forced convection. An example of *natural convection* in a home heating system is illustrated by the flow of air within a room created by the radiators and baseboard units. As air molecules within the room are heated by coming into contact with the heat distribution units, the air becomes less dense and rises to the ceiling. Cooler air from below moves toward the radiator to replace the warmer air that has risen. This creates a natural-convection airflow pattern, as the warm air is constantly rising and the cool air is moving in to take its place (Figure 1.3).

A similar convective air pattern is established by virtue of the construction of the finned baseboard heating unit, which comprises a great deal of surface area exposed to the passing air currents (Figure 1.4). Air enters the bottom of the baseboard unit and is heated as it passes over the fins of the unit. The fins are so constructed as to provide a great deal of surface area to the moving air, enhancing convective efficiency. The heated air then rises through the top of the baseboard and creates a moving current of air within the room.

The use of a circulating pump within a hot water heating system is an example of *forced-convection* heat transfer. Forced convection increases the efficiency of the heat transfer process since a moving fluid will give up its heat to the surrounding cooler surfaces more quickly than by natural means alone. Forced convection provides greater temperature differentials between the fluids within the heating system.

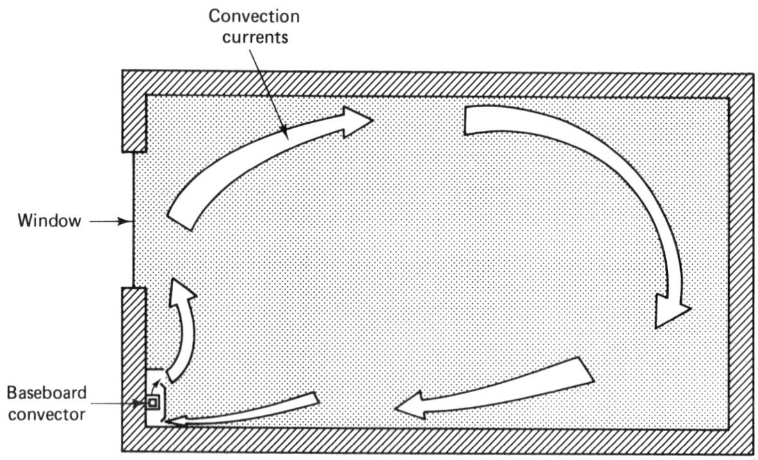

Figure 1.3 Natural-convection currents in a baseboard hot water heating system. (Courtesy of Martin L. Greenwald and Thomas K. McHugh, *Practical Solar Energy Technology*, © 1985, p. 14. Reprinted by permission of Prentice-Hall, Inc., Englewood Cliffs, NJ.)

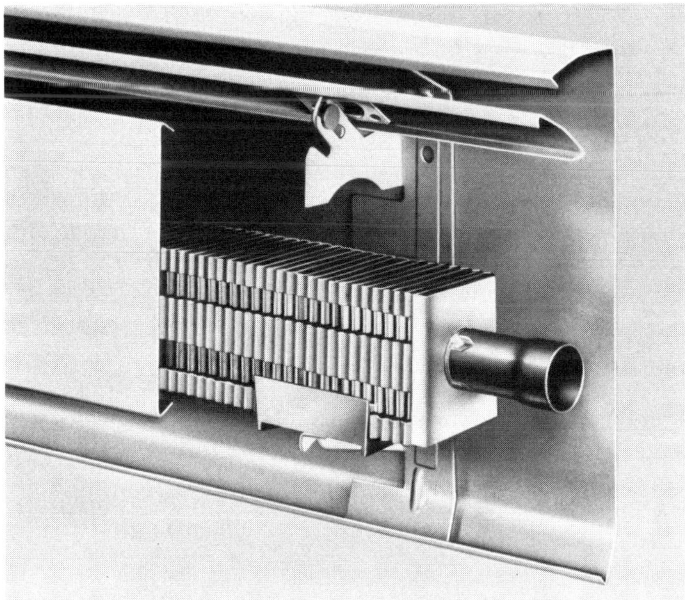

Figure 1.4 Construction details of finned baseboard radiation. (Courtesy of Slant/Fin Corp.)

The greater the temperature differential between the fluid within the heating system and the temperature of the burner mechanism, the greater will be the efficiency of the heating system. A finned baseboard heating unit therefore utilizes a combination of forced and natural convection: forced convection for movement of the heat transfer fluid from the boiler to the heat distribution units; natural convection in the movement of air currents passing over the heating unit to achieve interior heating levels. The quantity of heat energy released to the room is a function of the following factors:

1. The square footage of the heating surface exposed to air
2. The ambient air temperature passing over the heating element
3. The volume of air passing over the heating element
4. The temperature of the baseboard heating element
5. The amount of time that the temperature differential occurs

The volume of air passing over the heating element in a typical baseboard unit is affected by the depth and the inlet/outlet dimensions of the enclosure. The height also plays an important part in increasing air volume since the higher the enclosure, the greater will be the chimney effect, resulting in greater airflow. Ratings of radiation are expressed in Btu per linear foot per hour (Btu/ft/hr) at various fluid temperatures. For test and rating purposes, room air temperature is assumed to be 65°F.

PIPING SYSTEMS

Hot water heating systems are usually classified according to the type of plumbing circuitry involved in the heating loop. Many modifications of piping arrangements exist and are often dictated by the size and complexity of the heating system as well as the amount of money that is available to install the system.

Gravity Supply

The gravity supply system is the simplest of all hot water heating system designs (Figure 1.5). The gravity system operates on the principle that the density of water

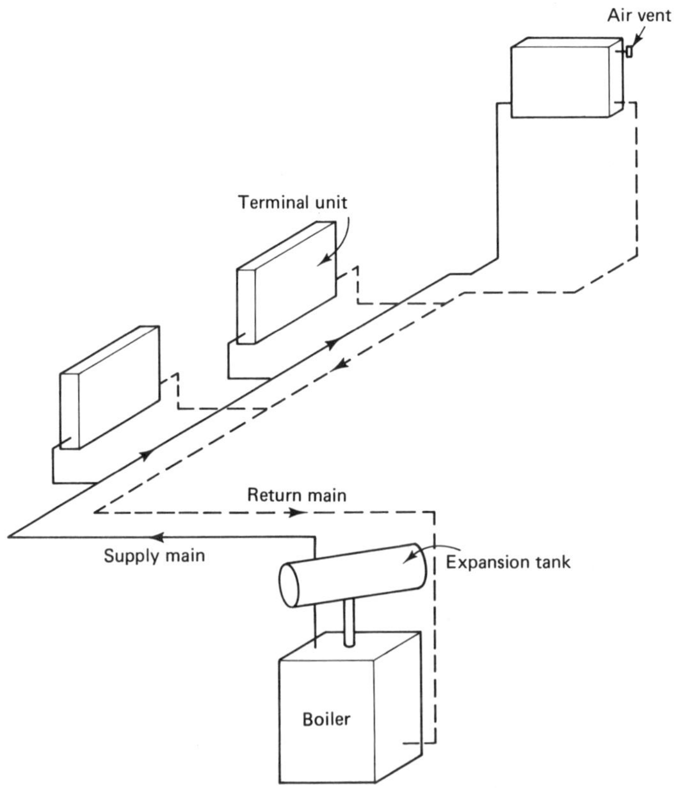

Figure 1.5 Gravity hot water heating system, direct return. Arrows show direction of water flow. (Reprinted by permission of Small Homes Council, Building Research Council, University of Illinois, Technical Note 13.)

decreases as it is heated. Thus hot water from the boiler rises up through the heating distribution units. As heat is transferred from the water to the surrounding air, the water cools, becomes heavier, and returns to the boiler via the return piping system. This operating principle is known as a *thermosiphon* and occurs without the use of a circulating pump to move the water throughout the piping circuitry. In gravity heating designs, the boiler is always located at the lowest point of the system. Since movement of the water is dependent on natural rather than forced convection, pipe sizes in gravity systems are large to minimize frictional losses. Gravity systems usually employ two pipes: one for hot water supply to the heat distribution units, the other for return water from the heat distribution units. These piping arrangements can be either direct return, as in Figure 1.5, or reverse return, illustrated in Figure 1.6.

In the direct-return piping arrangement, the heat distribution unit closest to the boiler has the shortest overall pipe run. It is the first to receive hot water from the boiler feed line and is the closest to the boiler return. This unit will also run

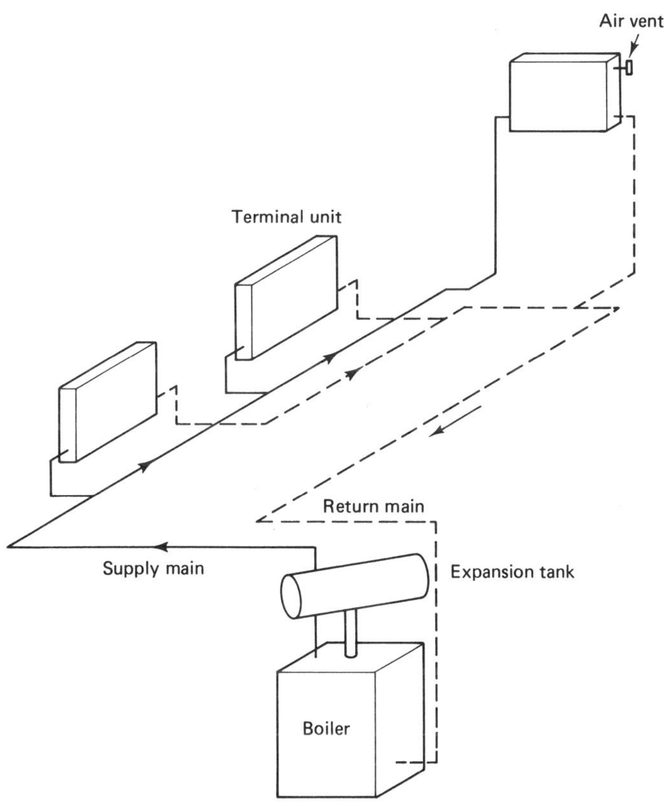

Figure 1.6 Gravity hot water heating system, reverse return. Arrows show direction of water flow. (Reprinted by permission of Small Homes Council, Building Research Council, University of Illinois, Technical Note 13.)

hotter than the other distribution units. The farther away from the boiler the heating units are, the greater will be their frictional pipe losses. For this reason, gravity systems usually employ reverse-return piping illustrated in Figure 1.6. Reverse-return design incorporates approximately the same total length of pipe run for each heat distribution unit. This allows for equal frictional losses throughout the system, enabling balanced fluid flow to produce equal heat output in each of the terminal heating units.

Forced-Circulation Series Loop

In a forced-circulation series-loop system, each of the baseboard heating units is connected in series (Figure 1.7). The heating circuit starts at the boiler, runs through each of the heating units in sequence, and terminates at the boiler. Since forced circulation by means of a low-head circulating pump is employed, pipe sizes can be smaller than in gravity heating systems; pipe of nominal diameter between $\frac{1}{2}$ and $1\frac{1}{4}$ in. is often used. This type of system is both the simplest and least expensive hot water baseboard system to install. One drawback to this type of arrangement is the absence of individual control over the heat distribution units, since restricting water flow through one unit would cause a significant reduction in the overall output of the remainder of the loop.

Radiant Slab Heating

Another type of radiant heating configuration is known as either radiant slab or radiant floor heating. In this type of system, the piping circuit is embedded in the floor of the structure. In most instances, the floor is made of concrete and is well insulated to prevent heat loss to the surrounding ground surfaces. This type of system is illustrated in Figure 1.8. In this type of radiant heating, lower water temperatures than those generally encountered in series-loop baseboard configurations can be used, since the floor of the structure acts as the heat distribution unit. Once the temperature of the floor reaches approximately 90°F, even heating of the residence will result. This system is highly resistant to temperature fluctuations due to boiler cycling since the thermal mass of the flooring tends to even out interior room temperatures, although initial pickup of the system takes much longer than does a standard baseboard configuration.

Supply Loop: Forced Circulation

The supply-loop system is a modification of the series-loop design. In the supply-loop arrangement, a main supply line, or trunk, runs around the perimeter of the building. Individual heat distribution units tap off the main trunk, as illustrated in Figure 1.9. Each heat distribution unit diverts a small portion of hot water from the trunk line. The water passes through the heating unit and is returned to the main trunk line. A special flow-diversion fitting is generally employed on either the heat-

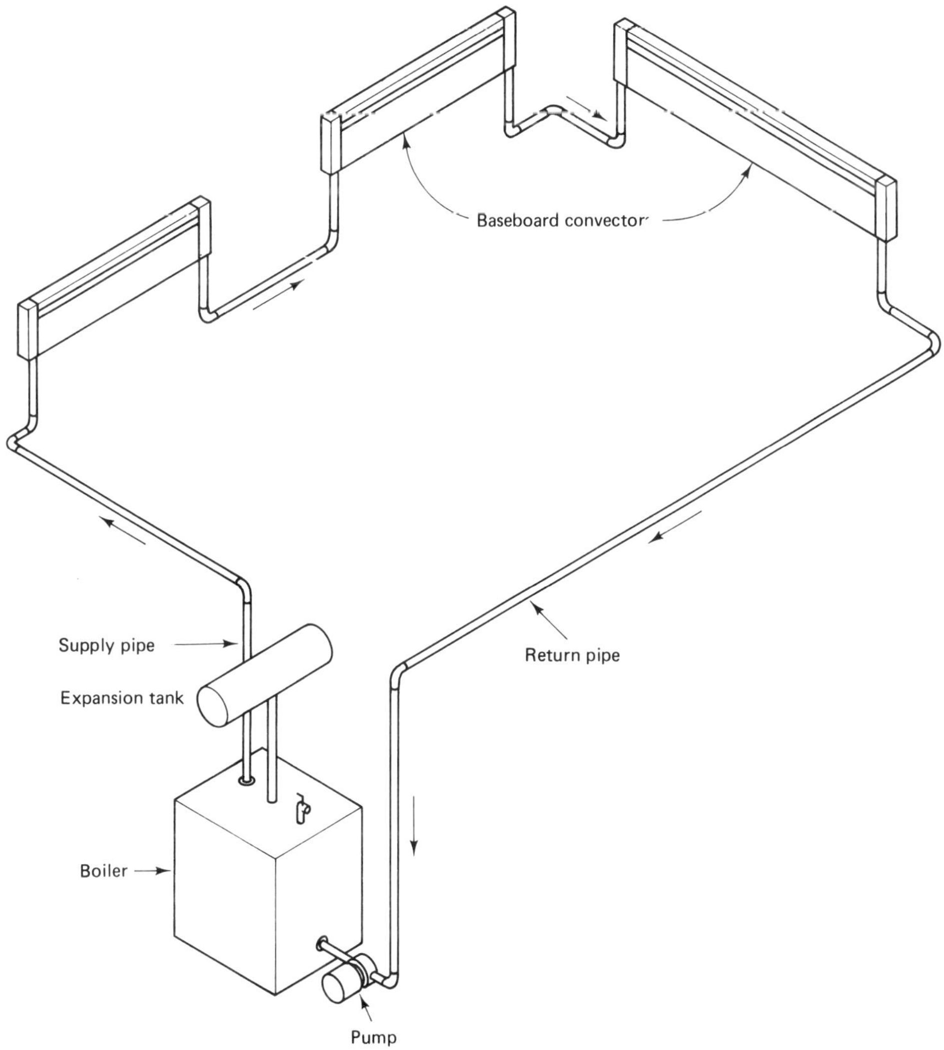

Figure 1.7 Single-circuit series-loop forced-circulation hot water heating system. (Courtesy of Martin L. Greenwald and Thomas K. McHugh, *Practical Solar Energy Technology*, © 1985, p. 135. Reprinted by permission of Prentice-Hall, Inc., Englewood Cliffs, NJ.)

ing unit feed or the return line to ensure the proper volume of hot water within the heating unit. One type of fitting for this purpose with a built-in venturi is illustrated in Figure 1.10. The supply-loop arrangement, sometimes referred to as a *one-pipe system,* has the advantage of offering heat output control over each of the individual heat distribution units. By adjusting the volume of hot water entering each unit, full zoning characteristics are achieved.

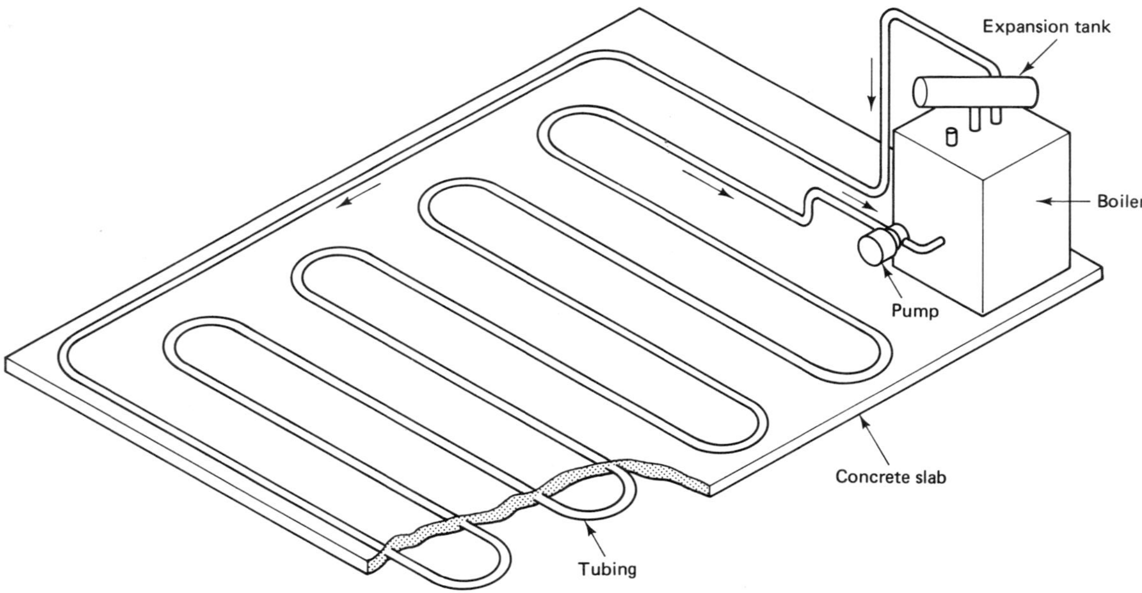

Figure 1.8 Radiant slab hydronic system. (Courtesy of Martin L. Greenwald and Thomas K. McHugh, *Practical Solar Energy Technology,* © 1985, p. 138. Reprinted by permission of Prentice-Hall, Inc., Englewood Cliffs, NJ.)

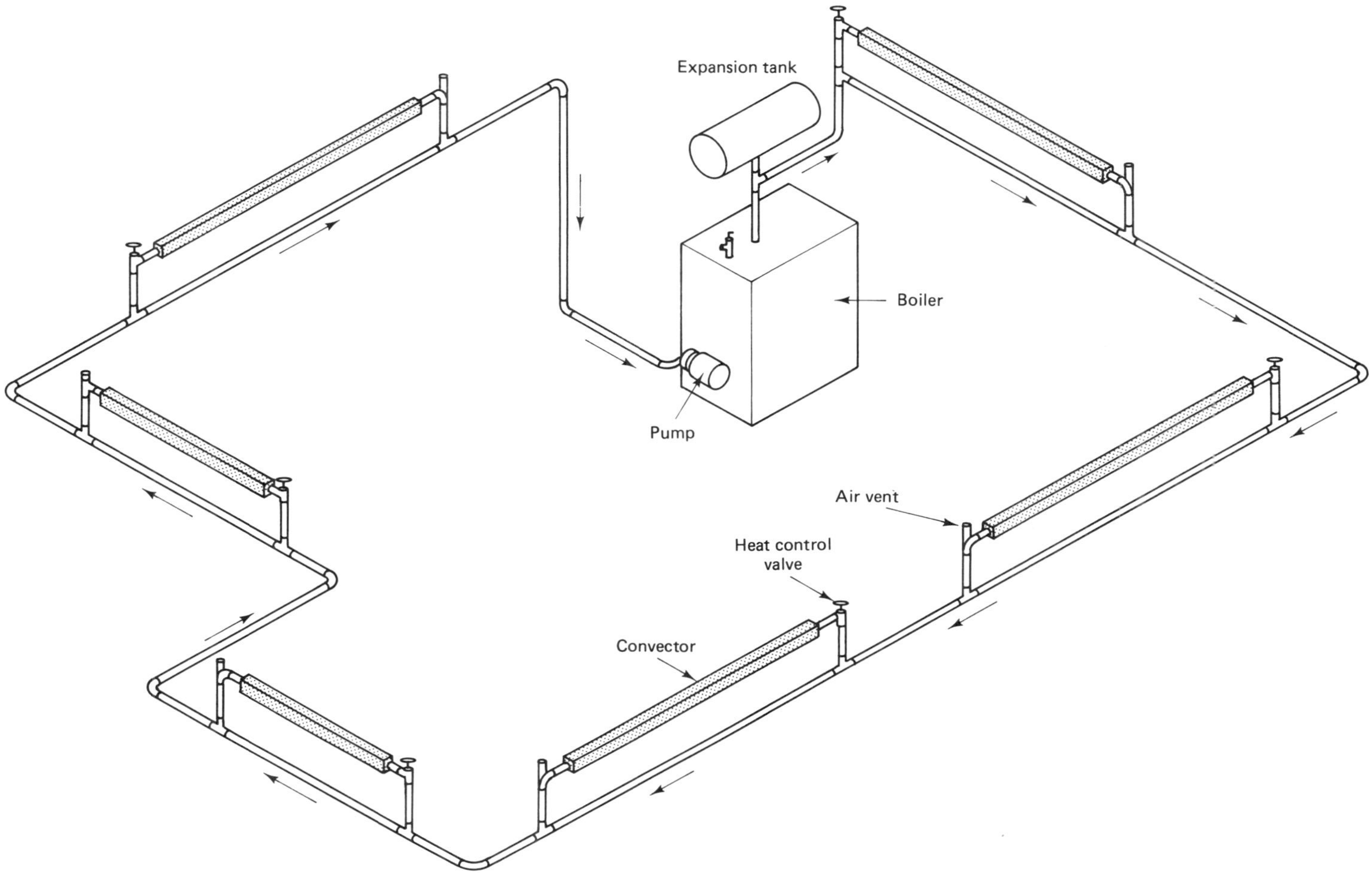

Figure 1.9 Single-zone supply-loop baseboard heating system. (Courtesy of Martin L. Greenwald and Thomas K. McHugh, *Practical Solar Energy Technology*, © 1985, p. 134. Reprinted by permission of Prentice-Hall, Inc., Englewood Cliffs, NJ.)

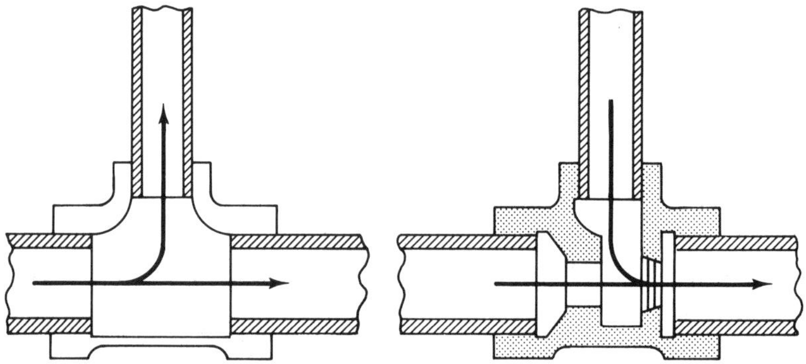

Figure 1.10 Cross section of typical flow-diversion fittings for use in supply-loop hydronic systems showing water flow through main and riser connections. (Courtesy of Taco, Inc.)

Supply Loop: Two-Pipe Direct Return

In this piping arrangement, a main feed or trunk line delivers water to each of the heat distribution units. A delivery pipe returns the cooled water to the boiler (Figure 1.11). Note that the heat distribution unit in closest proximity to the boiler has the shortest overall pipe run on both the hot water feed and the cold water return line. This piping arrangement can sometimes cause difficulty in system balancing or heat output because the internal frictional pipe losses of each heating unit are different, based on their overall distance from the boiler. For this reason, two-pipe direct-return systems are not generally recommended for standard residential applications.

Supply Loop: Two-Pipe Reverse Return

In a reverse-return piping configuration, the heat distribution unit closest to the boiler feed line is connected to the return line at the greatest distance from the boiler (Figure 1.12). In this manner, approximately equal piping runs, hence internal friction, allow for effective balancing of the heating system.

Steam Systems

In a typical steam heating system, steam rises through the pipes to a heating radiator. As the steam condenses against the interior surfaces of the radiator, the latent heat of condensation of the steam is released. The relatively large surface areas of the radiators release this heat into the room by radiant heat transfer. The condensate within the radiator then travels back down the piping system to the boiler. A typical steam heating system is illustrated in Figure 1.13. Steam piping configurations are classified as either one- or two-pipe and are discussed in greater detail in Chapter 6.

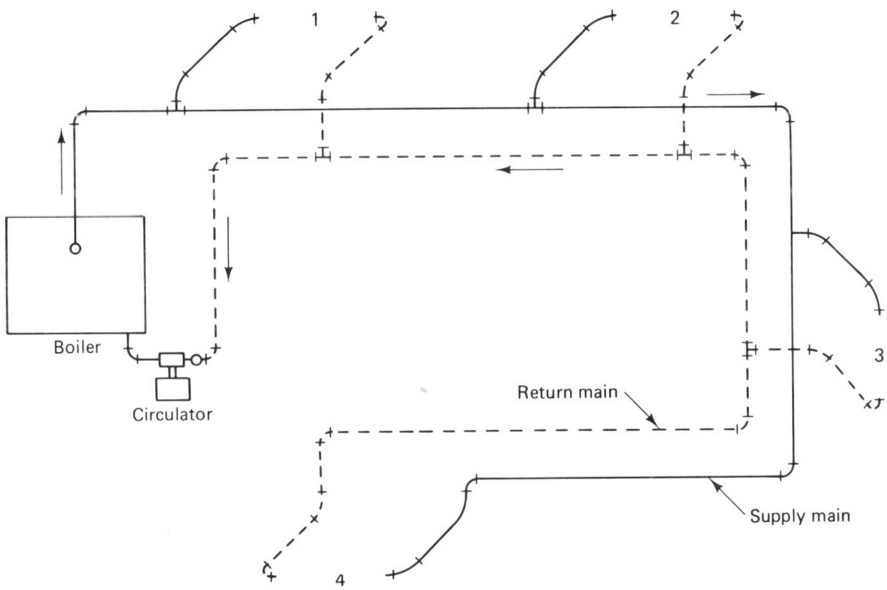

Figure 1.11 Supply-loop system, direct return. (Courtesy of the Hydronics Institute.)

Zoned Heating Systems

A zoned heating system features separate piping circuits that serve individual portions of a residence. Each of the heating circuits, or zones, are separate from one another but share the heating boiler as the hot water source. This allows for a great deal of flexibility in heating system operation. Heat is delivered to the zones of the house only where and when it is needed. Substantial fuel savings can accrue from this type of system versus the single-zone series-loop arrangement. Zoning is usually accomplished in one of two ways: by the use of motorized zone valves or by the use of separate circulators serving each heating zone.

Zone valves. A zone valve is an electrically operated water valve that opens and closes in response to a standard room thermostat. A typical zone valve is illustrated in Figure 1.14. A two-zone circuit arrangement is illustrated in Figure 1.15. Note that the zone valves are generally installed on the hot water feed lines.

The operational sequence of the system is as follows: When the zone thermostat calls for heat, the zone valve automatically opens, clearing the piping circuit for hot water flow. The normal zone valve is a low-voltage (24 V ac) device which has an end switch contact that closes when the zone valve opens. This completes a low-voltage circuit in a separate relay, which in turn provides line voltage to the hot water circulator. In most zone valve arrangements, one circulator provides all the

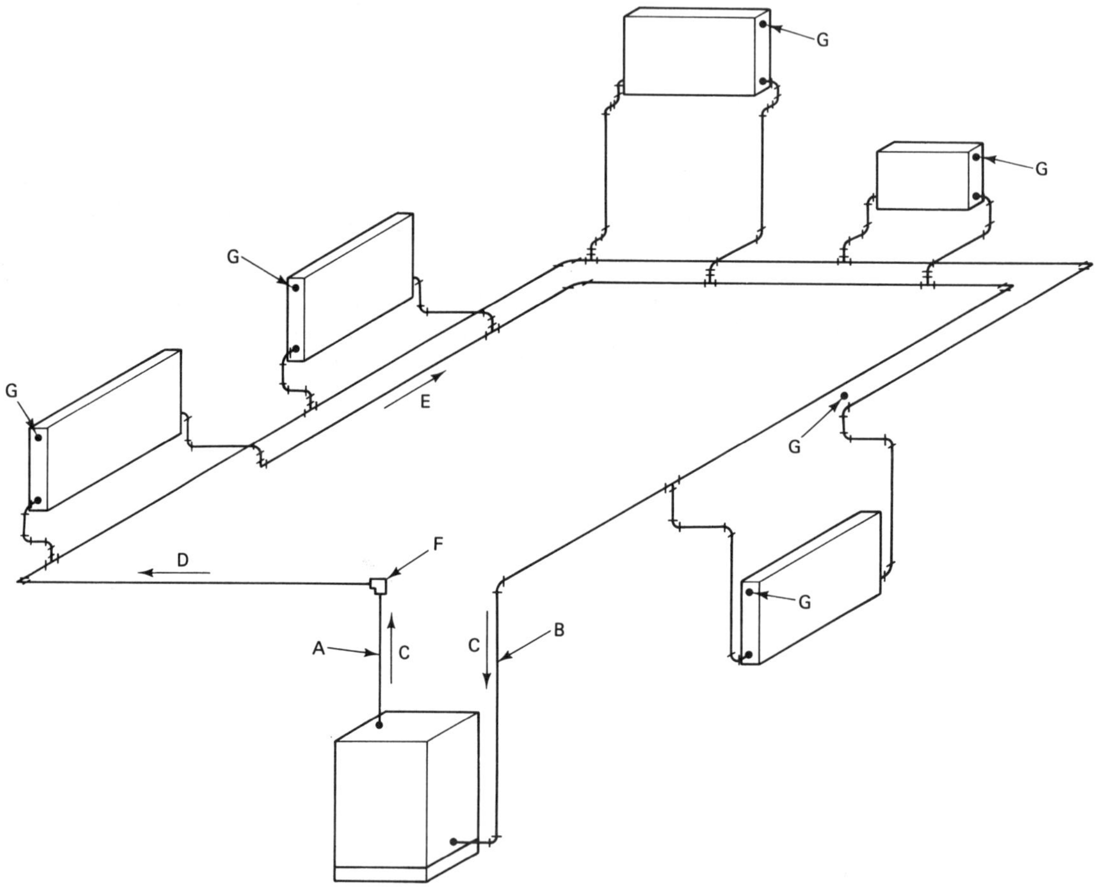

Figure 1.12 Supply-loop two-pipe reverse-return system. A, Supply pipe; B, return pipe; C, direction of flow of water; D, pitch supply pipe upward; E, pitch return pipe downward; F, flow control valve required if an indirect water heater is used and optional if an indirect water heater is not used; G, air vent on each unit and at end of supply main. (Courtesy of the Hydronics Institute.)

water circulation within the heating system. For a full schematic representation of a zone valve system, including aquastat connections, see Chapter 3.

Zone valve systems are employed in homes, apartments, and small commercial applications. The system circulator is sized to supply an adequate flow of hot water based on the total gallons per minute required if all the heating loops operate simultaneously. Hydronic circulators are available in a wide variety of capacities suitable for any heating situation. For a further discussion of circulator sizing and zoning requirements, consult Chapter 2.

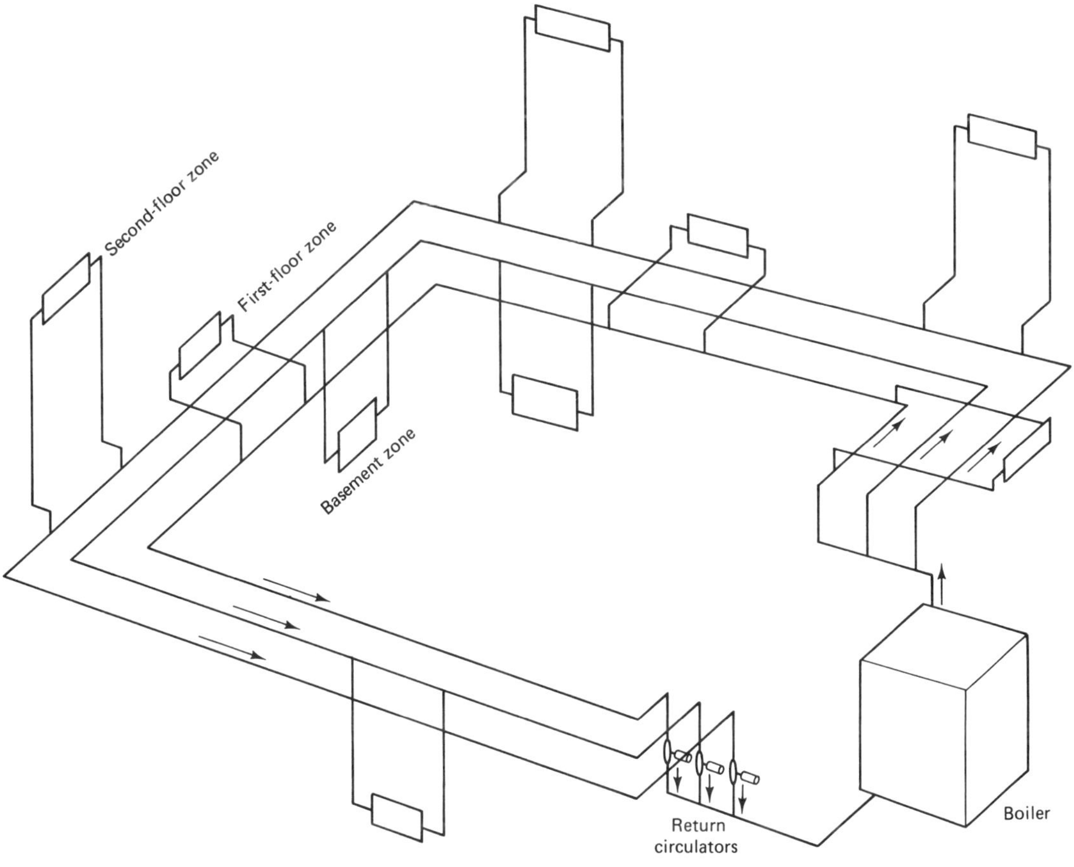

Figure 1.13 Three-circuit heating system with pumped return. (From *Solid Fuel Furnaces and Boilers* by John W. Bartok, Garden Way Publishing Co., Charlotte, Vt., 1982. Available from Storey Communications, Pownal, Vt., 05261.)

Circulator zoning. Separate heating circuits can be controlled by the use of individual circulators to move water within each heating zone. This arrangement is illustrated in Figure 1.16. In this arrangement the heating thermostat for each zone controls the operation of a specific circulator. Note that flow-control valves are placed on the hot water feed line exiting the boiler to each zone. A typical flow-control valve contains a weight that is designed to open only when the circulating pump is energized. A flow-control valve commonly used in hydronic systems is illustrated in Figure 1.17.

Most flow-control valves feature a manual adjustment that permits the valve to be opened to allow for gravity circulation of the hot water from the boiler. This

Figure 1.14 Electric zone valve. (Courtesy of Flair Mfg. Corp.)

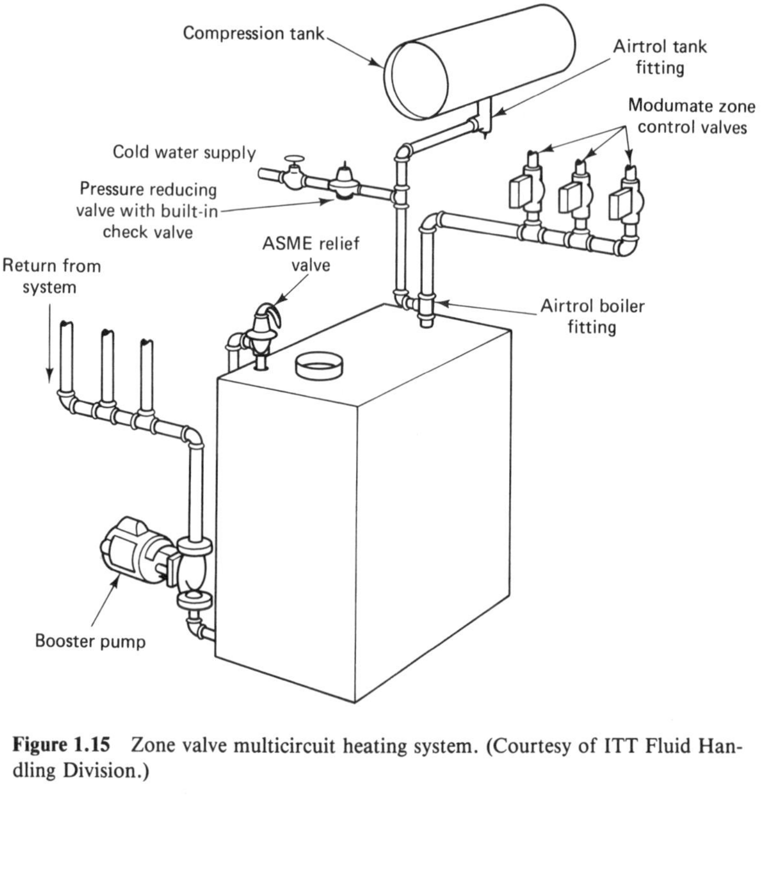

Figure 1.15 Zone valve multicircuit heating system. (Courtesy of ITT Fluid Handling Division.)

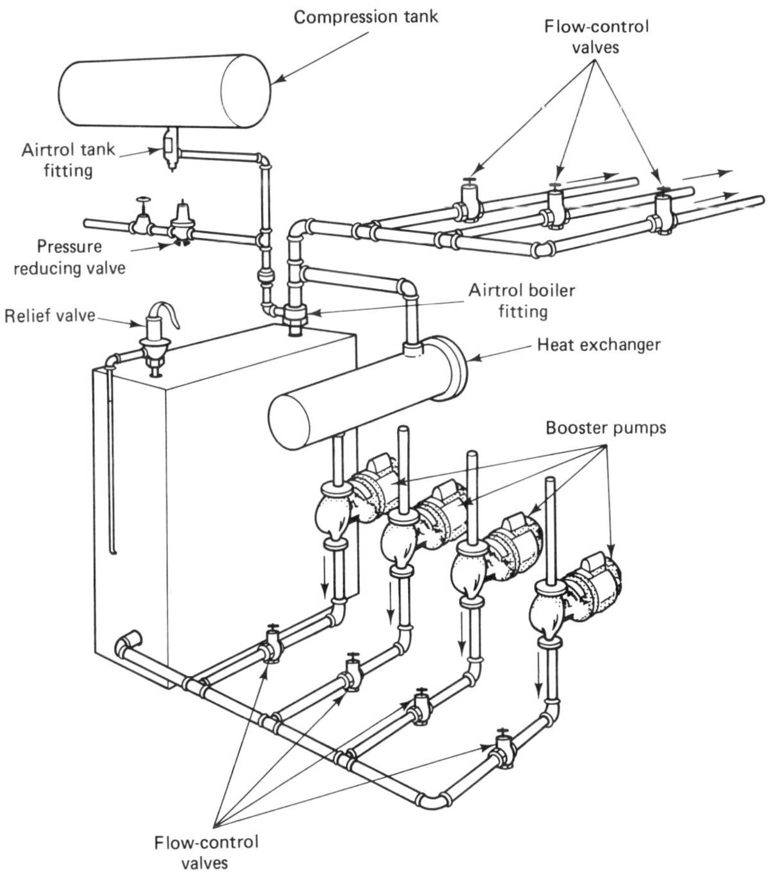

Figure 1.16 Zone control employing separate circulators. (Courtesy of ITT Fluid Handling Division.)

Figure 1.17 Flow-check valve. (Courtesy of Taco, Inc.)

adjustment would be engaged during a power outage, for example, when the circulator would be unable to pump water from the boiler. Under normal operating conditions, the valve will prevent gravity circulation of water through the zone when the circulator is not operating.

INDUSTRY APPLIANCE RATINGS

The effectiveness of the home heating system depends on efficient heat transfer between the heat developed from the combustion of fuel to the surrounding water within the confines of the boiler. Figure 1.18 illustrates the basic design configuration of a modern hot water boiler, highlighting the travel of the flue gases from the combustion chamber through the heat exchange surfaces within the appliance.

Most modern boilers yield high operating efficiencies and are designed and built to extremely high standards of quality. The testing and certification of heating boilers as well as their operating efficiency falls under the auspices of a number of

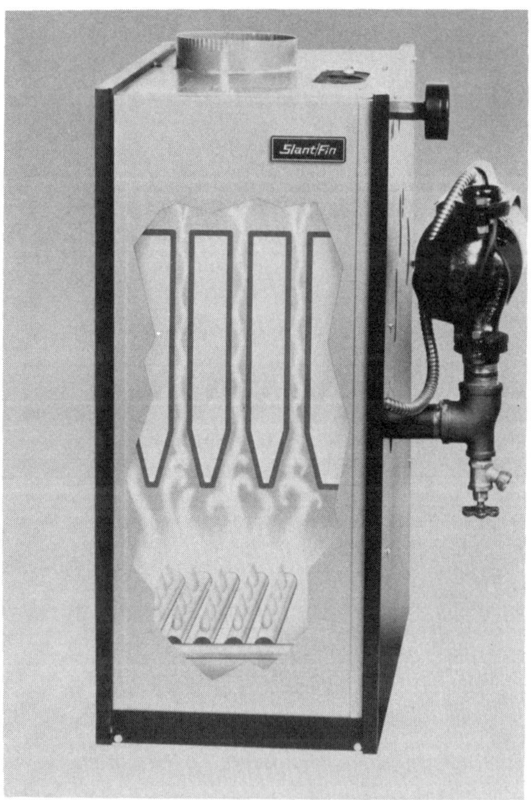

Figure 1.18 Cross-sectional design configuration of modern hot water boiler. (Courtesy of Slant/Fin Corp.)

agencies: the Hydronics Institute, the American Gas Association, the American Society of Mechanical Engineers, and the U.S. Department of Energy. Each of these agencies plays a different, although interrelated, role within the hydronics industry.

Hydronics Institute

The Hydronics Institute is an umbrella agency composed of two predecessor organizations: the Institute of Boiler and Radiator Manufacturers (IBR) and the Steel Boiler Institute (SBI). An independent organization supported by manufacturers of hydronic heating equipment and accessories, the institute conducts testing and certifies the efficiency ratings of a variety of hot water heating boilers. It also tests baseboard and commercial finned radiation equipment.

The IBR and SBI ratings for cast iron and steel boilers, as well as their annual fuel utilization efficiencies, are published in a volume entitled *Boiler Ratings and Efficiencies*. This publication is updated periodically with addenda and is available from the Hydronics Institute, 35 Russo Place, Berkeley Heights, NJ 07922. The ratings for baseboard and commercial radiation are published in a volume entitled *IBR Ratings for Baseboard Radiation and Finned Tube (Commercial) Radiation,* available from the institute at the address above. In addition to equipment testing and certification, the institute also publishes several booklets and pamphlets used in educational curricula within the hydronics industry.

American Gas Association

The American Gas Association (AGA) is the agency responsible for the testing and certification of gas appliances and controls sold in the United States. Boilers that have been tested and certified by the AGA are included in the IBR and SBI ratings published by the Hydronics Institute. The AGA maintains an extensive research library and offers many publications relating to gas appliances' manufacture and rating standards.

The AGA publication devoted to the safe operation, durable construction, and acceptable performance of gas-fired steam and hot water boilers is entitled *American National Standard for Gas-Fired Low-Pressure Steam and Hot Water Boilers,* available from the AGA, 1515 Wilson Boulevard, Arlington, VA 22209. This volume, printed by the AGA, is the American National Standards Institute (ANSI) rating Z21.13 and is updated periodically. The ANSI committee responsible for updating this code represents a coalition of members of the AGA, public utilities, consumer advocates, insurance industry representatives, and representatives from the appliance manufacturers and heating trades.

Of interest to those involved in the gas appliance industry is the *National Fuel Gas Code,* ANSI Standard Z223.1, carrying the National Fire Protection Association (NFPA) Code 54. This publication is available from both the AGA and from the NFPA in Quincy, Massachusetts.

Figure 1.19 ASME certification stamp. (Courtesy of American Society of Mechanical Engineers.)

American Society of Mechanical Engineers

The American Society of Mechanical Engineers (ASME) is responsible for setting forth the design and construction features of heating boilers in the latest edition of *Section IV of the ASME Boiler and Pressure Vessel Code.* This document is available from ASME, 345 East 47th Street, New York, NY 10017. ASME also maintains an extensive engineering library on the premises where *Section IV* and other documents are available for examination. All ASME-certified boilers carry an "H" stamped within a cloverleaf pattern on the boiler nameplate (Figure 1.19).

U.S. Department of Energy

The Department of Energy (DOE) sets the standards and test procedure to be followed in which boiler and appliance manufacturers certify the efficiency levels of their heating equipment. These test procedures have undergone continual upgrading as an integral part of the national energy conservation program and are examined later in further detail.

BOILER TESTING PROCEDURES AND OPERATING DEFINITIONS

The current testing procedures for residential heating boilers under DOE guidelines are carried out by the individual equipment manufacturers, who must certify their test results to the Federal Trade Commission (FTC). Testing by the Hydronics Institute serves as objective verification of the manufacturers' figures. This procedure applies to hot water boilers under 300,000 Btu/hr input. Larger boilers are tested independently by the institute. To understand and interpret the testing procedures fully, the following working definitions should be understood.

Steady-State Efficiency

When a boiler is operating under steady-state conditions, the hot water output of the unit is constant. Testing of the boiler at this point includes measuring the temperature and properties of the flue gases, electrical consumption of the unit, and firing rate and fuel consumption of the burner. These measurements yield a steady-state efficiency rating. This rating is defined as 100% minus the heat losses from the unit up the chimney.

DOE Heating Capacity

Multiplying the steady-state combustion efficiency of the boiler by the Btu/hr fuel input yields the heating capacity of the boiler. This figure represents the rate of useful heat output delivered by the boiler as hot water available for space heating. The heat transfer through the boiler jacket is not considered a loss, since it is assumed that this heat will be of use in maintaining interior temperatures.

Gross Output

Gross output is the amount of heat energy at the outlet of the boiler, expressed in Btu/hr, delivered under standard test conditions. The gross output figures account for all operating losses of the system, including boiler jacket losses.

Net Rating

The net rating of a boiler, in Btu/hr, is the result of the gross output or DOE heating capacity minus an allowance that is made for piping and pickup losses in the heating system. Thus it represents the heat energy available to the space to be heated. This figure, the net rating, should be matched to the calculated heat loss of the residence when sizing the boiler.

Piping and Pickup Losses

Most hydronic and steam heating systems have piping and pickup losses associated with them, and these allowances are based on assumptions that may or may not be correct on a system-by-system basis. On steam boilers, a deduction from the boiler output of 25% is made to allow for the pickup and piping to arrive at the net rating of the system. With the advent of forced hot water systems and series-loop piping systems featuring nonferrous baseboard, pipe losses were substantially reduced. For this reason, the current rating system reflects only a 13% allowance for these losses on hot water boilers. Should the system not fall into this category, additional allowances should be considered in the selection of the boiler. Conversely, the boiler may in some cases be selected on straight output ratings. For example, in a heating system in which all the piping is contained within the heated area and this area is kept at a constant temperature without any setback, piping losses are nonexistent and space heating can be satisfied by the output of the boiler.

Annual Fuel Utilization Efficiency

The annual fuel utilization efficiency (AFUE) applies to all heating boilers under 300,000 Btu/hr input tested under DOE specifications. AFUE is calculated as 100% minus stack losses during normal system cycling, in addition to heat loss within the dwelling due to infiltration of outside air that is used by the boiler for combustion and draft control within the flue.

Minimum standards for both the AGA and the Hydronics Institute list a steady-state efficiency of 75% in order for a boiler to carry either AGA or IBR certification. A graphic representation of the definitions pertaining to DOE testing procedures are illustrated in Figures 1.20 and 1.21.

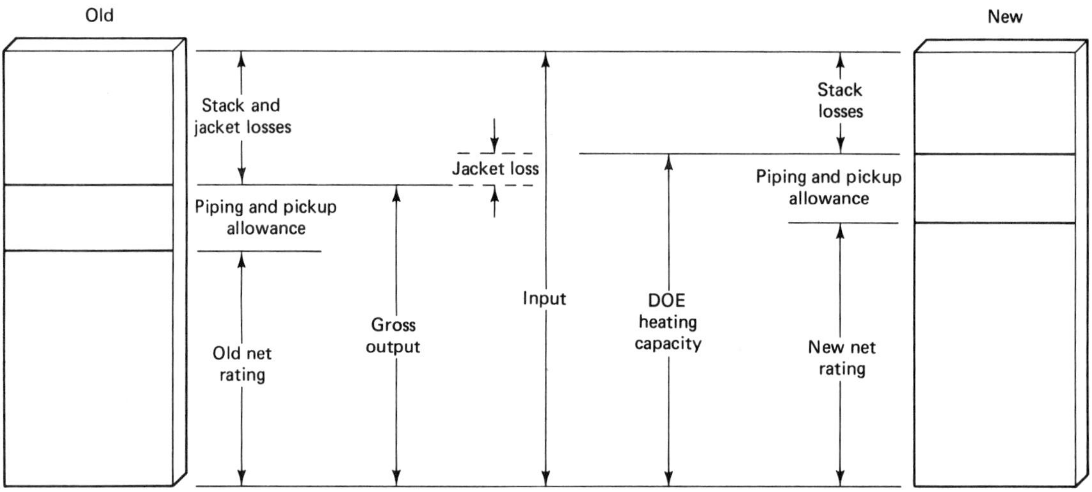

Figure 1.20 Boiler rating methods for oil-fired units up to 300,000 Btu/hr. (Courtesy of the Hydronics Institute.)

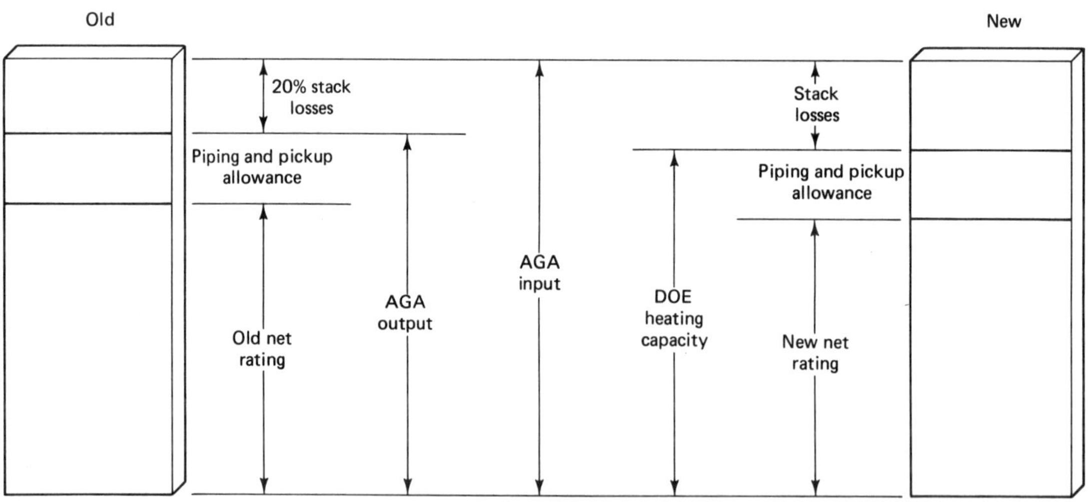

Figure 1.21 Boiler ratings for gas-fired units up to 300,000 Btu/hr. (Courtesy of the Hydronics Institute.)

DOE Test Procedures

The DOE test procedures include a steady-state efficiency test, a cool-down test, and a subsequent heat-up test. A profile of the test pattern is illustrated in Figure 1.22. The boiler is tested in a fully instrumental laboratory designed to record all temperatures, fuel and electrical consumption, and water qualities of the boiler during the test procedures. The fuel burner is adjusted to within 2% of the rated input value along with the manufacturer's recommended CO_2 value at a maximum of No. 1 smoke. Water output from the boiler is adjusted until an exit temperature of 200° ± 5°F is achieved. Once these conditions have been met, the steady-state efficiency test is conducted. These conditions are illustrated in Figure 1.23. The steady-state efficiency test is conducted over a period of 30 minutes. During this time the fuel and electrical consumption, stack temperature, and draft condition within the flue are all carefully monitored.

Cool-down test. To establish a cool-down curve for the boiler after completion of the steady-state efficiency test, fuel ignition within the boiler and water flow are shut off. Flue temperature readings are taken at 3:45, 22:30, and 45 minutes after boiler shutdown. The cool-down test provides the basis for calculating off-cycle losses of the boiler. This figure is one component of the AFUE.

Heat-up test. Fifty minutes after shutdown, the burner and rated water flow in the boiler are reestablished. Flue temperature readings are taken at 1 minute and 5:30 seconds after this secondary startup. These figures establish a heat-up curve for the boiler, again one a component of the cycling losses included in the AFUE calculations.

Test assumptions. The annual fuel utilization tests (AFUEs) are an attempt to analyze boiler operation under a "typical" set of operating conditions. A computer program developed for this purpose therefore contains certain assumptions as to what constitutes an "average" or "typical" operating system. These assumptions are as follows:

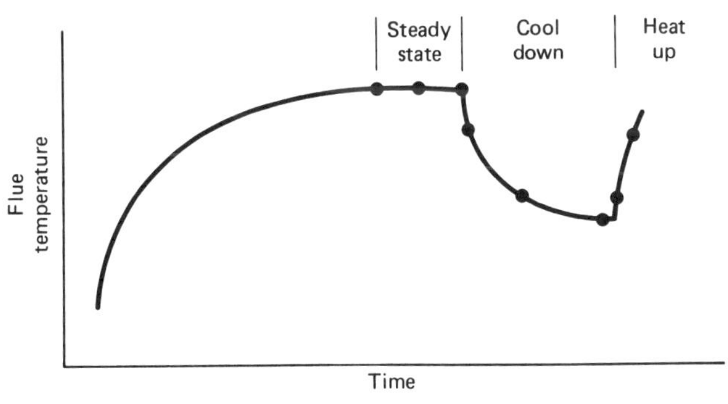

Figure 1.22 Profile of flue temperatures during Department of Energy (DOE) test. (Courtesy of the Hydronics Institute.)

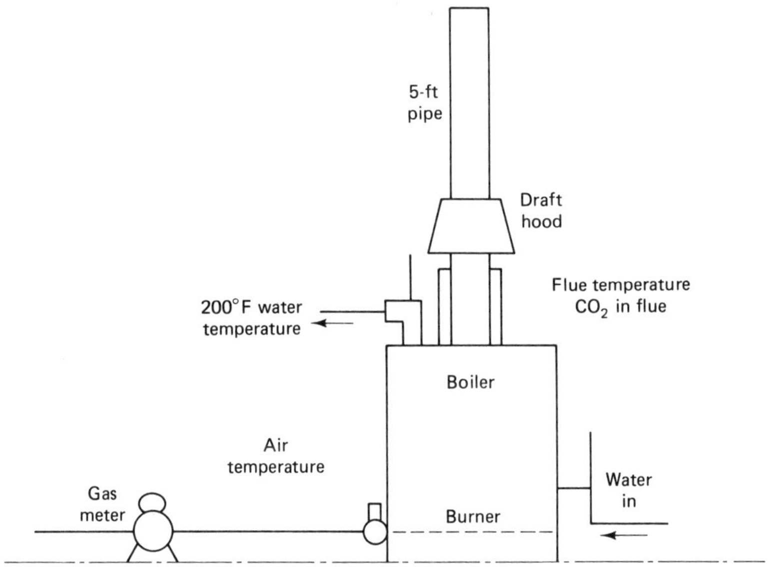

Figure 1.23 Gas boiler setup for DOE efficiency test. (Courtesy of the Hydronics Institute.)

1. The house is located in Pittsburgh, Pennsylvania, with an average outside temperature of 42°F. At this temperature, the boiler is oversized for the heating load by approximately 70%.
2. The boiler cycles twice each hour. Each cycle runs for 9.68 minutes.
3. The operating boiler water temperature is 200°F.
4. All boiler heat loss through the jacket constitutes a contribution of heat to the building.
5. Combustion air for the fuel burner, and air for draft control through the barometric damper, is assumed to have been heated by the home heating system and replaced by normal air infiltration.

Vent dampers. The AFUE calculations include provisions for calculating the efficiency of boilers with the use of automatic stack dampers. Automatic dampers reduce off-cycle heat losses by closing off the chimney passage when the burner mechanism is not operating. In gas units, provision is made within the damper to allow sufficient draft to remove the products of combustion from standing pilot ignition systems. The inclusion of a vent damper can therefore significantly reduce off-cycle losses of the boiler, raising the efficiency levels of the unit. A typical set of efficiency figures illustrating the inclusion of vent dampers appears in Table 1.1

It should be remembered that the AFUE is similar in nature to the EPA mileage figures for automobiles. As such, the AFUE can be used to compare the op-

TABLE 1.1 TYPICAL NATURAL AND PROPANE BOILER RATINGS AND DOE EFFICIENCY WITH AND WITHOUT VENT DAMPERS

Boiler number	AGA and CGA input (MBtu/hr)[a]	DOE heating capacity (MBtu/hr)	IBR net rating (MBtu/hr)[b]	Net rating water (ft^2)[c]	CGA gross output (MBtu/hr)	Recommended chimney size[d] round diameter (in. × ft)	DOE annual efficiency (%)			
							Standing pilot	Standing pilot and vent damper	El	El and vent damper
203	62	51	44.3	295	49.6	4 × 15	69.1	74.7	72.1	79.6
204	96	79	68.7	458	76.8	5 × 15	69.4	75.4	72.1	79.8
205	130	107	93.0	620	104.0	6 × 15	69.6	76.1	72.1	79.9
206	164	135	117.4	783	131.2	6 × 15	69.9	76.8	72.1	80.0
207	198	162	140.9	939	158.4	7 × 15	70.2	77.4	72.1	80.2
208	232	190	165.2	1101	185.6	7 × 15	70.4	78.1	72.1	80.3
209	266	218	189.6	1264	212.8	8 × 15	70.7	78.8	72.1	80.5
210	299	245	213.0	1420	239.2	8 × 15	70.9	79.5	72.1	80.6

Source: Burnham Corp.

Note: DOE heating capacity and annual efficiency are based on U.S. government standard tests. Propane available for standing pilot only.

[a]AGA ratings shown are for installations at sea level and elevations up to 2000 ft. For elevations above 2000 ft, AGA ratings should be reduced at the rate of 4% for each 1000 ft above sea level.

[b]IBR ratings shown are based on normal IBR piping and pickup allowance of 1.15. Consult the manufacturer for installations having unusual piping and pickup requirements such as intermittent system of operation, extensive piping systems, etc.

[c]Based on 170°F average water temperature in radiators (heat emission rate of 150 Btu/hr/ft^2). For higher water temperatures, select boiler on basis of net ratings in Btu/hr.

[d]Recommended chimney for rectangular application is 8 in. × 8 in. × 15 ft for sizes 203 through 209, and 8 in. × 12 in. × 15 ft for size 210 (15 ft. chimney height is from bottom of draft hood opening to top of chimney).

eration of two different units under the same set of conditions to help determine the difference in operating costs between the different boilers. These figures should not be used to predict exact fuel costs in a specific heating situation, since the assumptions used to calculate AFUE will rarely be identical to any specific residential heating application. For an up-to-date list of AFUE ratings, consult the *Boiler Rating and Efficiencies* manual published by the Hydronics Institute.

2

Sizing the Heating System

There are many different methods in use for sizing residential heating systems. These vary from instinct to the performance of detailed computerized analyses.

To obtain reliable information regarding what size boiler and how much radiation is required in a particular installation, a series of sequential design steps should be taken that will result in a heating system engineered to deliver topnotch performance in a given situation. These design steps are as follows:

1. Determine the heat loss of the building, on a room-by-room basis, including an overall heat loss of the structure at a design temperature dictated by local geographical conditions.
2. Size the heating boiler and associated fittings based on the maximum net heat delivery to be supplied by the heating system within operating design temperatures.
3. Select the type and size of heat distribution units required to supply the necessary Btu/hr.
4. Make a drawing, to scale, of the proposed heating system. Inherent in the drawing phase is a decision as to the type of heating system to be installed (series loop, two-pipe reverse return, etc.), together with any zoning configurations to be used.
5. Size all trunk lines and peripheral components such as circulators and expansion tanks to ensure ample heat delivery throughout all heating circuits.

Each of these design phases will be discussed, together with simple calculations to be performed, that will clarify all the procedures presented.

DETERMINING BUILDING HEAT LOSS

Determination of the building heat loss is the first step in designing a hot water or steam heating system. In this phase the Btu requirements of the boiler, as well as the required amount of baseboard or steam radiators, are determined.

There are three ways to determine building heat loss. These are:

1. Utilization of past fuel consumption records in an existing home.
2. Performance of detailed heat-loss calculations for each room of the residence.
3. Assumption of Btu heat-loss figures per square foot of living space based on construction techniques utilized in the home, and multiplying these factors by the total square footage of the residence.

Let us examine each of these methods in a bit more detail.

When working in retrofit situations, a system installer can use method 1, utilization of past fuel consumption records, to determine heat loss in an existing home. This method involves comparison of past fuel consumption in the home with degree-day records for a specific geographical area.* If done properly, this procedure is the most accurate of all the methods listed because it is based on actual energy requirements of the house, not on theoretical assumptions as to heat loss.

It is important to note that when following the methods outlined in this procedure, the size of an existing boiler or other heating appliance should be disregarded, since building contractors often oversize a heating system to be on the safe side in providing adequate heat to the residence. Fuel consumption figures take into account the life-style of the family and other conditions that cannot be included in a theoretically calculated heat loss but which do have a significant impact on the heating load. To illustrate how this method works, consider the following example.

Example 2.1

A family of four living near Philadelphia, Pennsylvania, burns No. 2 fuel oil for both space heat and hot water. They have a 15-year-old boiler with a separate oil-fired domestic hot water heater. The water heater is 5 years old with a high-speed flame-retention burner. The hot water temperature in the water heater is set at 130°F. The oil equipment is serviced yearly. The service record on the unit identifies the efficiency of the boiler at 70%. Approximately 1000 gal of No. 2 fuel oil was used during the past 12 months. The degree-day total for that area (see Appendix D) was 5400. What is the heat loss of the home?

*Consult Appendix D for degree-day listings of major U.S. cities.

Solution Before the space-heating load of the home can be determined, the amount of oil used for the production of domestic hot water must be calculated and subtracted from the total amount of fuel oil consumed. This can be done by using the following simple calculations:

$$\text{number of people} \times 20 \text{ gal} \times 8.3 \text{ lb/gal} \times \text{set temperature} - \text{incoming temperature} = \text{number of Btu/day}$$

$$\frac{\text{Btu/day}}{\text{efficiency of unit}} \div \text{Btu/gal} \times \text{days used} = \text{DWH oil consumption}$$

Therefore,

$$4 \text{ people} \times 20 \text{ gal} \times 8.3 \text{ lb/gal} \times (130°\text{F} - 55°\text{F}) = 49{,}980 \text{ Btu/day}$$

and

$$\frac{49{,}980 \text{ Btu}}{0.60 \text{ efficiency}} \div 140{,}000 \text{ Btu/gal No. 2 fuel oil} \times 365 \text{ days/yr} = 217 \text{ gal oil/per year for domestic hot water production}$$

Since 217 gal of fuel oil was used to heat the domestic hot water, 783 gal of oil was used for home heating. The heat loss of the home can now be calculated as follows:

$$\frac{\text{gal of oil consumed} \times \text{Btu/gal of fuel} \times \text{efficiency of unit}}{\text{degree-days (DD) for area}} = \text{heat loss in Btu/DD}$$

Substituting the oil consumption figure from our previous calculations, we get

$$\frac{783 \text{ gal No. 2 fuel oil} \times 140{,}000 \text{ Btu/gal} \times 0.55 \text{ efficiency}^*}{5400 \text{ DD}} = 11{,}165 \text{ Btu/DD} \div 24 \text{ hr} = 465 \text{ Btu/°F-hr}$$

To use the fuel consumption method for calculating building heat loss, the Btu content of various fuels and the overall efficiency of their respective heating units must be known. Also, if the family in Example 2.1 used a tankless water heater instead of a separately fired water heater, oil consumption would be affected. In this instance, two separate calculations would need to be made to determine fuel oil consumption for domestic hot water production due to the difference in effi-

*The overall efficiency of the heating unit is lower than the burner combustion efficiency, due to flue and boiler jacket losses. Similar reductions in efficiency are apparent in hot air heating systems as well.

ciency of the tankless coil between the winter heating months and the summer non-heating months. Whereas the efficiency of a typical tankless coil is in the neighborhood of 55% during the winter months when the boiler is providing residential space heating, this figure drops to an average of 15% during the summer months since the entire boiler must be kept hot 24 hours a day solely to provide for domestic hot water production capability. Under these circumstances, the hot water fuel oil consumption calculations would be performed as follows:

Example 2.2

49,980 Btu required per day ÷ 0.55 efficiency × 275-day heating season ÷ 140,000 Btu/gal = 179 gal of oil used for domestic hot water during the nine-month heating season

49,980 Btu required per day ÷ 0.15 efficiency × 90-day heating season ÷ 140,000 Btu/gal = 214 gal of fuel oil used for domestic hot water production during the nonheating season of three months

From the figures above, it is easy to see why the present hot water and space-heating system capability of the family should be considered when setting up the design parameters of the heating system. The second heat-loss calculation method is more complicated than the past-fuel-consumption methodology, but in essence is rather simple.

In new construction, the architect or heating contractor will normally have performed a heat-loss analysis in order to size the heating system. If attempts to obtain this information fail or are not available, a calculated building heat loss must be performed. The Btu rating of any heating equipment lists should be disregarded since the boiler might have been oversized to allow for construction errors or poor heating system performance.

The heat loss of the building, as we have seen, is normally expressed in Btu/hr and is based on some temperature differential between outside ambient air and indoor comfort levels. For example, in Philadelphia, this design temperature differential is 70°F, while in Albany it is 85°F. The design temperature differential is usually based on an indoor temperature of 70°F minus the normally coldest outside air temperature (0°F in Philadelphia and −15°F in Albany). The Btu/hr figure can be converted as follows:*

$$50{,}000 \text{ Btu/hr with a } 70°\text{F } \Delta T = \frac{50{,}000 \text{ Btu/hr}}{70°\text{F}} = 714 \text{ Btu/°F-hr}$$

To predict monthly or yearly fuel requirements, the heat-loss figure can be converted to Btu/DD, as illustrated in the following example.

*The Greek capital letter delta (Δ) is used to denote temperature differentials.

Example 2.3

A new home is being built in Boston, Massachusetts, which has a calculated heat loss of 60,000 Btu/hr with an 85°F design temperature differential. The boiler to be installed in the home will burn natural gas priced at $1 per therm. The average yearly number of degree-days near the construction site is 5634. The expected overall efficiency of the new space heating system is 65%. Assuming an average winter with steady gas prices, how much will it cost to heat this home the first year?

Solution We first find the heat loss of the house in Btu/DD using the following formula:

$$\frac{\text{heat loss of building per hour}}{\text{design temperature differential}} \times 24 \text{ hr} = \text{heat loss, Btu/DD}$$

Therefore,

$$\frac{60{,}000 \text{ Btu/hr}}{85°\text{F}} \times 24 \text{ hr} = 16{,}941 \text{ Btu/DD}$$

To convert from heat loss in Btu/DD to amount of energy required per year to heat the home, we use the following formula:

$$\frac{\text{heat-loss Btu/DD} \times \text{degree-days per year}}{\text{efficiency of heating system} \times \text{Btu/therm of gas}} = \text{therms required per year}$$

$$\frac{16{,}941 \text{ Btu/DD} \times 5634 \text{ DD/yr}}{0.65 \text{ efficiency} \times 100{,}000 \text{ Btu/therm}} = 1468 \text{ therms required per year}$$

$$1468 \text{ therms} \times \$1 \text{ per therm} = \$1468 \text{ first-year heating costs}$$

In this example, we assumed that the heat-loss calculations were available for use. In many instances, however, they are not and the building heat loss must be calculated before the figures above can be obtained. As discussed in Chapter 1, *R* values, the resistance to the flow of heat of various objects, are the mathematical reciprocals of *K* values, a material's thermal conductivity. When performing building heat-loss calculations, the cumulative conductive properties of many different building materials must be derived. The term used to describe how well a wall, window, or floor conducts heat is the overall coefficient of heat transmission, the *U* value. A smaller *U* value denotes a higher *R* value and therefore less heat loss. *U* and *R* values are expressed in Btu/hr-ft^2/ΔT(°F). The *R* values for common building materials are listed in Table 2.1.

Consider the use of this information in the following example.

TABLE 2.1 *R* VALUES OF COMMON INSULATION AND BUILDING MATERIALS

Material description	*R* value, per inch
Fiberglass insulation	3.7
Cellulose insulation	3.7
Vermiculite	3.5
Polystyrene	4.0
Urethane foam	7.5
Building brick	0.20
Wood siding	0.85
Concrete block	0.20
Asphalt shingles	0.40 (common thickness)
Wood shingles	0.94 (common thickness)
Insulating sheathing board	2.64
Gypsum board (sheetrock)	0.45 (standard $\frac{1}{2}$ in.)
Plywood	0.63 (standard $\frac{1}{2}$ in.)
Plaster	0.20
Cement	0.20
Ceiling tile	1.20 (standard $\frac{1}{2}$ in.)
Solid wood door	2.00 (standard $1\frac{3}{4}$ in.)
Combination storm windows	1.80

Source: Martin L. Greenwald and Thomas K. McHugh, *Practical Solar Energy Technology,* © 1985, p. 14. Reprinted by permission of Prentice-Hall, Inc., Englewood Cliffs, NJ.

Example 2.4

A 20-ft wall 9 ft high, which has no windows, has an *R* value of 20. What is the heat loss of the wall for a design temperature of 70°F?

Solution

$$20 \text{ ft} \times 9 \text{ ft} \times \frac{1}{20} \times 70 = 630 \text{ Btu/hr heat loss}$$

Let us take Example 2.4 one step further.

Example 2.5

The wall in Example 2.4 now has two windows added to it. The windows measure 3 ft × 4 ft and have a *U* value of 0.65. What is the heat loss of the wall and window units combined, based on a design temperature differential of 70°F?

Solution Heat loss of the walls:

$$(20 \text{ ft} \times 9 \text{ ft}) - (2 \text{ ft} \times 3 \text{ ft} \times 4 \text{ ft}) \times \frac{1}{20} \times 70 = 546 \text{ Btu/hr heat loss}$$

Heat loss of windows:

$$2 \times 3 \text{ ft} \times 4 \text{ ft} \times 0.65 \times 70 = 1092 \text{ Btu/hr}$$

Combined heat loss:

$$564 + 1092 = 1638 \text{ Btu/hr heat loss of wall/window combination}$$

Air Infiltration

Another significant source of heat loss from a building is air infiltration, the direct loss of heat due to the exchange of air into and out of the building. Warm air that leaks through the cracks around windows and doors must be replaced by new, unheated air from outside the building. The heat capacity, c, of air is 0.018 Btu/ft^3-°F. If a house is maintained at a temperature of 68°F while the outside air is 38°F, every cubic foot of air that is exchanged will result in a heat loss of 0.54 Btu. The exchange of air, Q, is measured in cubic feet per foot of crack per hour.

Half of the total crack length around all the windows and doors in the home should be used for these calculations, because for each location where cold air enters the home there must be a separate location where the warm air escapes, and the air exchange should not be counted twice. Table 2.2 lists Q values for selected wind speeds and types of windows. For doors, use the same chart but double the Q value. The perimeter of a window or door is normally used as the crack length, but if the frame is not tightly sealed to the wall of jamb, it must also be considered. Let us see how the values in Table 2.2 are used to determine air infiltration heat losses in an average setting.

Example 2.6

A room has two windows on the west wall and two windows on the east wall. All four windows have average fittings and weather stripping and measure 3 ft × 4 ft. The room is maintained at a temperature of 68°F and the outside air is 0°F. If there is a prevailing wind of 15 mph from the west, what is the infiltration heat loss for the room?

TABLE 2.2 AIR INFILTRATION THROUGH WINDOWS: Q VALUES (FT^3/HR-FT)

	Wind speed	
	15 mph	20 mph
Weather-stripped	25	35
Non-weather-stripped	40	60

Source: ASHRAE, *Handbook of Fundamentals,* 1981.
Note: For wood double-hung windows.

Solution To determine infiltration heat loss, use the formula

$$\text{length of crack} \times Q \text{ value} \times \Delta T \times C = \text{infiltration heat loss}$$

Substituting the values into our equation from the discussion above, we get

$$2 \times 3 \text{ ft} \times 4 \text{ ft} \times 24 \times 68°\text{F} \times 0.18 = 705 \text{ Btu/hr heat loss}$$

By using the *R* values of common building materials and air films that are listed in Table 2.1, the total heat loss from all surfaces of the residence can be calculated. After adding infiltration losses to all other heat-loss figures, the building heat-loss process is complete.

The third method for calculating the heat loss of a building involves the use of default figures, in Btu/ft^2, that are based on the construction characteristics of a particular home. This method, due to inherent inaccuracies of assuming a Btu/ft^2 heat loss for a structure, is not meant to provide a heating system designer with a set of heat-loss figures on which to make final determinations as to boiler sizing and heat distribution unit calculations. Rather, it gives the system designer a reference point with which to work.

The Btu heat-loss figures are based on the following characteristics of residential construction:

1. Older, poorly insulated homes that feature little or no insulation in the walls and only single-glazed windows
2. Average modern residential construction techniques that feature standard insulation in walls and ceilings and double-glazed windows
3. Extremely tight construction techniques and super insulation, such as that found in passive solar homes, with little air infiltration and tight-fitting, well-insulated window units

Once assumptions have been made as to the construction technique category into which a particular residence fits, Btu/ft^2 heat-loss figures can be assigned. These are as follows:

1. 75–80 Btu/ft^2 for poorly insulated homes
2. 40–50 Btu/ft^2 for average insulation, modern construction
3. 25–35 Btu/ft^2 for super-tight construction

To calculate the heat loss of the residence, the total square footage of living space to be heated is multiplied by the Btu/ft^2 heat-loss figure. For example, if a house measures 40 ft × 30 ft, is built on a concrete slab with no basement and only one level of living space, and meets average modern construction techniques as to window glazing and insulation characteristics, the heat-loss figure can assumed to be approximately 50 Btu/ft^2. The assumed heat loss for the residence will then be 60,000 Btu/hr.

The limitations of this method should be recognized. Note that the calculations do not include a particular design temperature differential, nor do they take into account any interior set temperature or specific design features of the house, such as insulation standards or window sizes and condition. However, the method does place the building heat loss into some kind of perspective.

SELECTING THE BOILER AND ACCESSORIES

Once the building heat loss has been completed and the calculations for each room have been performed, the size and type of heating boiler can be selected. Different criteria should be used for the selection of oil- or gas-fired units versus solid-fuel units. Solid-fuel unit selection is covered in Chapter 7.

Conventionally Fueled Boilers

For oil and gas boilers of both conventional and high-efficiency design, the net IBR ratings or DOE heating capacity listings of the boiler should be equal to, or slightly larger than, the heat loss of the building.

If the boiler is to house a tankless coil for domestic hot water, the Btu capacity of the unit should be increased accordingly to take into account the additional energy drain on the unit. Manufacturers' specifications should be consulted to determine the Btu capacity suggested, or calculations can be performed to determine the increase. The reader is directed to the calculations performed in Examples 2.1 and 2.2. Hot water consumption figures of 20 gal per person per day should be used together with a suggested hot water set temperature of 130°F.

Once all calculations have been performed, boiler selection can be made from a manufacturer's chart that lists several varieties and options available. For example, let us assume that a home has a heat loss of 110,000 Btu/hr, and a tankless coil is to be included in the boiler for domestic hot water. Consulting the manufacturer's catalog in Table 2.3 shows that boiler model 3 is sized to supply the heating requirements of the home.

If it is anticipated that additional heating zones will be needed due to an added bedroom, family room, or hot-tub application in the future, heat-loss calculations should be made to size the boiler for future expansion of the heating system.

When selecting a boiler, the customer must usually make a choice between a steel or a cast iron unit. Each has distinct advantages, and in fact some companies manufacture both units with almost identical warranties. No inherent advantages are suggested for either system. While cast iron has been the predominant material for boiler manufacture for many years and offers greater thermal mass than a steel unit, steel boilers transfer heat somewhat more quickly than cast iron and are generally less expensive. The customer should make a purchasing decision based on the

TABLE 2.3 MANUFACTURER'S BOILER SELECTION GUIDE (SLANT/FIN)

	Ratings steam or water[a]			Net IBR			
Model number[b]	AGA input (Btu/hr)	DOE capacity (Btu/hr)	Water (ft^2)	Water (Btu/hr)	Steam (Btu/hr)	Steam (ft^2)	Tankless heater capacity[c] (gal/min)
GX-100	100,000	80,000	460	69,600	60,000	250	$2\frac{1}{2}$
GX-125	125,000	100,000	580	87,000	75,000	310	3
GXS-150	150,000	120,000	695	104,300	90,000	375	$3\frac{1}{4}$
GX-175	175,000	140,000	810	121,700	105,000	440	$3\frac{1}{2}$
GX-200	200,000	159,000	920	138,300	119,000	495	4
GX-225	225,000	179,000	1040	155,700	134,300	560	4
GX-250	250,000	198,000	1150	172,200	148,500	620	$4\frac{1}{2}$
GX-275	275,000	218,000	1265	189,600	163,500	680	5
GX-300	300,000	240,000[d]	1390	208,700	180,000	750	$5\frac{1}{2}$

Source: Slant/Fin Corp.

Note: All boilers under 300,000 Btu input are tested and rated for capacity under the U.S. Department of Energy (DOE) test procedure for boilers.

[a] Net ratings for baseboard radiation are based on a piping and pickup allowance of 1.15 (hot water) and 1.333 (steam). Slant/Fin should be consulted before selecting a boiler for installation having unusual piping and pickup requirements. For elevations above 2000 ft, reduce ratings by 4% for each 1000 ft above sea level.

[b] Add suffix PT for packaged water boiler with tankless heater, PZ for steam boiler, PZT for packaged steam boiler with tankless heater. Add suffix E for intermittent pilot ignition system (available only with 24-V gas valve). Type of gas: After model number, specify gas by name: natural or propane.

[c] Tankless heater ratings based on intermittent draw.

[d] AGA gross output rating (Btu/hr).

merits of each unit, including applicable warranties, operating efficiencies, and burner and circulator options available.

Add-On Solid-Fuel Boilers

Although solid-fuel boilers are dealt with separately in Chapter 7, some time needs to be spent in outlining the sizing procedures to be used when add-on solid-fuel units are to be included in the heating system. While the sizing procedures for multifuel units are similar to those used for conventional oil- and gas-fired systems, somewhat different considerations should be taken into account for solid-fuel add-ons.

In general, add-on units can be sized with a smaller Btu rating than that for the conventionally fueled boiler, even though the add-on unit often functions as the main heat source, with the oil or gas boiler serving as the system backup. The reason that solid-fuel add-ons can be sized with a smaller Btu heating capacity than their conventionally fueled counterparts is because they do not have to supply all the central heating during times of extremely low temperatures.

Over the course of a typical heating season, outdoor temperatures will average a good bit higher than the lowest design temperatures, which are, in most instances, encountered only rarely. Let us assume that the design temperature in a particular geographical location is −10°F. Winter temperatures may most often be found to hover in the low 20s to 30s for a significant part of the winter heating season, while temperatures of −10°F will be encountered perhaps 7 to 10 days per year. Whereas the heat loss of a particular residence may be 100,000 Btu at −10°F, it will be significantly less at 20 to 30°F; in many instances, cut almost by half. For this reason, during a majority of the time the add-on boiler, operating with outdoor temperatures averaging 25°F, will only need to provide perhaps 50,000 to 60,000 Btu/hr to keep the interior of the home at 68°F. Sizing the add-on boiler in this manner to provide for the "majority" of the heat load to be encountered, rather than the "maximum" heating load to be encountered, saves the purchaser money in two ways: first, a smaller add-on unit will cost less than a larger one; second, the smaller unit will operate at higher efficiencies with less creosote and soot buildup during the heating season. Although there are no specific guidelines for sizing add-on units that are fail-safe, generally they may be sized to provide between 20 and 40% less net Btu heat output than that for the backup oil or gas boiler.

Expansion Tanks

There are two types of expansion tanks in general use today: air cushion and prepressurized diaphragm tanks. Regardless of the design of the expansion tank, specific tank sizes need to be correlated to maximum Btu boiler output. Techniques for selecting and sizing these tanks are detailed in Chapter 3.

SELECTING THE HEAT DISTRIBUTION UNITS

Types of Heat Distribution Units

The heat distribution units used in residential applications are of three basic types: baseboard radiation, convectors, or radiators.

Standard residential baseboard contains a copper pipe ranging in diameter from $\frac{1}{2}$ to 1 in. with aluminum fins firmly attached to the pipe, spaced closely together. This assembly is placed within an enclosure which is open at the bottom and top to allow for gravity circulation of room air (Figure 2.1).

Convectors are heat distribution units surrounded on all sides by a metal enclosure having an air inlet at the bottom and top to allow for gravity circulation of room air. Convectors are generally installed fully or partially recessed in the wall (Figure 2.2).

Radiators are heat distribution units composed of two or more tubular sections generally fabricated from cast iron (Figure 2.3). Room air circulating around the radiator is heated by radiation and convection currents that allow the warm air to rise and the cooler air to fall in an on-going cycle.

Ratings of Heat Distribution Units

To size accurately for the number of heat distribution units that are required for a given heating situation, standard testing and rating procedures have been adopted which simplify this stage of the system design procedure. The amount of heat that can be delivered by a heat distribution unit is dependent on three factors:

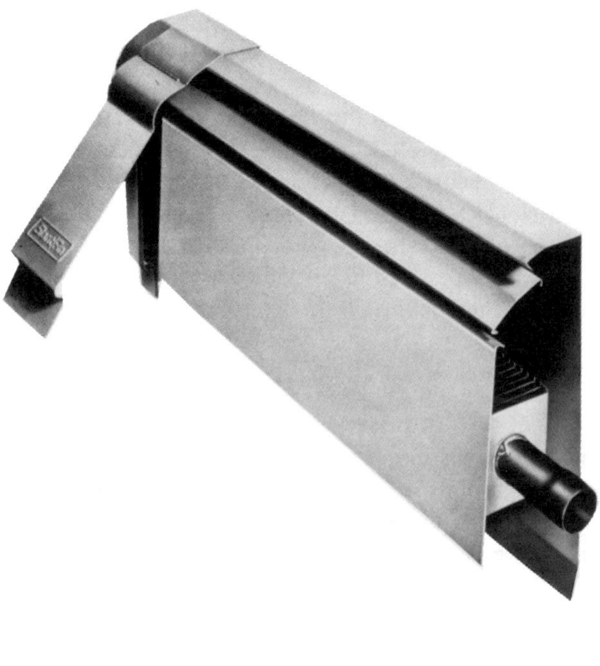

Figure 2.1 Baseboard radiation unit. (Courtesy of Slant/Fin Corp.)

Figure 2.2 Hot water convector for free-standing or partially recessed installation. (Courtesy of Burnham Corp.)

Figure 2.3 Cast iron radiator used in hot water and steam heating systems. (Courtesy of Burnham Corp.)

1. The basic design configuration of the unit, including the size of the pipe used, and fin type and configuration
2. The physical size of the unit
3. The temperature of the water circulating within the heat distribution unit

All these factors are combined into what is known as the *IBR ratings* for heat distribution units.

IBR ratings for finned tube radiation. The Hydronics Institute, formerly known as the Institute of Boiler and Radiator Manufacturers, publishes a list of approved ratings of residential and commercial finned tube radiation. The ratings, which are updated periodically, may be obtained from the Hydronics Institute, 35 Russo Place, Berkeley Heights, NJ 07922.

The ratings for finned tube radiation, the most commonly used heat distribution unit in residential applications, are based on a flow rate through the unit of 1 gal/min, equivalent to 500 lb of water per hour:

$$8.3 \text{ lb/gal of water} \times 60 \text{ min} = 500 \text{ (rounded)}$$

In some instances a manufacturer may request that a particular baseboard unit be tested at 2000 lb/hr equivalent to a flow rate of 4 gal/min. This higher flow rate is useful to reduce the length of baseboard required in a particular installation if the higher flow rate can be designed into the heating system.

The ratings for convectors and radiators as well as fan-coil heat distribution units are also available, either through the IBR rating charts or from the manufacturer of the specific unit. (Ratings for cast iron radiators are discussed in Chapter 6.)

Design Temperature and Baseboard Selection

Once the heat-loss figures for each room have been determined, selection of the heat distribution units is relatively simple. The IBR charts or manufacturer's specifications for the particular unit are consulted to determine the required length and/or size of the unit needed to supply a specified amount of heat. To illustrate how this procedure works, let us assume that a homeowner is to install $\frac{3}{4}$-in. finned baseboard radiation in one large room which has a heat loss of 65,000 Btu/hr. Assuming that a specific manufacturer has been selected, the homeowner consults the IBR ratings for the particular type of baseboard, illustrated in Table 2.4. Note that the Btu output of the baseboard is a function of the water temperature flowing through the radiation. For example, the heat output of model A radiation is 480 Btu/ft at 170°F and 680 Btu/ft at 200°F. To determine the number of linear feet of baseboard required, the person must know what the operating temperature of the water heating system will be. In general, operating temperatures between 180 and 200°F are the norm for hydronic systems employing finned $\frac{1}{2}$- and $\frac{3}{4}$-in. baseboard. If the operating temperature is to be 200°F with a corresponding baseboard

TABLE 2.4 IBR SIZING CHART FOR TYPICAL BASEBOARD MANUFACTURER

Element[a]	Water flow (gal/min)	Pressure drop	170°F 76.7°C	180°F 82.2°C	190°F 87.8°C	200°F 93.3°C	210°F 98.9°C	215°F 101.6°C	220°F 104°C	230°F 110°C	240°F 116°C
No. 15-75E baseboard with E-75 element; $\frac{3}{4}$-in. nominal copper tubing, $2\frac{5}{8}$-in. × $2\frac{1}{8}$-in. × 0.009-in. aluminum fins bent to $2\frac{19}{64}$-in. × $2\frac{1}{8}$-in., 55 fins/ft	1	47	480	550	620	680	750	780	820	880	950
	4	525	510	580	660	720	790	820	870	930	1000
No. 15-50 baseboard with E-50 element; $\frac{1}{2}$-in. nominal copper tubing, $2\frac{5}{8}$-in. × $2\frac{1}{8}$-in. × 0.009-in. aluminum fins bent to $2\frac{5}{16}$-in. × $2\frac{1}{8}$-in., 55 fins/ft	1	260	490	550	610	680	740	770	800	860	920
	4	2880	520	580	640	720	780	810	850	910	970

Source: Slant/Fin Corp.

Note: IBR-approved hot water ratings for Fine/Line 15 series [Btu/hr per linear foot with 65°F (18.3°C) entering air]. Ratings are for element installed with damper open, with expansion cradles. Ratings are based on active finned length (5 in. to 6 in. less than overall length) and include 15% heating effect factor. Use 4-gal/min ratings only when flow is known to be equal to or greater than 4 gal/min; otherwise, 1-gal/min ratings must be used.

[a]End fins are of tinned steel for extra ruggedness.

output of 500 Btu/ft, then 130 ft of $\frac{3}{4}$-in. baseboard is required to supply the 65,000-Btu/ft heat loss in the room:

$$500 \text{ Btu/ft} \div 65{,}000 \text{ Btu heat loss} = 130 \text{ ft } \tfrac{3}{4}\text{-in. baseboard required}$$

Btu Heating Circuit Design Limitations

While the calculations performed above illustrate that 130 ft of $\frac{3}{4}$-in. baseboard will supply the room heat loss of 65,000 Btu/hr, there are subtle limitations that must be imposed on the heating system if this sizing technique is to work effectively. These limitations are based on the temperature drop and number of gallons per minute that can circulate throughout the heating loop.

Temperature drop and gal/min calculations. The temperature drop in the heating system, that is, the difference in the temperature of the water leaving the heating boiler and that in the return line entering the boiler after traversing the heating system is designed to be 20°F. To provide a given number of Btu to a room to make up for the heat loss, two factors must be taken into account: Water must flow through the heating loop at a certain volume per unit of time, and a specific temperature drop must be designed into the system for which a specified water flow rate will work effectively.

To determine the gal/min requirements necessary to provide a specific amount of heat to a given area, the following equation is used:

$$\text{gal/min required} = \frac{\text{Btu/hr load}}{8.33 \times 60 \times \text{TD}}$$

where

8.33 = weight of 1 gal of water, lb/gal

60 = number of minutes per hour

TD = temperature drop of the heating circuit (assumed to be 20°F)

Btu/hr load = heat loss to be met by the heating system

To calculate the heat delivered by 1 gal/min of circulating water, the following equation is used:

$$\text{heat (Btu/hr)} = 1 \text{ gal/min} \times 8.33 \text{ lb/gal} \times 60 \text{ min/hr} \times \frac{1 \text{ Btu}}{\text{lb°F}} \times 20\text{°F}$$

$$= 10{,}000 \text{ Btu/hr}$$

This figure, 10,000 Btu/hr, is the industry standard that is applied to determine the gal/min requirements for low-temperature residential hydronic systems.

Coupled with this figure are the limitations imposed by the pipe sizes in each heating loop, which affect the ability of the heating circuit to deliver an adequate supply of heat to each room in the system.

If we return to our example of a room with a heat loss of 65,000 Btu/hr sized for 130 ft of $\frac{3}{4}$-in. baseboard, the design limitations are brought out when the recommended flow rates for given sizes of pipe used in residential installations listed in Table 2.5 are examined. If we enter the chart at the nominal pipe size of $\frac{3}{4}$ in. and move to the right side under "copper tubing," note under the column listing "maximum gal/min" that $\frac{3}{4}$-in. pipe is limited to a flow rate of 3.5 gal/min. If we multiply 3.5 gal/min by the standard figure of 10,000 Btu/hr per gallon of water, it becomes clear that $\frac{3}{4}$-in. pipe can adequately carry only approximately 35,000 Btu/hr. In most design practices, therefore, Btu capacities for single-circuit heating zones are limited to between 15,000 and 20,000 Btu/hr for $\frac{1}{2}$-in. pipe, 40,000 Btu/hr for $\frac{3}{4}$-in. pipe, and 80,000 Btu/hr for 1-in. pipe. In order to use $\frac{3}{4}$-in. finned baseboard in the single-room heating zone in this example, the system must be broken up into two separate circuits, with one main feed pipe branching into two circuits on a single zone, illustrated in Figure 2.4. For a complete list of the maximum recommended Btu for series loop or piping circuit capacities using standard baseboard radiation, see Appendix C.

Taking all the foregoing factors into account, let us apply this design criterion to a single-family residence with a total heat loss of 89,000 Btu/hr at the specified design temperature to determine both the total gal/min required to supply the entire

TABLE 2.5 RECOMMENDED FLOW RATES (GAL/MIN) FOR SPECIFIC SIZES OF PIPE

	Steel pipe				Copper tubing			
Nominal pipe or tubing size (in.)	Minimum (gal/min)	Velocity (ft/sec)	Maximum (gal/min)	Velocity (ft/sec)	Minimum (gal/min)	Velocity (ft/sec)	Maximum (gal/min)	Velocity (ft/sec)
$\frac{1}{2}$	—	—	1.8	1.9	—	—	1.5	2.1
$\frac{3}{4}$	1.8	1.1	4.0	2.4	1.5	1.0	3.5	2.3
1	4.0	1.5	7.2	2.7	3.5	1.4	7.5	3.0
$1\frac{1}{4}$	7.2	1.6	16	3.5	7	1.9	13	3.5
$1\frac{1}{2}$	14	2.2	23	3.7	12	2.1	20	3.6
2	23	2.3	45	4.6	20	2.1	40	4.1
$2\frac{1}{2}$	40	2.7	70	4.7	40	2.7	75	5.1
3	70	3.0	120	5.2	65	3.0	110	5.2
$3\frac{1}{2}$	100	3.2	170	5.4	90	3.1	150	5.2
4	140	3.5	230	5.8	130	3.5	210	5.6
5	230	3.7	400	6.4	—	—	—	—
6	350	3.8	610	6.6	—	—	—	—
8	600	3.8	1200	7.6	—	—	—	—
10	1000	4.1	1800	7.4	—	—	—	—
12	1500	4.3	2800	8.1	—	—	—	—

Source: Taco, Inc.

Note: For all piping except branch connections on Taco Venture systems.

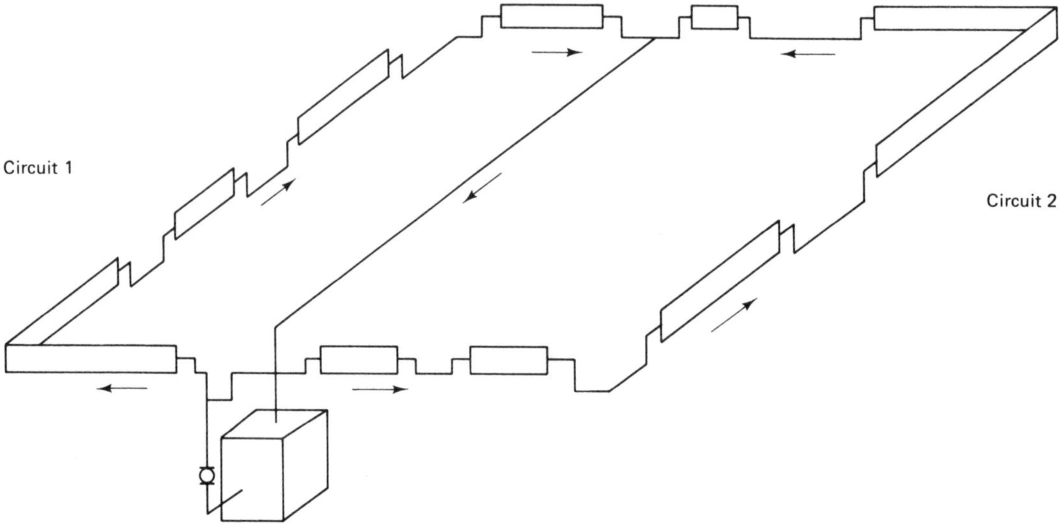

Figure 2.4 Double-circuit single-zone hydronic system designed to overcome pipe size limitations imposed by single-circuit design. Circuit 1: 35,000 Btu/hr, 70-ft baseboard; circuit 2: 30,000 Btu/hr, 60-ft baseboard. (Courtesy of the Hydronics Institute.)

heating load of the building, as well as the individual gal/min requirements of the individual rooms in the home.

The first step in this process is to determine the total gal/min required to supply the 89,000 Btu/hr heat load required by the building. Substituting the appropriate figures into our equation, we find the following:

$$\text{gal/min} = \frac{\text{heating load in Btu/hr}}{8.33 \times 60 \times 20} = \frac{89{,}000}{10{,}000} = 8.9 \text{ gal/min required}$$

The total delivered gal/min by the heating system must be 8.9 gal in order to supply the heat loss of 89,000 Btu.

We can now apply this formula to determine the gal/min requirements for each room within the heating system. The heat losses of the individual rooms in the home are given in Table 2.6. Performing the calculations for the building configuration above, the following flow rates are required: 1.5 gal/min for room 1, 2.0 gal/min for room 2, 3.0 gal/min for room 3, and 2.4 gal/min for room 4. The total number of gal/min required for the house, 8.9, is determined either by adding up the individual room gal/min figures or by using the total heat-loss figures for the residence. In actual practice, individual room requirements are more accurate, although only minor variations will be encountered in either method.

Also, keeping in mind the Btu limitations per circuit based on the maximum flow rates for the piping utilized, the home in this example will have to be broken into separate heating circuits so as not to exceed the maximum Btu limitations for the type and size of baseboard chosen.

TABLE 2.6 INDIVIDUAL ROOM HEAT-LOSS CALCULATIONS

Room number	Calculated heat loss (Btu/hr)
1	15,000
2	20,000
3	30,000
4	24,000

LAYING OUT THE HEATING CIRCUIT

Making a scale drawing of the proposed heating circuit is prerequisite to any subsequent installation work. There are three steps involved in this process:

1. Preliminary sketches of the system design, showing approximate locations and spatial relationships of the various heating system components.
2. Determination of the system configuration. This includes the specification of zoning requirements, type of circuit layout to be used (series loop, supply loop, etc.), and the individual circuit limitations imposed by the restrictions of the pipe sizing and gal/min requirements.
3. Production of a scale drawing of the system, including the exact location of all components, placement of baseboard, and all applicable pipe sizes used in the circuitry.

We will now examine each of these design phases in more detail.

Preliminary Sketches

Making preliminary sketches of the heating system, sometimes known as "thumbnail" sketches, is useful for any project. This step allows thoughts about the system to be expressed graphically without fear of making mistakes or changes in the system configuration. When ideas are initially expressed in this fashion, some features that may at first have appeared workable may, in fact, turn out not to be, and vice versa. Carefully prepared drawings, at least in the preliminary stages of system design tend to inhibit changes in system features, if for no other reason than the implied necessity of redoing the drawings. Preliminary rough sketches help free the mind to access a variety of different ideas and will, through a process of elimination, offer a workable plan for the job. All preliminary sketches should be saved even though the finished drawings have been prepared. Often information penciled in on the rough drawing that may be necessary for later stages in system planning or installation is not transferred to the final draft of the heating system and must be retrieved from the rough sketches.

Determining the System Configuration

Determination of system configuration involves two basic decisions: the first is deciding on the best type of piping circuit to use; the second concerns the decision to include one or more heating zones in the system. In Chapter 1, a variety of piping circuits were examined. Of all the types of circuitry available, most residential heating systems employ one-pipe systems in either a series or supply loop forced-circulation configuration. Each of these arrangements is illustrated in Figure 2.5. While each layout is highly efficient and effective, the one-pipe series-loop configuration

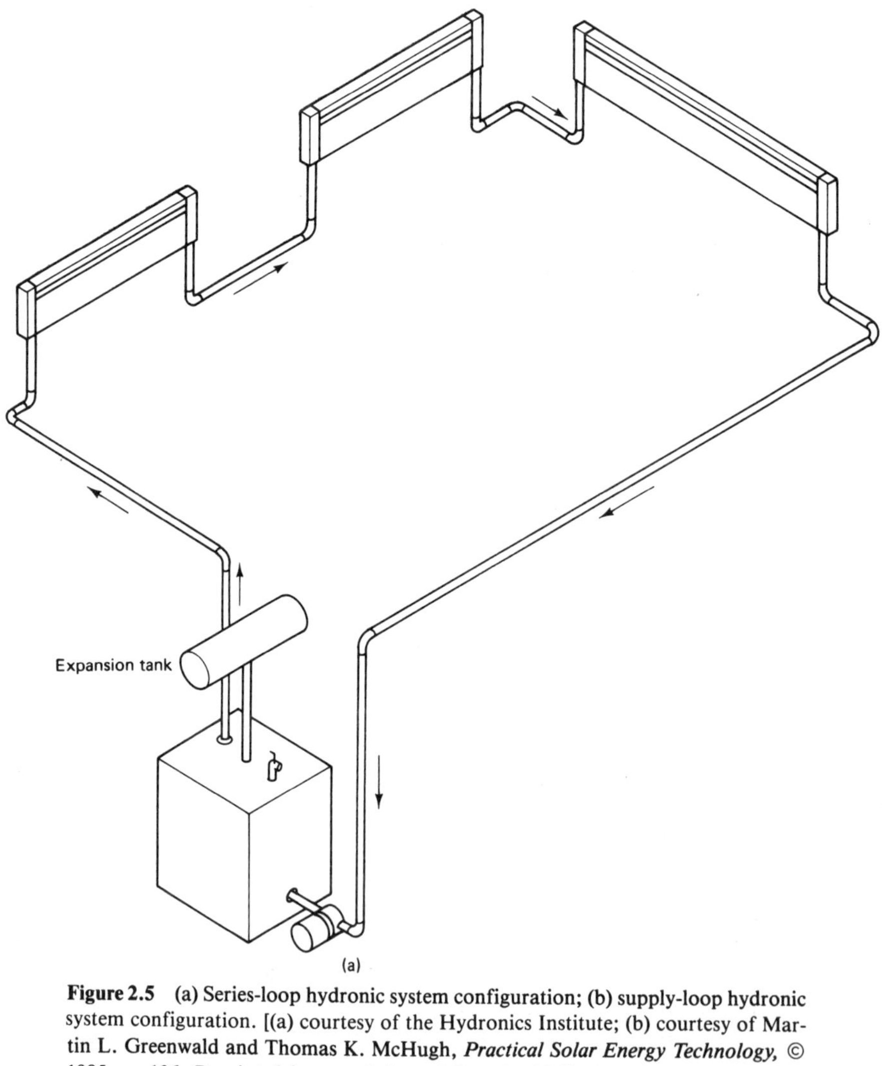

Figure 2.5 (a) Series-loop hydronic system configuration; (b) supply-loop hydronic system configuration. [(a) courtesy of the Hydronics Institute; (b) courtesy of Martin L. Greenwald and Thomas K. McHugh, *Practical Solar Energy Technology,* © 1985, p. 136. Reprinted by permission of Prentice-Hall, Inc., Englewood Cliffs, NJ.]

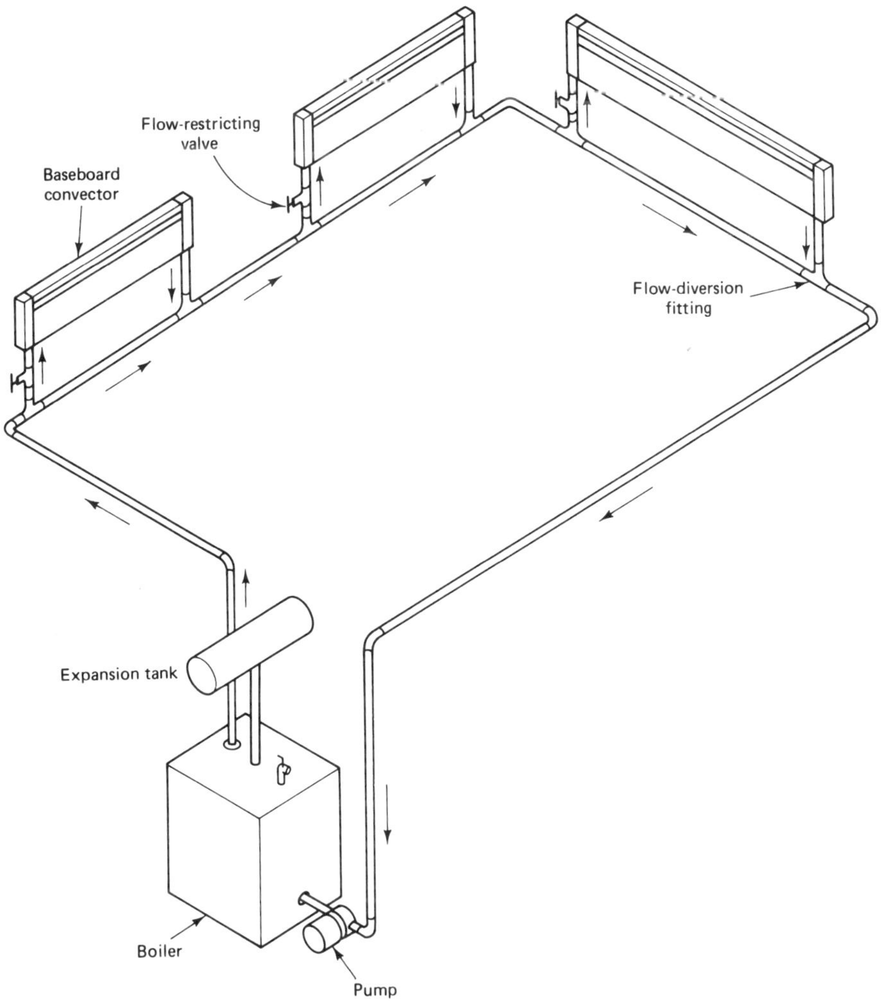

(b)

Figure 2.5 *(continued)*

is the least expensive in terms of materials and labor to install. Also, at this stage of system planning, the decision as to whether or not to zone the system should be examined.

While zoning modifications increase the initial cost of the installation somewhat, savings accrue once operation of the system begins. Heat can be increased, reduced, or eliminated where required by the dictates of the living patterns of the family. The Btu output of the boiler, hence fuel consumption, will always be less with a well-designed zoned system than with an equivalent system without benefit of zoning.

Drawing the System to Scale

After all decisions regarding component size, placement, and piping configurations have been made, a final scale drawing of the heating system should be prepared. Figures 2.6 and 2.7 illustrate drawings typical of this type that have been prepared for both a single- and a multiple-zone layout for the same single-family residence. Note that the drawings indicate the gal/min and length of baseboard requirements for each room in the home. Also, all pipe sizes and lengths of pipe runs have been placed on the drawings for quick reference.

SIZING THE CIRCULATOR AND TRUNK LINES

The size of all trunk lines as well as the capacity of the system circulator(s) must be calculated before the installation can begin. In most commercial and industrial heating systems, the pressure-drop calculations for all fittings, valves, and piping must be carefully prepared. However, in a majority of residential applications, sizing tables have been developed that take into account frictional losses through an average number of fittings and valves used in typical heating systems. The procedures outlined in this section are referenced to series-loop heating systems. Appendixes A and B contain the tables to be consulted for sizing supply-loop and two-pipe reverse-return systems.

Sizing the Circulating Pump

The function of the circulating pump in a forced-convection hydronic heating system is to move the water, which is under system pressure, from the boiler throughout the heat distribution units and back into the boiler in a closed-loop cycle. This cycle continues until the heating demand has been satisfied. During the heating cycle, the water will flow through a variety of valves, fittings, and pipe, all of which offer a certain amount of resistance to the flow of water. Therefore, a specified amount of pressure must be developed by the circulating pump that is sufficient to overcome all internal friction within the system. A heating system designer must know how much pressure a pump can deliver for a given required gal/min capacity. This information is given by each pump manufacturer in its descriptive literature.

The ratings of circulating pumps are described in terms of "feet of head" or "head pressure." To illustrate this concept, if a circulating pump is connected to a supply of water of the intake side of the pump, and the discharge side is connected into a vertical pipe that is 5 ft high, and the unit delivers a 12 gal/min flow out of the pipe, we can say that the pump delivers 12 gal/min at 5 ft of head. If the height of the vertical pipe is increased from 5 ft to 8 ft, and we observe that the volume

---

Figure 2.6 (*opposite page*) Single-zone series-loop system scale drawing. Total length of circuit, 144 ft; gal/min requirement, 3.9 (Courtesy of the Hydronics Institute.)

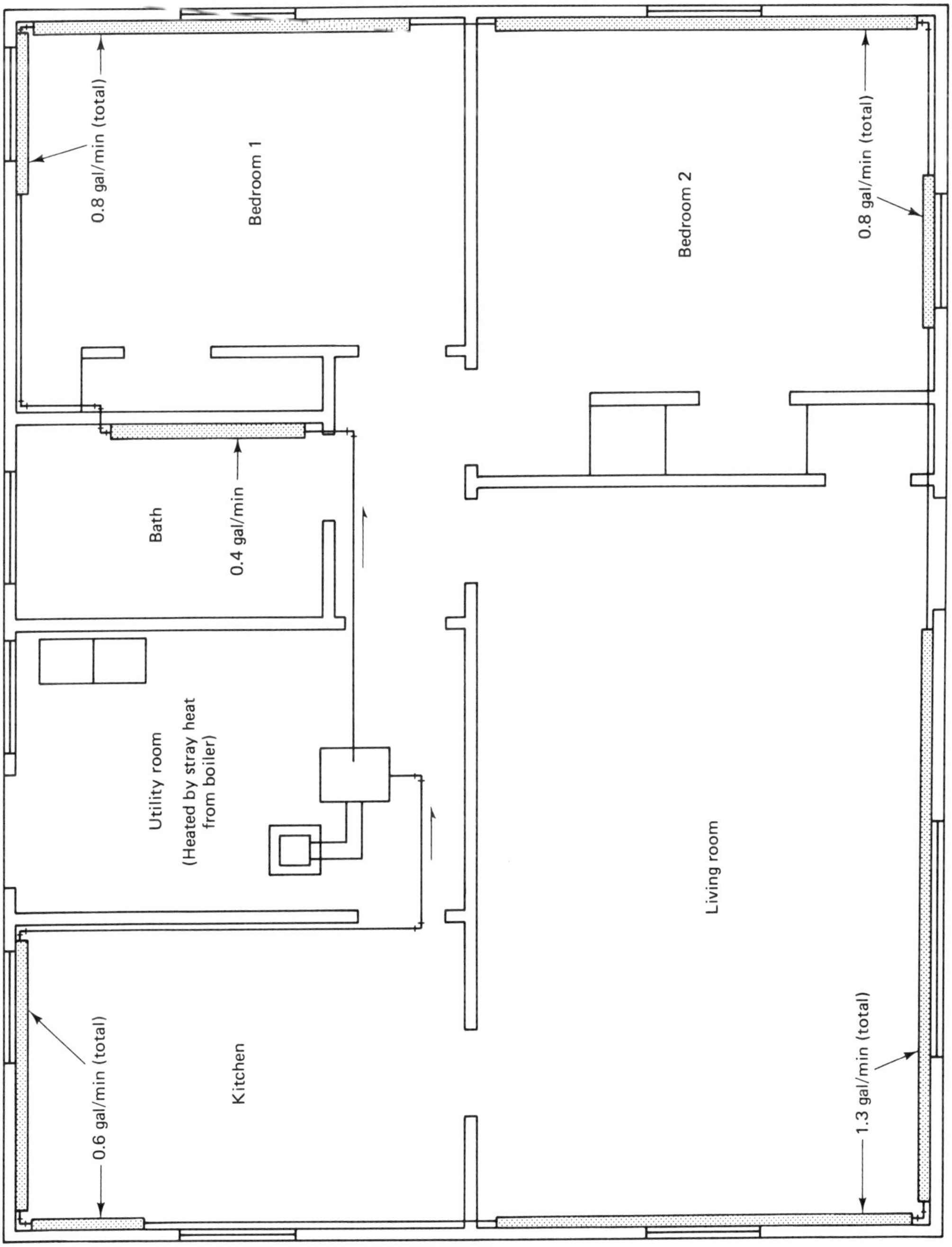

0.8 gal/min (total)
Bedroom 1
Bedroom 2
0.8 gal/min (total)
Bath
0.4 gal/min
Utility room
(Heated by stray heat
from boiler)
Living room
Kitchen
0.6 gal/min (total)
1.3 gal/min (total)

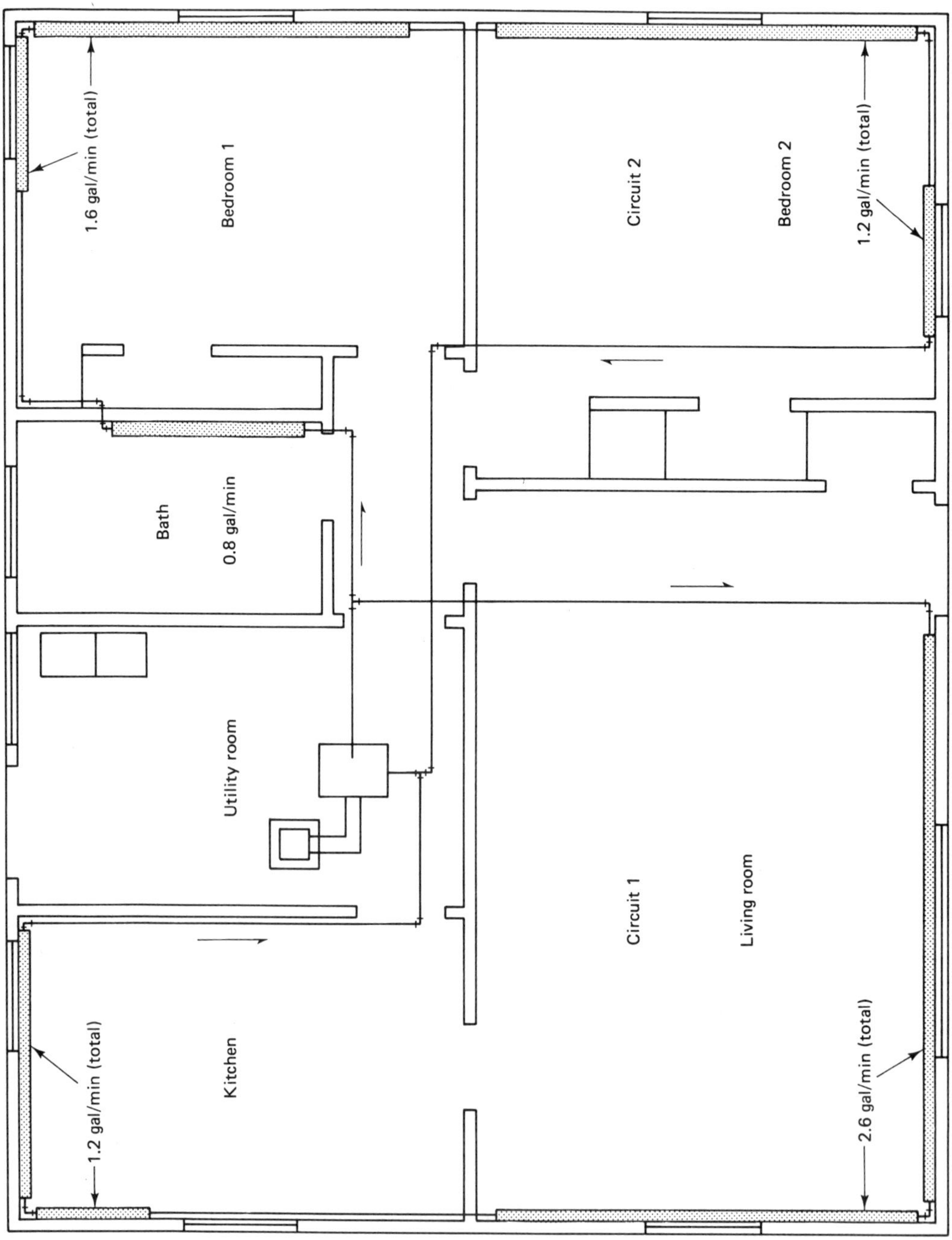

1.6 gal/min (total)
Bedroom 1
Circuit 2
Bedroom 2
1.2 gal/min (total)
Bath
0.8 gal/min
Utility room
Circuit 1
Living room
Kitchen
1.2 gal/min (total)
2.6 gal/min (total)

of discharged water drops from 12 gal/min to 6 gal/min, it can now be stated that the pump delivers 6 gal/min at 8 ft of head pressure. The circulators used in most hydronic systems are low-head units. These pumps differ from standard shallow and deep-well water pumps, which can develop considerable head pressure and pumping capacity. Since relatively small volumes of water are circulated in hydronic heating systems, and the system pressure on both the feed and discharge side of the pump is about the same, low-head circulators are more efficient than their high-head counterparts in this application. Also, most hydronic circulators are driven by fractional-horsepower motors under $\frac{1}{12}$ hp, whereas standard water pumps start $\frac{1}{3}$ hp and larger.

Circulating pump manufacturers measure the volumetric capacity of their pumps under a variety of head pressures in order to derive a performance curve for a particular pump. A typical performance curve is illustrated in Figure 2.8. To illustrate the use of this chart, if we wish to determine the head pressure developed by a particular circulating pump at a capacity of 4 gal-min, move across the chart until the 4-gal/min line intersects the pump curve. Thus, in this example, the pump in question can deliver almost 12 ft of head pressure at a capacity of 4 gal/min.

Sizing the Pump Head Pressure

All design steps lead to the final sizing of the circulator based on the length of all the system circuits and the gal/min requirements. Once the performance curves of the circulator under consideration have been examined, Table 2.7 should be consulted for final circulator sizing. For example, the house heating system depicted in Figure 2.6 has a total calculated heat loss of 39,000 Btu/hr, with a heating circuit length of 144 ft, using $\frac{3}{4}$-in. baseboard and interconnecting pipe for the single-circuit series-loop system. The total gal/min requirements are 3.9, based on a delivered Btu content of 10,000 Btu/gal of heated water within the heating system. The system has been designed to provide the proper amount of heated water to provide for the heat loss of each room within the residence. To determine the pump capacity required for this system, we enter Table 2.7 at the $\frac{3}{4}$-in. circuit size and read to the left of the chart to find the figure equal to, or slightly larger than, the 3.9 gal/min figure. In Table 2.7, this figure is 4.1. At this point, we read down the column into the lower portion of the table (total length of circuit) to find the figure that is closest to 144 ft, the circuit length in our proposed heating system. The figure in this instance is 140. Moving across the column to the head figures indicate that a pump with 10 ft of head is required for the system. Thus the pump selected for this system must be able to develop a head pressure of 10 ft at a delivered volumetric capacity of 3.9 gal/min.

Figure 2.7 (*opposite page*) Two-zone series-loop scale drawing. Circuit 1: total length, 97 ft; required gal/min, 3.8. Circuit 2: total length, 104 ft; required gal/min, 3.6. (Courtesy of the Hydronics Institute.)

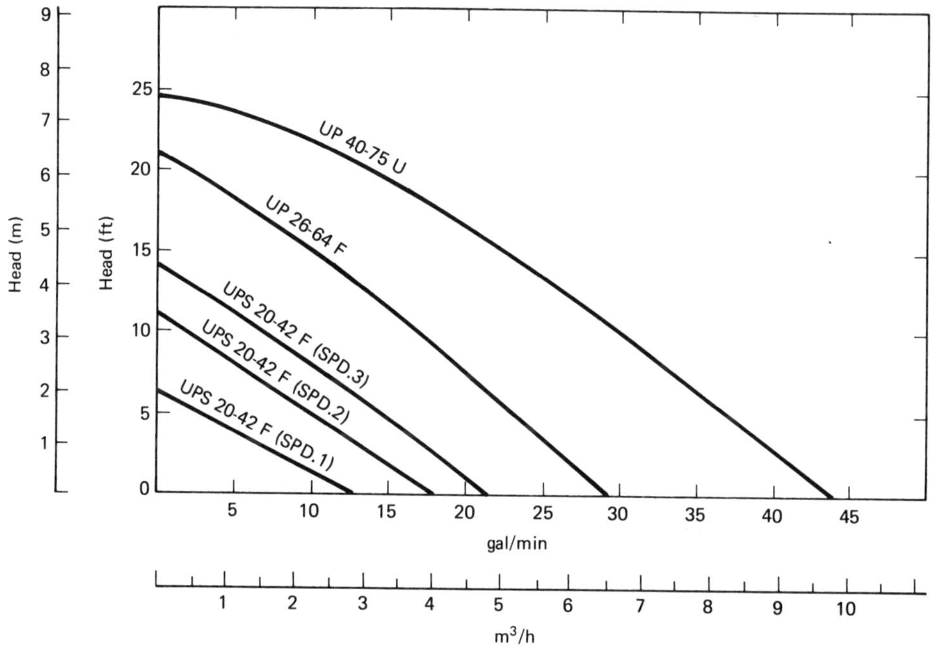

Figure 2.8 Pump performance curve. (Courtesy of Grundfos Pump Corp.)

Sizing the Pipe Circuits

When selecting circulating pumps for multiple-circuit systems, the head pressures of each circuit must be calculated prior to selecting the pump. To illustrate how this procedure works, refer to the system design illustrated in Figure 2.7. In this instance, the residence has a total designed heat loss of 74,000 Btu/hr and has been divided into two separate heating circuits, each utilizing a $\frac{3}{4}$-in. series circuit in each loop.

Circuit 1 has a total length of 97 ft, with a gal/min requirement of 3.8 (based on 10,000 Btu/gal of delivered water throughout the heating system). To determine the pump requirements for this circuit, we consult the pipe sizing chart in Table 2.7. Entering the chart at a circuit size of $\frac{3}{4}$ in. we find that the gal/min figure closest to our requirements is 3.8. Moving down this column into the lower section of the table, we find that the total length of the circuit closest to our requirements of 97 ft is 110 ft. Moving across to the left column of the chart, it is found that the pump must develop a head pressure of 4 ft.

Circuit 2 has a total length of 104 ft and a total gal/min of 3.6. Entering the table at the $\frac{3}{4}$-in. circuit size, we find the closest gal/min figure for our purpose is 3.6. Moving down into the lower portion of the table, the circuit length closest to our requirements is 110 ft. Moving to the left column of the table, it is shown that a circulating pump with a head pressure of 6 ft is required for this heating circuit.

TABLE 2.7 IBR CIRCULATOR SIZING GUIDE FOR SERIES-LOOP BASEBOARD SYSTEM

Circuit or trunk pipe size (in.)	Capacity of circuit (gal/min)															
$\frac{1}{2}$	2.3	2.0	1.9	1.8	1.7	1.7	1.6	1.5	1.5	1.4	1.3	1.2	1.2	1.1	0.9	0.8
$\frac{3}{4}$	5.0	4.3	4.1	3.8	3.7	3.6	3.4	3.2	3.1	2.9	2.8	2.6	2.4	2.2	2.0	1.8
1	9.6	8.3	7.7	7.3	7.0	6.8	6.5	6.3	5.9	5.7	5.5	5.0	4.6	4.3	3.8	3.4
$1\frac{1}{4}$		18	17	16	15	15	14	14	13	12	11	11	9.7	9.0	8.3	7.3
Available head (ft of water)	**Total length of circuit (as measured on piping layout)**															
4	35	45	50	60	65	70	75	80	90	100	110	130	150	180	220	290
5	45	60	65	70	80	90	95	100	120	130	140	160	190	230	290	360
6	55	70	80	90	100	110	120	130	140	160	180	200	240	290	350	450
7	65	90	100	110	120	130	140	150	170	190	210	240	290	340	420	540
8	75	100	110	130	140	150	160	180	200	220	250	290	330	400	490	620
9	85	110	130	150	160	170	190	200	230	250	290	330	380	450	560	710
10	100	130	140	170	180	190	210	230	260	290	320	370	430	510	620	790
11	110	140	160	190	200	220	240	260	290	320	360	410	480	570	690	880
12	120	160	180	200	220	240	260	290	320	350	400	450	530	620	760	960
14	150	190	210	250	260	290	310	340	380	420	470	540	620	730	900	1120

Source: The Hydronics Institute.

To use this table:

1. *Pump selection:* Enter the upper left-hand column at the circuit or baseboard size. Read to the right to the gal/min capacity of the circuit equal to or greater than required. Read down this column into the lower portion of the table to the figure closest the total length of the circuit. Read across to the left-hand column to determine the pump head pressure required. Follow this procedure for each circuit. Enter the pump manufacturer's curve at the total gal/min of the system and select a pump with a head pressure equal to or greater than the highest head pressure required. If separate pumps are used for each circuit, select each pump based on the circuit gal/min, circuit length, and pipe size.
2. *Trunk selection:* Enter the lower left-hand column at the head pressure of the pump selected. Read to the right to the figure closest to the total length of the *longest* circuit. Read up to the gal/min capacity equal to or greater than required. Read across to the left-hand column to determine the trunk pipe size required. If separate pumps are used for each circuit, the trunk is sized based on the lowest pump head, the length of the longest circuit, and the total gal/min of the system.

Once the head pressure for each heating circuit has been calculated, the circuit with the highest head pressure is used for selecting the pump, in this instance 6 ft. The gal/min figures are simply added together for each circuit to make a final determination of pump capacity. Therefore, in this example, a circulating pump must be selected that can deliver 7.4 gal/min at a head pressure of 6 ft.

To gain experience in making these calculations, the reader is urged to do several practice calculations in order to become familiar with the procedures prior to undertaking the design and sizing of a specific hydronic system. For sizing supply-loop and reverse-return heating systems, consult Appendixes A and B for the appropriate sizing tables.

3

Selecting the Peripheral Components

Peripheral system components in a hydronic heating system include all those accessories other than the boiler and terminal heating distribution units. The selection of any system component must be made carefully, since all must be engineered to work together.

THERMOSTATS

The thermostat is responsible for activating the heating system and is triggered by a drop in room temperature. When the room temperature drops to the thermostat set point, an electrical circuit within the thermostat is closed, which in turn triggers a relay within the boiler control to begin heating system operation.

Locating the Thermostat

The best location for a room thermostat is approximately 5 ft above the floor, where average-temperature room air can freely circulate over the unit. Thermostats should not be located on exterior building walls, or where hot or cold drafts from heating units or home appliances can affect the accurate room-temperature readings. For proper operation, thermostats should be installed perfectly level.

Single-Stage Thermostats

A thermostat that responds by opening and closing at only one set point is referred to as a single-stage thermostat (Figure 3.1). The operation of the thermostat relies on a coiled bimetallic spring and mercury switch (Figure 3.2). A drop in room temperature causes the coiled bimetallic spring to contract, rotating the mercury-filled capsule. Upon tilting, mercury within the capsule falls against two switch terminals, completing the electrical circuit within the thermostat. This low-voltage circuit then activates a relay within the boiler aquastat to begin burner operation and water circulation. This type of switching action is referred to as single-pole, single-throw switching: One pole and one switch of an electrical circuit are controlled by the thermostat.

Single-stage thermostats are also designed to operate both a heating and cooling system. In this design configuration, the thermostat would activate the heating system on a drop in building temperature and turn on a central cooling system when the building temperature rises. This electrical switching circuitry is referred to as single-pole double-throw: Two separate relays or switching systems are activated at two separate thermostat set points. These thermostats are generally mounted on subbases that incorporate the additional wiring and terminal arrangements to achieve the proper switching. A schematic representation of a double-throw thermostat for control of both heating and cooling is illustrated in Figure 3.3.

Heat Anticipators

The heat anticipator is a device located within the thermostat that is designed to create artificial heat by the use of a resistor, anticipating the arrival of heat from the heating system that has not yet penetrated the cover of the thermostat (Figure 3.4). Without the heat anticipator circuit, room temperature would exceed the thermostat setting. Overheating a house in this manner occurs due to the residual heat contained in the baseboard or radiators, which continues to be transferred to the

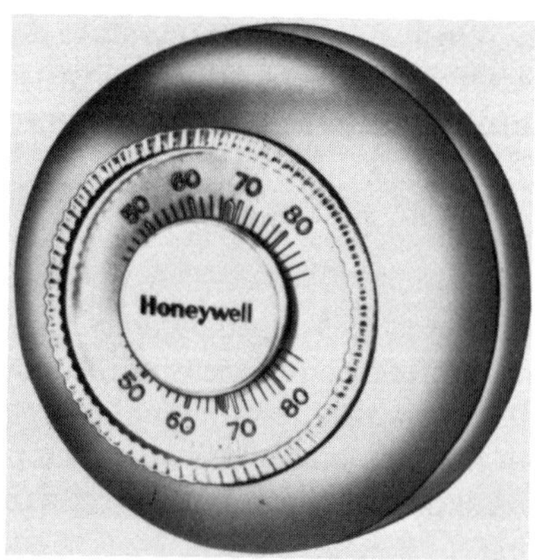

Figure 3.1 Single-stage thermostat. (Courtesy of Honeywell, Inc.)

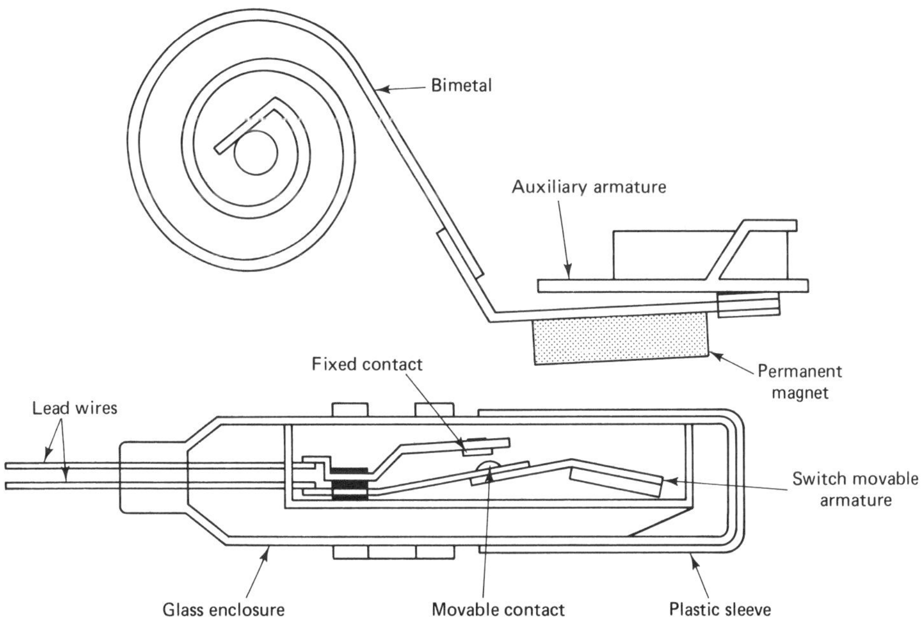

Figure 3.2 Bimetallic spring coil and switch contact in typical house thermostat. (Courtesy of Robertshaw Controls Company.)

living space after the thermostat setting has been satisfied. Since the amount of heat generated by the anticipator will be proportional to the current draw in the thermostat electrical circuit, an adjustment is provided by the thermostat manufacturer for proper calibration of the anticipator. The most accurate way to make this adjustment is to measure actual current draw in the thermostat circuit. If this is not possible, the relay should be set to the current draw listed on the appliance relay or other device to which the thermostat is connected. Failure to make these adjustments can result not only in unsatisfactory comfort levels within the home, but also in increased number of run cycles per hour of the heating system, which can have a significant effect on overall heating system efficiency. See Chapter 8 for a more detailed discussion on heat anticipator setting procedures.

Automatic Setback Thermostats

The ability to lower thermostat settings automatically when less heat is required in a home on a predictable basis can result in significant fuel savings during the heating season. The savings that can be realized by the use of setback thermostats in a variety of geographical locations within the United States is examined in greater detail in Chapter 8. A typical solid-state automatic setback thermostat is illustrated in Figure 3.5.

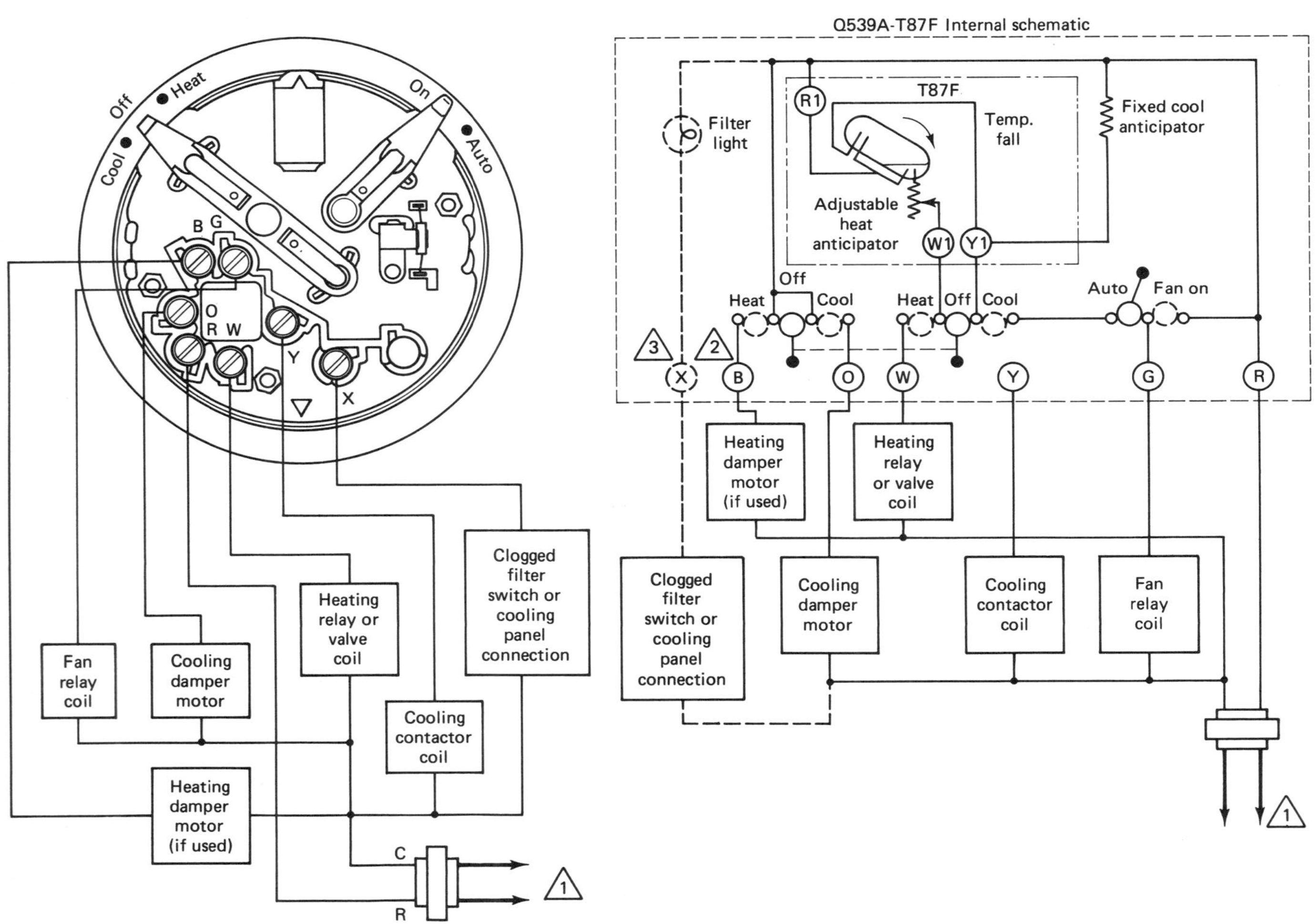

Figure 3.3 Typical hookup and internal schematic for multistage heating and cooling thermostat. 1, Power supply; provide overload protection and disconnect means as required. 2, No. 4 terminal is the same as B terminal. 3, X terminal used on Q539 models with factory-installed malfunction light. (Courtesy of Honeywell, Inc.)

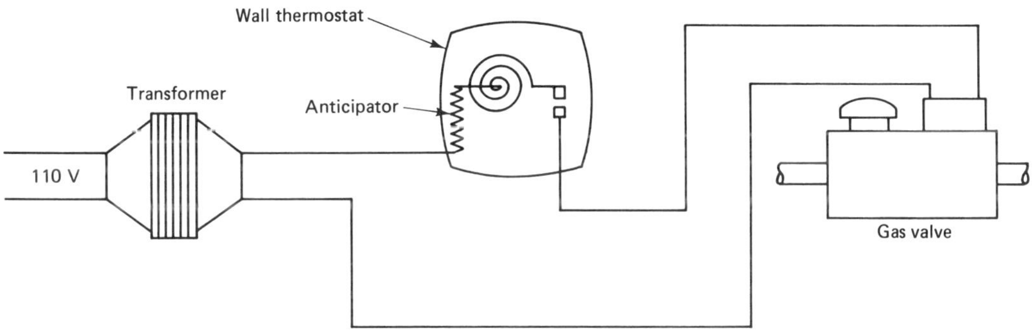

Figure 3.4 Heat anticipator circuit in thermostat. This circuit helps to control cycling of the heating system to provide for even interior temperatures. (Courtesy of Robertshaw Controls Company.)

Multistage Thermostats

Multistage thermostats incorporate two or more switching poles to control complex heating and cooling functions from one centralized control (Figure 3.6). Multistage thermostats are generally designed so that there is a 2 to 3° isolation between the heating stages of the unit. A two-stage thermostat in a hot water heating system would operate as follows. The first stage of the thermostat (highest temperature set point) would open a zone valve and commence circulator operation; water from the boiler would flow throughout the heating circuit, regardless of boiler-water tem-

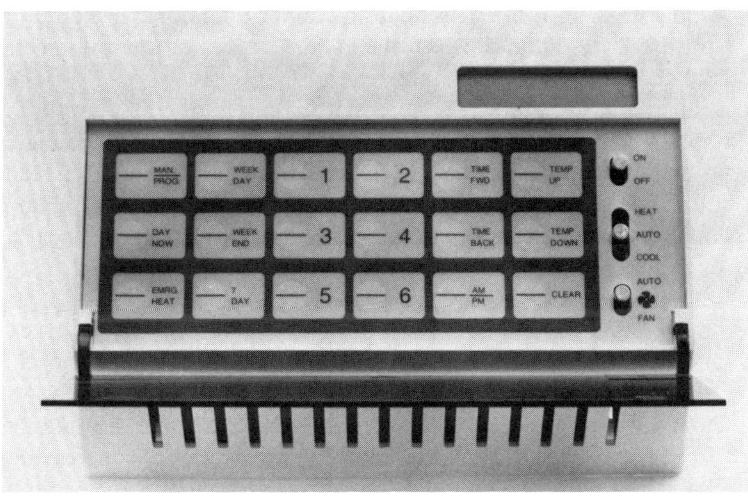

Figure 3.5 Automatic setback thermostat. (Courtesy of American Stabilis.)

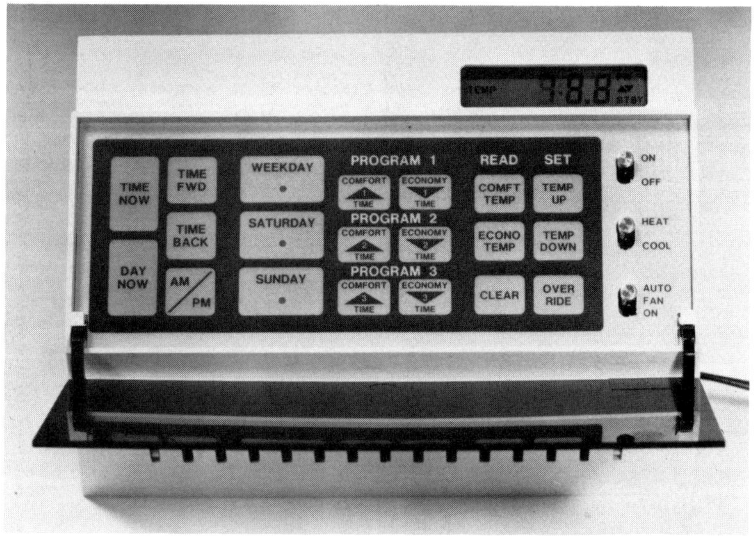

Figure 3.6 Multistage programmable thermostat. (Courtesy of American Stabilis.)

perature. Should the heat within the boiler water be sufficient to maintain room temperature and prevent the thermostat from reaching the second-stage setting (lower temperature set point), the burner mechanism would not be activated. If, on the other hand, the temperature within the heating zone dropped 2 to 3° and energized the second-stage thermostat circuitry, the burner mechanism would be activated to provide additional heat. This two-stage arrangement has the advantage of utilizing all available heat from within the boiler water and correlates operation of the burner to room heating needs rather than boiler-water temperature. The temperature differential between the heating stages on the thermostat are usually field adjustable. A schematic of the operational sequence of two-stage hot water heating is illustrated in Figure 3.7. Note that the pump operates continuously unless system water temperature falls below the setting of the reverse-acting control because of no calls for heat. Further, if the boiler-water temperature reaches the high-limit control setting, the gas burner valve is shut off. A drop of approximately 20° in boiler-water temperature will start the gas burner again. Local ordinances may require a second high-limit control (set 10°F higher).

CIRCULATORS

Circulating pumps are used in residential hot water heating systems to promote forced convection of the heat transfer process, thereby increasing system efficiency. Chapter 2 examined the selection of a circulator for a specific heating application based on the gal/min requirements of the heating circuit and the feet of head re-

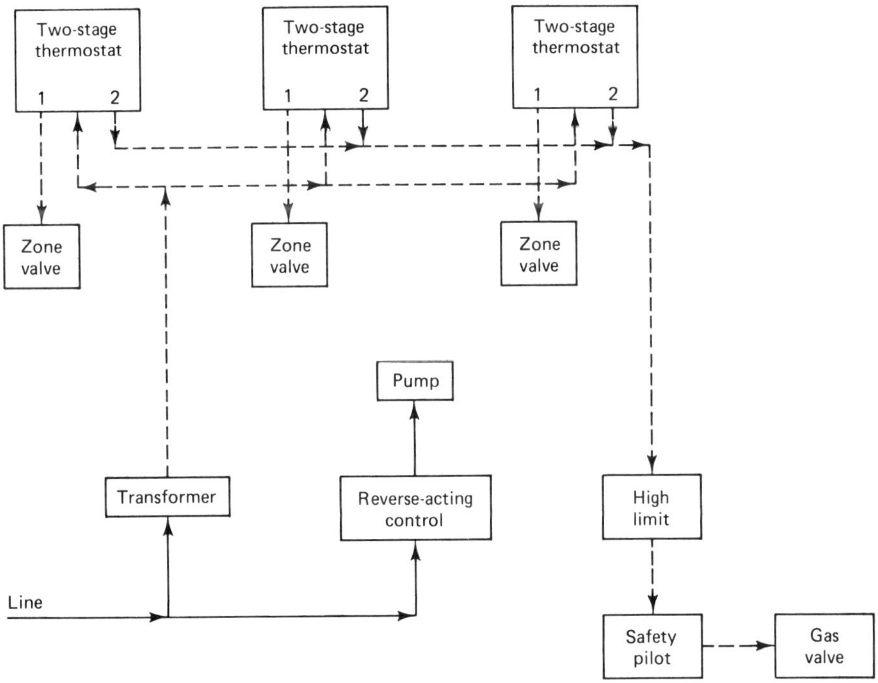

Figure 3.7 Two-stage circuitry for residential hot water heating system. (Courtesy of the Hydronics Institute.)

quired to overcome frictional resistance within the heating loop. There are a variety of circulators available for residential heating systems, and a knowledge of basic pump design is helpful in understanding overall heating system operation.

Basic Pump Design Features

Single-stage centrifugal pumps are most commonly used in residential hydronic applications. A single-stage pump assembly consists of one impeller within a housing that contains water inlet and outlet channels. The impeller contains a series of vanes that are arranged in such a manner so that when the impeller is rotated at high speed, water within the vanes is flung out into the discharge port of the pump. As the water is discharged, incoming water takes its place entering the pump housing at the inlet port. This arrangement is illustrated in Figure 3.8.

In-Line Pump Design

The in-line pump is the most common design configuration utilized in small hydronic systems up to 1 hp in power rating (Figure 3.9). These units operate at speeds between 1750 and 3450 revolutions per minute (rpm), are quiet, and develop head

Figure 3.8 Basic component arrangement of hydronic heating system circulator. Note position of impeller within the intake and exhaust flange assembly. (Courtesy of Grundfos Pump Corp.)

Figure 3.9 In-line centrifugal pump. (Courtesy of ITT Fluid Handling Division.)

pressures sufficient to handle most residential and small commercial applications. The in-line design features the impeller housing connected into the piping circuit, which supports the entire pump assembly. Two variations of in-line pumps are commonly used: oil lubricated and water lubricated. An oil-lubricated pump is illustrated in Figure 3.10. These pumps feature mechanical seals where the motor shaft penetrates the stuffing box to connect to the impeller. These mechanical seals are very durable and in most instances are field repairable. These units generally require semiannual lubrication to ensure adequate oil levels for seal and bearing lubrication.

Figure 3.10 Oil-lubricated centrifugal circulating pump. (Courtesy of Taco, Inc.)

A water-lubricated circulating pump (Figure 3.11) utilizes the circulating water within the heating system for pump lubrication and cooling purposes. Rather than standard sleeve bearings in the motor, water-lubricated pumps generally feature a porous ceramic bearing and stainless steel pump components for efficient lubrication. The output or performance curve for a typical in-line centrifugal pump is shown in Figure 3.12.

Figure 3.11 Water-lubricated centrifugal circulating pump. (Courtesy of Grundfos Pump Corp.)

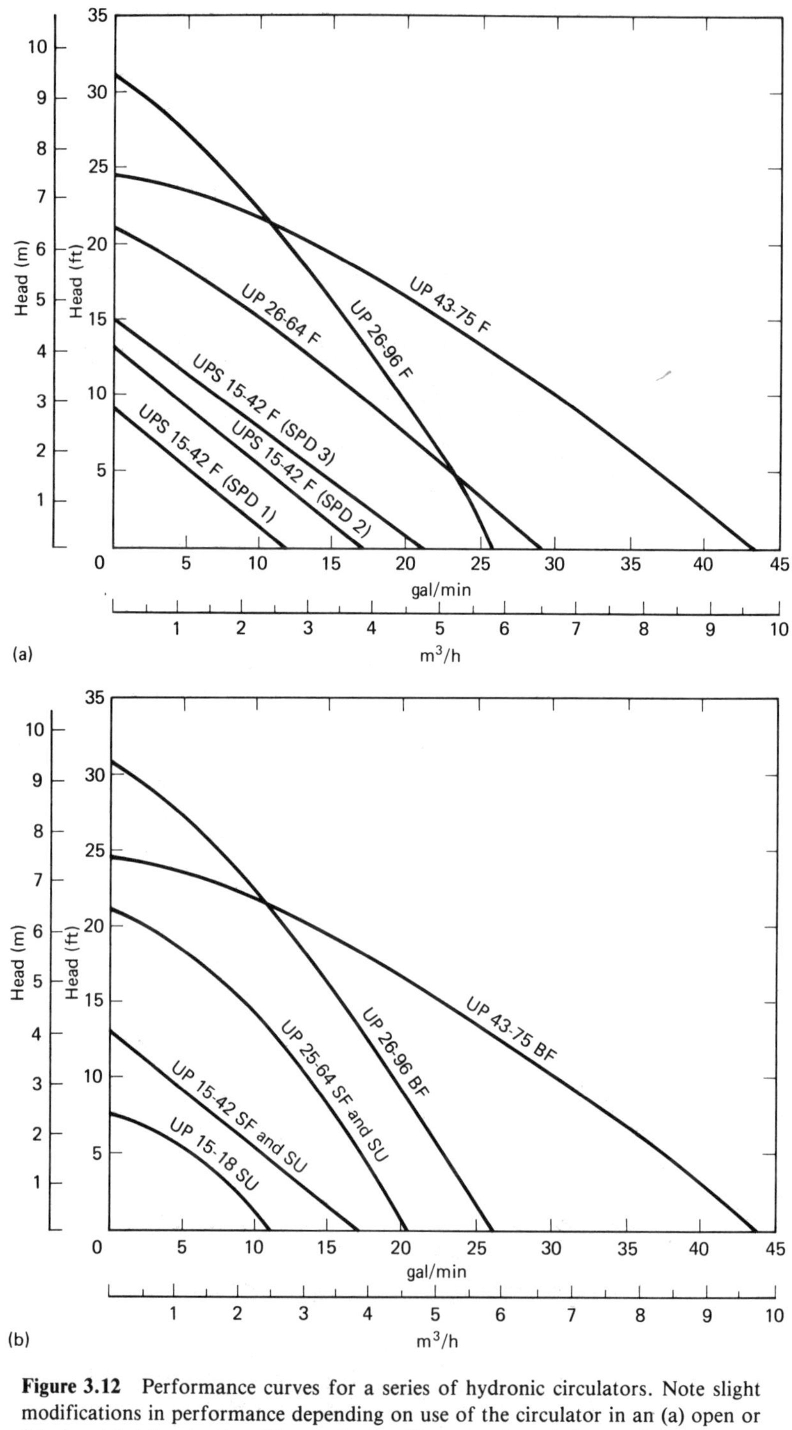

Figure 3.12 Performance curves for a series of hydronic circulators. Note slight modifications in performance depending on use of the circulator in an (a) open or (b) closed heating system. The product suffixes are: S, stainless steel; F, flanged; B, bronze; U, union. (Courtesy of Grundfos Pump Corp.)

Isolation Valving

When installing circulator pumps, it is good piping practice to incorporate isolation valving on either side of the circulator flanges. This allows for removal of the circulator from the plumbing circuit without the need to drain the entire heating system. Gate valves or isolation ball valves are generally used for this purpose. A typical isolation valving arrangement is illustrated in Figure 3.13.

Locating the Circulating Pump

Proper location of the circulating pump within the plumbing circuitry has an important bearing on overall system performance. The location of the expansion tank is the point of no pressure change within the heating system. Placement of the circulating pump with the expansion tank located on either the suction or discharge side of the circulating pump can cause different pressure fluctuations within the system. Each of these circulator hookups is illustrated in Figures 3.14 and 3.15.

For proper operation of the heating system, a positive pressure must be maintained at all times at the highest point in the piping array. The equations given in Table 3.1 should be used to determine the minimum system pressure to be main-

Figure 3.13 Isolation valving of system circulator. This type of configuration allows the circulator to be repaired without the need to drain either the heating system or the boiler.

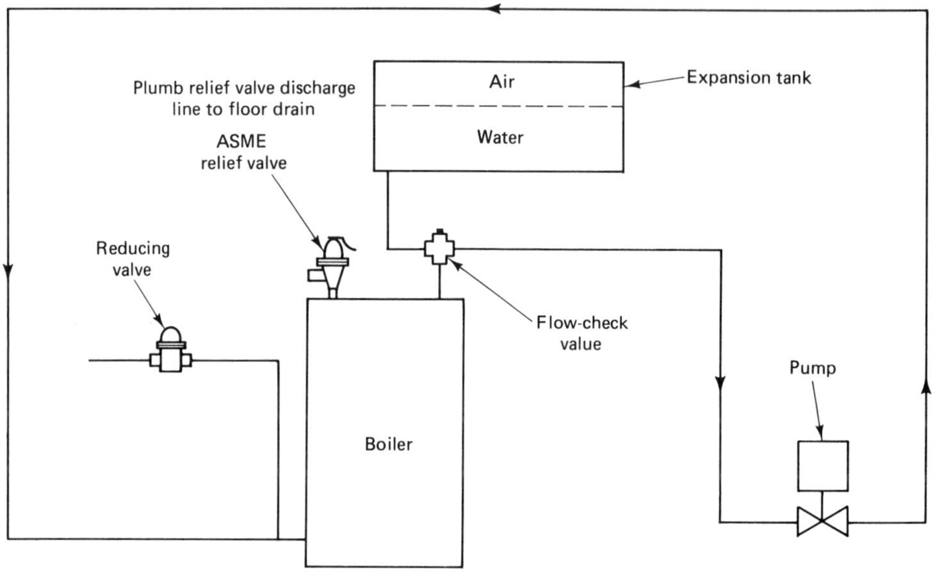

Figure 3.14 Location of circulating pump when expansion tank is connected to the suction side of the pump. (Courtesy of Taco, Inc.)

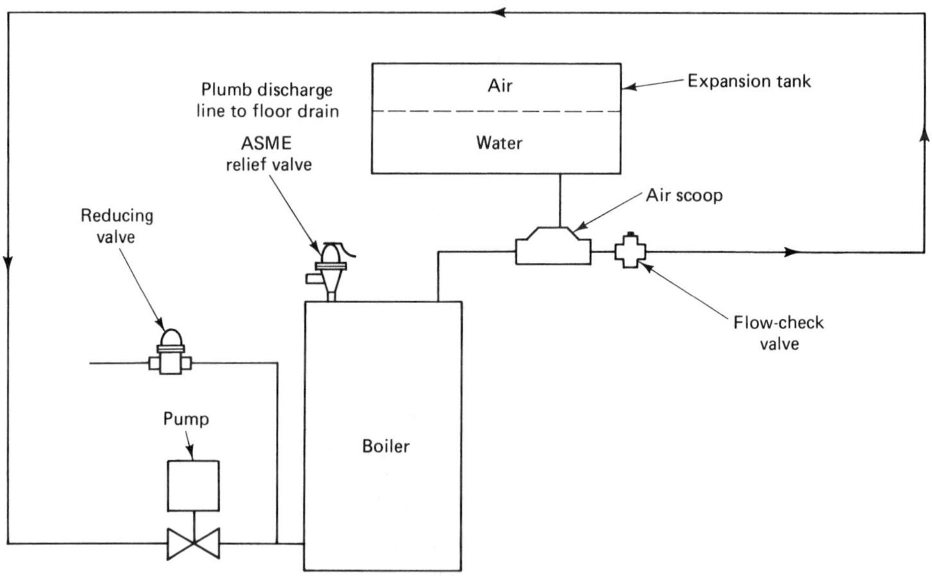

Figure 3.15 Location of circulating pump when expansion tank is connected to the discharge side of the pump. (Courtesy of Taco, Inc.)

TABLE 3.1 EQUATIONS FOR ESTABLISHING HEATING SYSTEM PRESSURE BASED ON THE LOCATION OF THE CIRCULATOR

(a) Systems with expansion tank on suction side of pump:

$$SH = \frac{RH + 4}{2.31}$$

(b) Systems with expansion tank on discharge side of pump:

$$SH = \frac{RH + 4 + PH}{2.31}$$

where SH = minimum static head in system
RH = distance from top of boiler to highest point in system in feet
PH = operating head of pump in feet of water

Note that in the latter equation it is necessary to add the operating head of the pump because this head appears as a lowered system pressure on the suction side of the pump.

Source: Taco, Inc.

tained based on the circulator location illustrated in Figure 3.14 or 3.15. Note in the recommended circulator configurations shown in Figures 3.14 and 3.15 that the circulator should *never* be installed between the pressure relief valve and/or the feedwater pressure-reducing valve and the boiler. To do so would create low pressure at the reducing valve when the pump is in operation, possibly causing feedwater to be admitted to the heating system in excess of that required. This will cause high operating system pressures and continual discharging of the pressure relief valve.

In a normal residential heating application requiring pumping head pressures of 10 ft or less, the location of the circulator is not critical. It should be noted, however, that some manufacturers of packaged boilers have designed their equipment predicated on a predetermined fluid flow pattern. Any deviation from this flow should be undertaken only after consultation with the equipment manufacturer.

Larger heating systems that require in excess of 10 ft of pump head pressure capacity usually perform best when the expansion tank is on the suction side of the circulating pump. As the reader will note from Table 3.1, locating the pump on the return lines requires an increase in the fill pressure which may not be possible given the height of the system piping and maximum operating pressures of the boiler.

AQUASTATS

Aquastats are switching devices that are designed to react to changes in water temperature. A typical aquastat control is illustrated in Figure 3.16. Aquastats are designed in two basic configurations: either immersion or surface-mounting types.

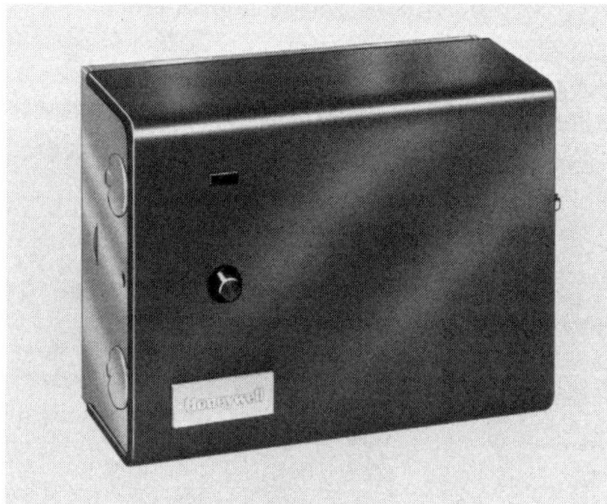

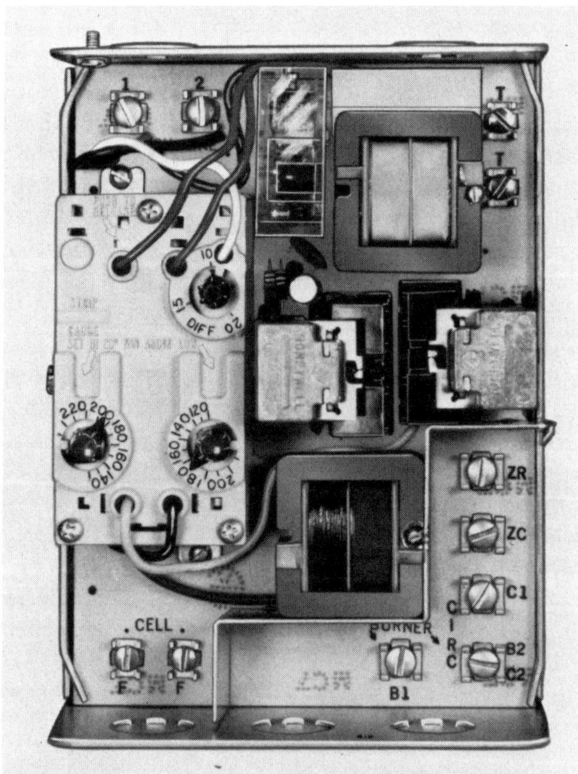

Figure 3.16 Aquastat control used in hydronic heating systems. This control features a high and a low limit as well as an adjustable differential. (Courtesy of Honeywell, Inc.)

Immersion Aquastats

A typical aquastat consists of a gas-filled bulb that is connected to a switching mechanism. As water temperature rises, gas within the bulb is heated and expands, either closing or opening an electrical circuit within the aquastat control. In most designs, the expansion of the gas is transmitted through a capillary tube, exerting pressure against a bellows that can operate both snap switches and relays (Figure 3.17).

The temperature-sensitive bulb is installed in a water-tight well mounted in a tapping provided in the housing of the boiler. In this way, the gas-filled bulb senses internal boiler water temperature. For precise control of the heating system, it is critical that the bulb "bottom out" and make contact with the wall of the immersion well. A heat-conductive grease can be placed within the well to ensure effective conductive heat transfer. A typical immersion well fitting is illustrated in Figure 3.18.

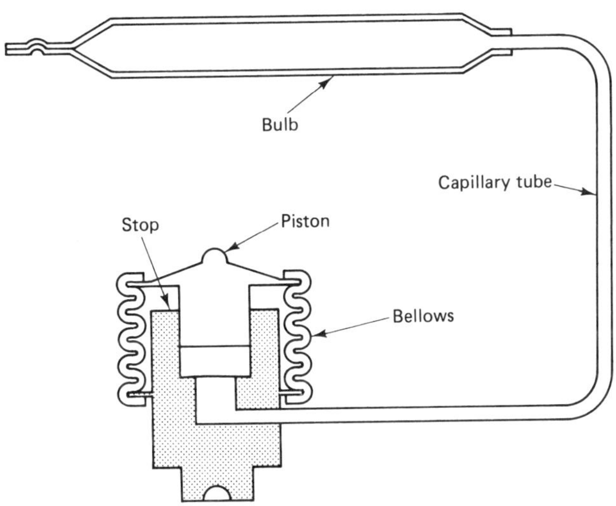

Figure 3.17 Sensing bulb, capillary tube, and bellows mechanism typical of many aquastat switching mechanisms. (Courtesy of Robertshaw Controls Company.)

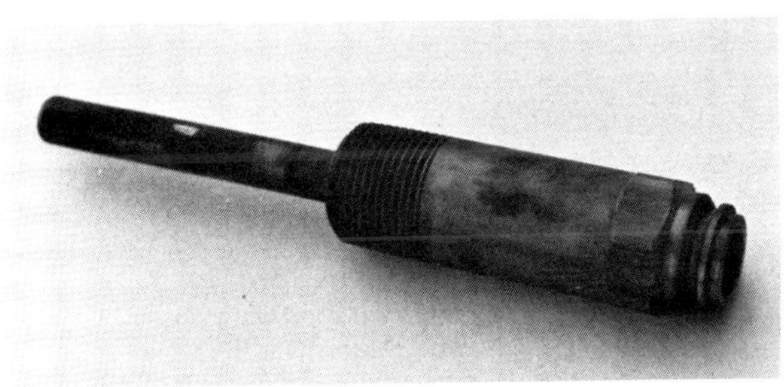

Figure 3.18 Immersion well used to hold aquastat sensing bulb. (Courtesy of Honeywell, Inc.)

Aquastats are usually designed to be "fail-safe," in that a rupture of the sensing bulb or capillary tube will cause the electrical contacts to open, deenergizing critical circuits for safe operation. Some aquastats are designed for direct immersion; that is, they are provided with a rigid sensing bulb that is part of the aquastat assembly. These aquastats are inserted through a special compression fitting directly into the boiler water. Although direct-immersion devices provide more sensitive responses to temperature changes than do indirect-immersion units, the boiler must be drained in the event that the control must be replaced. This disadvantage has made the use of this type of control more limited than the immersion well type.

Surface-Mounted Aquastats

Surface-mounted aquastats are mounted directly on exterior pipe or heat exchanger surfaces for temperature monitoring. A typical surface mount aquastat is illustrated in Figure 3.19. While ease of installation is one advantage of this type of design, one obvious disadvantage is that the aquastat reacts to changes in surface temperature of the object rather than actual internal water temperature. The result of this is a variation in temperature control from actual system water temperature on both heat-up and cool-down boiler cycling. The degree of these variations will differ depending on the conductivity of the surface on which the aquastat is mounted and the actual location of the control. These controls can also be affected by ambient air temperature that surrounds the device. For these reasons, the use of surface-mounted aquastats is usually limited to applications where precise temperature control is not essential.

Figure 3.19 Surface-mounted aquastat. These units clamp onto the surface of the pipe and sense temperature by conduction of heat from the pipe to a hidden sensing bulb within the clamp. (Courtesy of White Rodgers Division, Emerson Electric Co.)

Switching Action

Aquastats are used for many different functions and are available in a variety of switching configurations for most applications. Most aquastats are designed to either open or close on a temperature rise or fall and can thereby handle a wide variety of switching applications. One of the most common uses of aquastat controllers is for high-limit protection. In this mode, the aquastat operates a single-pole, single-throw (SPST) electrical switch whose contacts are normally closed. Used in this manner, electrical current is directed to the inlet side of the switch when there is a call for heat from a room thermostat. If the water temperature within the boiler is at a level below the set point of the aquastat control, the electrical contacts remain closed, allowing current to flow through the switch to the burner mechanism. If the temperature should exceed the set point of the aquastat, however, expansion of the gas within the sensing bulb causes the electrical contacts to open, preventing the flow of electricity to the burner, terminating the ignition sequence of the boiler. A typical high-limit aquastat control is illustrated in Figure 3.20.

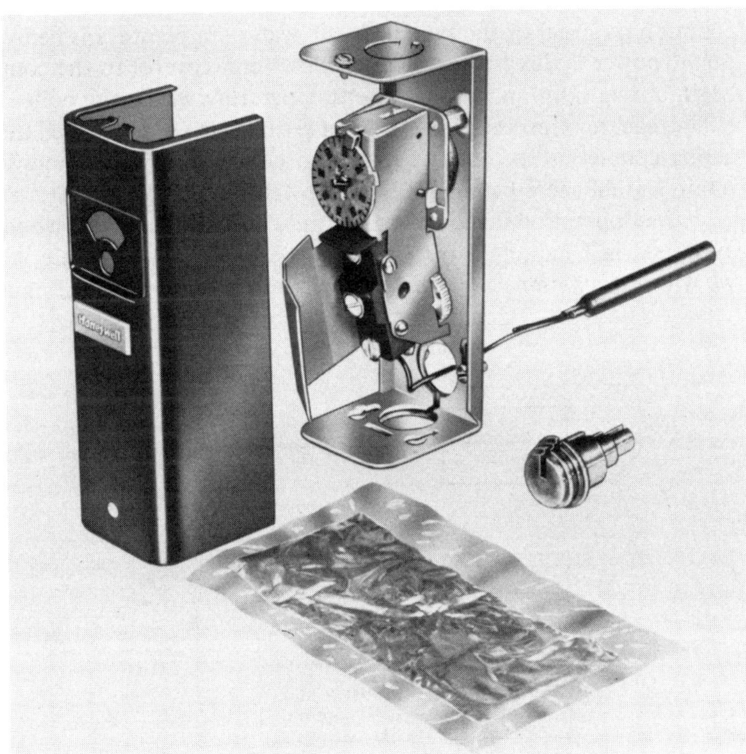

Figure 3.20 A high-limit aquastat controller opens an electrical switch on temperature rise, interrupting operation of the burner mechanism or circulator. (Courtesy of Honeywell, Inc.)

This same device, wired in a different manner, can also function as a low-limit or operating control. Used in this way the control is set at the water temperature that is to be maintained within the boiler. Any time that the water temperature falls below the control setting, contraction of the gas within the sensing bulb will cause the electrical switching contacts to close, completing the electrical circuit to the burner. In this arrangement, the control is usually wired in parallel with the room thermostat. Thus a call for heat from the thermostat is not necessary for operation of the burner mechanism. Only the high-limit control can override the requirements of the low-limit-switching aquastat, since power is delivered to the low-limit control in series with the high limit. One use for this control is in providing domestic hot water via a tankless coil in a steam heating system. The operating schematic of this type of control configuration is illustrated in Figure 3.21.

A number of control functions are often combined into one master aquastat control unit. One such combination is a low-limit, reverse-acting control. The purpose of this aquastat is to provide minimum temperatures within the boiler and also prevent operation of the circulator until a minimum low temperature within the boiler has been reached. To accomplish these control functions, a single-pole, double-throw (SPDT) switch is used within the switching relay. Power flows into the common terminal of the switch. If the water temperature is below the low-limit set point, power is directed through the low-limit control to the combustion device to begin burner ignition. As the water temperature within the boiler begins to rise and either reaches or exceeds the low-limit setting, power is diverted through the reverse-acting contacts to a separate relay and power is now available for the circulating pump should there be a call for heat by the room thermostat.

The operational sequences for oil-, gas-, steam-, and wood-fired boilers are examined in Chapters 4 through 7, respectively. To put these sequences in focus, we will now examine typical operating modes for reverse-acting aquastats.

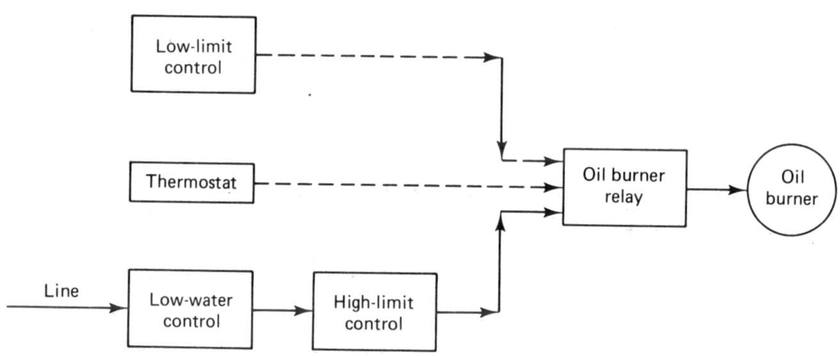

Figure 3.21 Low-limit control in steam system for domestic hot water production. (Courtesy of the Hydronics Institute.)

Operational Sequences

There are a variety of operational sequences used in aquastat controls. The switching sequence illustrated in Figure 3.22 describes the operation of a typical reverse-acting boiler aquastat relay. The high-limit control opens on a temperature rise to turn off the burner mechanism when the boiler water reaches the high-limit set point. On a 10°F temperature drop from high limit, the burner switch makes contact. This switching action will reignite the burner if the thermostat is still calling for heat. If the thermostat setting is satisfied, the burner will not ignite even though the switching contacts have closed.

The low-limit control performs two functions simultaneously: control of the low-temperature water within the boiler and operation of the circulating pump. If the boiler-water temperature falls below the low-limit set point, the contacts controlling circulator operation open automatically. This prevents the circulator from operating until the water temperature has risen to the low-limit setting to ensure that only heated water will be circulated throughout the heating system. Also, in tankless coil installations, the burner will automatically ignite 10°F below the set point, even though the thermostat is not calling for heat. This switching action ensures that the boiler-water temperature is sufficient for indirect water heating with a tankless coil or other external domestic hot water heat exchanger.

An adjustable differential is sometimes incorporated into the aquastat switch-

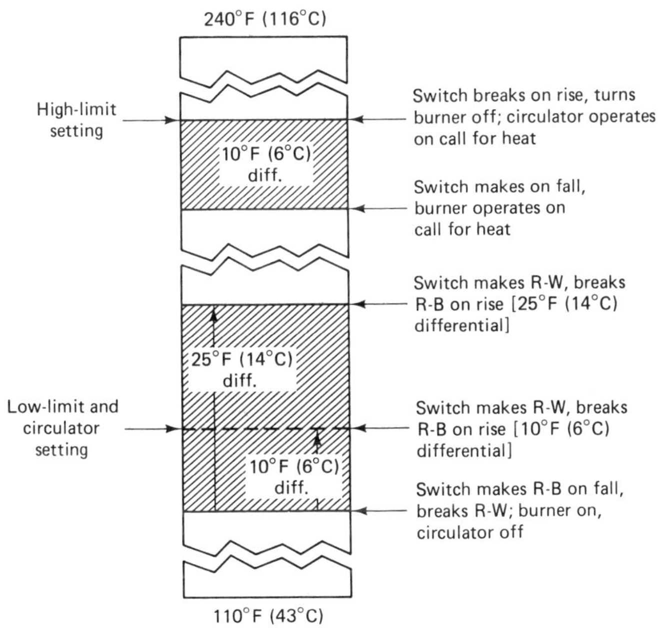

Figure 3.22 Aquastat switching sequence. (Courtesy of Honeywell, Inc.)

ing design. When the boiler water drops 10°F below the low-limit setting, the burner automatically ignites. At this temperature, the low-limit electrical circuit to the circulator is open. When the burner has brought the boiler-water temperature back up to the low-limit setting, the circulator contacts will close. The adjustable differential controls water temperature *above* the low-limit setting and is measured from the fixed turn-on point, which is 10°F *below* the low-limit setting. For example, an adjustable differential setting of 20°F allows a total temperature difference of 20°, which is measured from 10° below the low-limit setting to 10° above the low-limit setting. Thus, if the low-limit setting on the aquastat control is 140°F, the burner will begin operation when the boiler water drops to 130°F. The circulator contacts will close to allow circulator operation when the boiler water reaches 140°F, and the burner will shut off when the water temperature reaches 150°F, assuming that there is no call for space heating by the room thermostat. Also, it should be noted that the 10° differential below the low-limit setting at which point the burner turns on is usually a fixed differential.

Figure 3.23 illustrates the system schematic for a typical reverse-acting aquastat. The operating sequence is as follows. Upon a call for heat by the room thermostat, relays 1K and 2K are energized to turn on the fuel burner. The safety heater is also activated at this time. Assuming that the burner ignites within the limits of the safety timing switch allowance (usually between 20 and 30 seconds), the safety heater circuit is bypassed. This switching action allows burner ignition to continue. Contact R/W in the aquastat controls the operation of the low-limit circulator setting. This operation will continue until the thermostat setting has been satisfied.

Should the boiler-water temperature drop 10°F below the low-limit setting, relay 2K will be energized to begin burner operation. If the thermostat is not calling for heat, the circulator will not operate and the burner will shut off when the water temperature reaches the differential programmed setting above the low-limit cut-in point. Note that Figure 3.23 illustrates a two-zone heating system with separate circulators for each heating zone. Each additional zone within the system requires a separate circulator and relay control. The circulator zone relay draws its power from the main boiler feed line and is wired through terminals ZC and ZR for low-limit control and burner operation when the second zone is calling for heat. Wiring the zone relays independently of the main boiler aquastat eliminates low-limit and burner mechanism control and is therefore avoided unless these control functions are not important to the proper operation of the system.

The use of zone valves for multiple heating circuit control utilizing a single circulating pump is illustrated in Figure 3.24. All thermostats in this arrangement are low voltage (24 V ac) and are wired in parallel with the TT terminals on the main boiler aquastat. A call for heat by any of the zone thermostats opens the motorized zone valve. When the valve is fully open, the pump begins operation. In this type of setup, the low-limit control functions independently to turn on the burner when there is a drop in water temperature just as in a single-zone, single-circulator system.

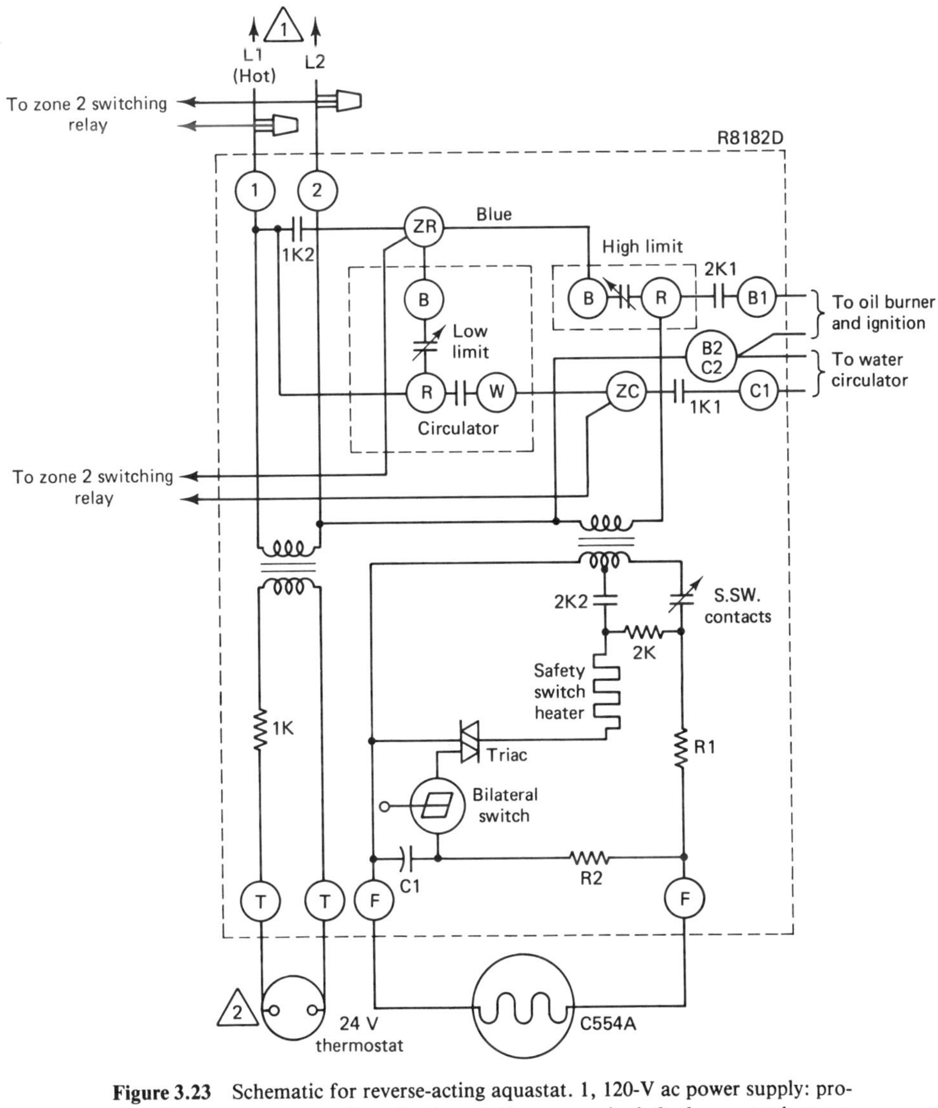

Figure 3.23 Schematic for reverse-acting aquastat. 1, 120-V ac power supply: provide disconnect means and overload protection as required; 2, thermostat heat anticipator setting, 0.2 A for R8182D; 3, thermostat heat anticipator setting, 0.4 A for R845A; 4, control case must be connected to earth ground, use grounding screw provided. (Courtesy of Honeywell, Inc.)

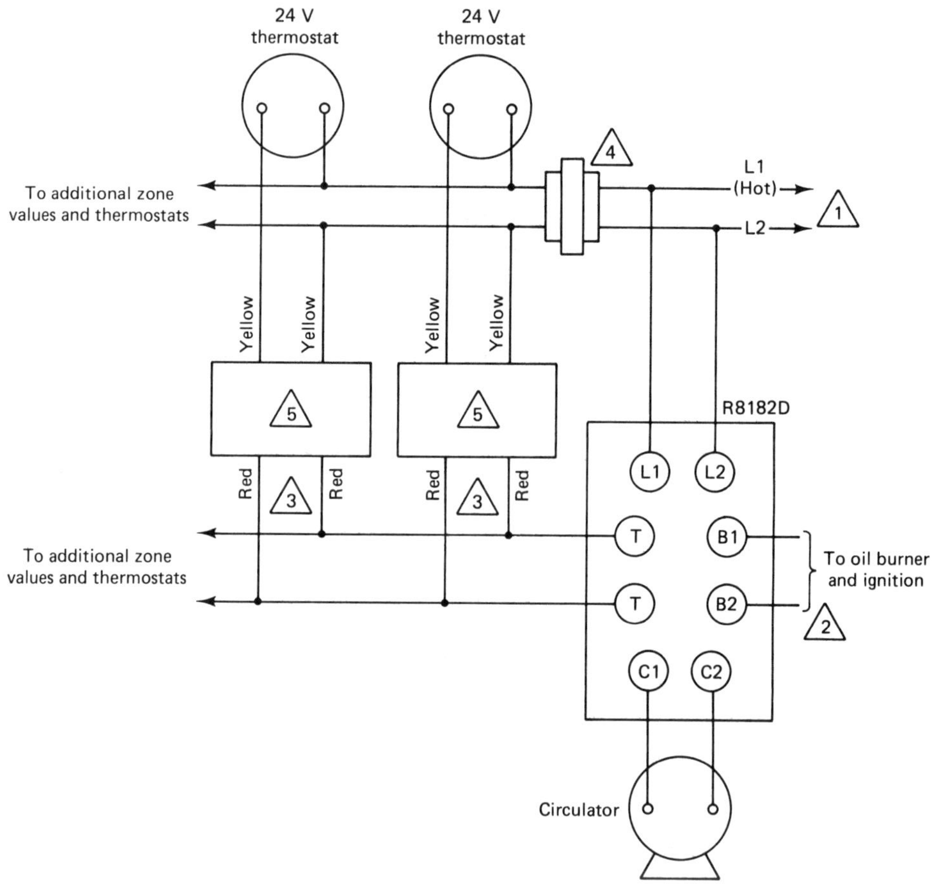

Figure 3.24 Schematic for multiple zone control with zone valves. 1, Power supply: provide disconnect means and overload protection as required; 2, R8182H has white and orange lead wires for oil burner and ignition connections; 3, red lead wires provided only on models with end switch; 4, choose AT72, AT87, or AT88 transformer to match maximum system load; 5, use V8043, V8044, or V8343 zone valves with auxiliary end switch only. (Courtesy of Honeywell, Inc.)

RELAYS

A relay is a device whose contacts are activated by energizing an electromagnet on demand by another electrical device. Relays are generally used in situations where it is necessary to control the activation of one electrical circuit by a device in another circuit whose current or voltage rating is different. In hydronic heating systems, the most common use of a relay is for control of the system circulator pump, which operates on 110-V electrical power by the room thermostat operating on a 24-V circuit. Since it is not possible to intermix the two differing voltages directly to

Figure 3.25 Hydronic switching relay control. These relays are generally used to control additional heating zone circulators. (Courtesy of Honeywell, Inc.)

achieve the desired control of the circulator, a relay is employed for this control function. A typical relay control is illustrated in Figure 3.25.

When the contacts close in the room thermostat, they complete an electrical circuit that provides 24 V to the coil of a small electromagnet. The magnetic action closes the contacts of the relay, which allows 110 V to pass to the circulator motor, activating the unit. As with aquastats, relays are available in a variety of switching actions, such as SPST or SPDT, both with normally open and closed contacts. Care should be exercised that the contacts of the relay to be used in a particular situation are rated for the current present in that particular electrical circuit to be controlled.

FLAME DETECTION SAFETY CONTROLS

All oil- and gas-fired boilers must incorporate into the control circuitry a system for flame detection and emergency system shutdown should a failure occur within the burner ignition system. Although the specific methods used for flame detection and automatic shutdown are different in oil- and gas-fired systems due to the differences in the nature of the fuels, the logic of both systems is identical. The safety sequence is designed to prevent the buildup of potentially explosive amounts of fuel within the ignition chamber of the boiler should the burner fail to achieve stable ignition.

Oil-Fired Flame Detection

Pyrostats. A flame detection control mechanism in use for many years within the heating industry incorporates an extended probe covering a bimetallic coil, commonly referred to as a *pyrostat* or *stack switch* (Figure 3.26). The pyrostat is installed on the flue pipe in close proximity to the boiler. When the burner is activated on a call for heat from the room thermostat, a safety heater switch within the control is energized. This safety switch features a built-in time delay mechanism, usually between 30 and 40 seconds. In normal operation, when ignition is established by the burner, a great deal of heat is generated in the process and begins to heat up the flue pipe. As the bimetallic coil of the pyrostat is heated, it rotates and upon sufficient temperature will lock out the safety switch circuitry. However, should ignition fail, the coil will rotate in the opposite direction, and the safety heater circuitry will be engaged to shut down the burner.

Cad cells. A cad cell primary control system consists of a relay and a light-sensitive cell mounted so as to view the flame. This cell is referred to as a *cad cell* since it is manufactured from cadmium sulfide and is illustrated in Figure 3.27. In darkness, cadmium sulfide has a very high resistance to the passage of electrical current. In the presence of light, its electrical resistance becomes very low and current is allowed to pass. On a call for heat, power is provided to the burner in a timed safety circuit to try for ignition. If, during this try period, resistance in the cad cell lowers, allowing the flow of current, another contact that maintains power to the burner closes after the safety circuit electrically opens. If, however, electrical

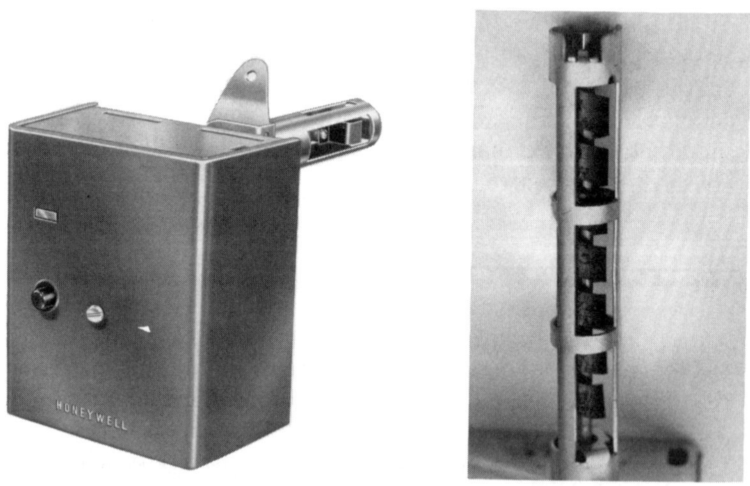

Figure 3.26 Pyrostat relay control used as a safety device to prove burner ignition. The heat-sensing element is made from a bimetallic helical coil that rotates in response to heat from the flue gases. (Courtesy of Honeywell, Inc.)

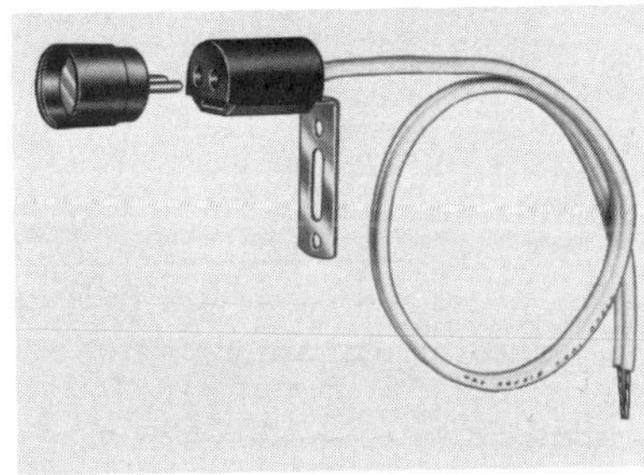

Figure 3.27 Light-sensitive cad cell used to detect flame presence in the safety circuit of oil-fired appliances and cad cell relay. (Courtesy of Honeywell, Inc.)

resistance in the cad cell remains high, indicating that a flame is not present, power to the burner is interrupted, and the control must be reset for another ignition try.

Note from the illustration in Figure 3.28 that the cad cell must be located in line-of-sight with the ignition electrode of the burner in order to "see" the flame during ignition. If the burner is adjusted properly, the resistance of the cad cell will be in the range 300 to 1000 ohms (Ω) when the burner is operating.* To ensure continued reliable burner operation, the resistance of the cell should be checked periodically and should never be above 1600 Ω during the burner "on" cycle. If cell resistance is greater than this figure, the cell surface should be cleaned and its line-of-sight adjusted.

Cad cells also provide an additional safety feature, in that power will not be provided to start the burner for an initial ignition try if the cell senses light prior to a call for heat. Thus, if the cad cell is out of the burner, or if the burner is open, exposing the cell to light prior to ignition, the primary control will not allow delivery of electrical power to the ignition components. The fast response time of this control, usually within 15 to 30 seconds, eliminates the time lag that is often found in bimetallic sensors and has made this system the standard in modern residential oil-fired heating equipment.

*The ohm is the basic unit of electrical measurement and is used to measure electrical resistance within a circuit.

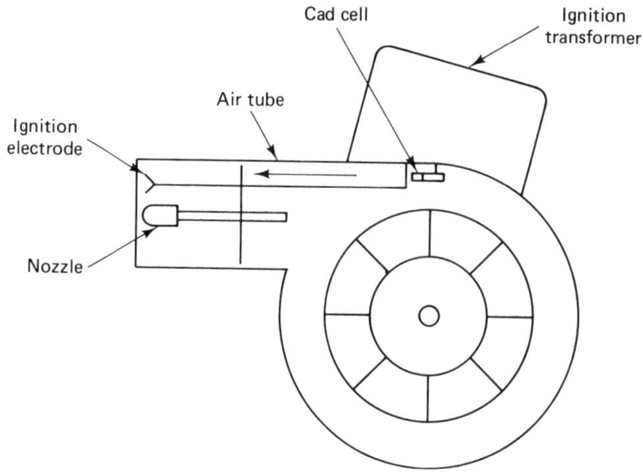

Figure 3.28 Location of cad cell in oil burner housing. The cell must be mounted in direct line-of-sight of the ignition electrodes and nozzle assembly for best results.

Gas-Fired Flame Detection

Standing pilot ignition. Modern gas heating systems are equipped with either a standing pilot or intermittent electronic ignition system. Each of these mechanisms is discussed in greater detail in Chapter 5; however, the basic configuration of a standing pilot gas control system is illustrated in Figure 3.29. The pilot light is lo-

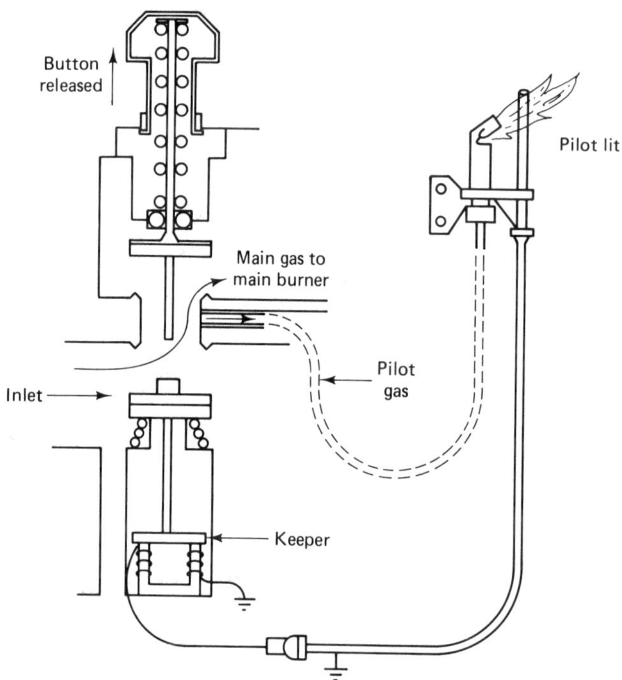

Figure 3.29 Standing pilot ignition system. (Courtesy of Robertshaw Controls Company.)

cated so that its flame is in direct contact with a thermocouple and burns constantly during normal boiler operation. The thermocouple is made of dissimilar metals which when heated produce a small electrical current. This current, in turn, operates a solenoid valve which opens and closes the gas feed line. Should the pilot light go out for any reason, the flow of electricity from the thermocouple ceases, which closes the gas solenoid valve, stopping the flow of all gas to the unit. Gas will not flow until the pilot light has been relit. Pilot light safety controls are found in either 100% shutdown or non-100% shutdown types. In the event of a pilot failure, the 100% shutdown system eliminates the flow of all gas, including the pilot gas. The non-100% system will stop the flow of gas to the main burners but *will not* stop the flow of pilot gas.

Electronic ignition. Intermittent ignition systems differ from standing pilot ignition in that the pilot flame is established only on a call for heat within the home. The pilot is ignited by a solid-state igniter and then lights the main burners. Should the pilot light go out for any reason, automatic shutdown of the gas system is quite fast—usually in 1 second, compared to thermocouple response times of between 30 to 60 seconds. Once the thermostat has been satisfied, both the pilot and main burners shut off until the next heating cycle. This design is illustrated in Figure 3.30.

In addition to the sensing systems on the pilot light, most gas boilers contain other backup safety systems, which will be discussed in greater detail in Chapter 5. Safety shutdown within the gas system can occur for many reasons: excessive water temperature levels in the boiler, pilot light failure, power failure, and main burner failure. Redundancy in safety control design, together with highly refined electrical circuitry engineering, virtually eliminates safety system failure except in the case of tampering or improperly maintained system components.

AIR PURGING

Air Scoops and Float Vents

When a heating system is initially filled with water or a combination of water and antifreeze, air must be purged from within the system in order to begin operation. Although removal of air from within the boiler and piping will allow for initial operation, provision must be made to remove additional air that is dissolved within the water on a continual basis. As this water is heated during normal boiler cycling, air will be released. Failure to vent this air can result in a buildup of air pockets that can cause a blockage in the piping, preventing water circulation, and can cause noisy operation of the heating system. The specific procedures for initial air purging of the heating system are discussed in greater detail in Chapter 4. The method of ongoing removal of trapped air during normal operation of the heating system involves the use of specially designed air scoops and float vents. An air scoop and float vent assembly is illustrated in Figure 3.31.

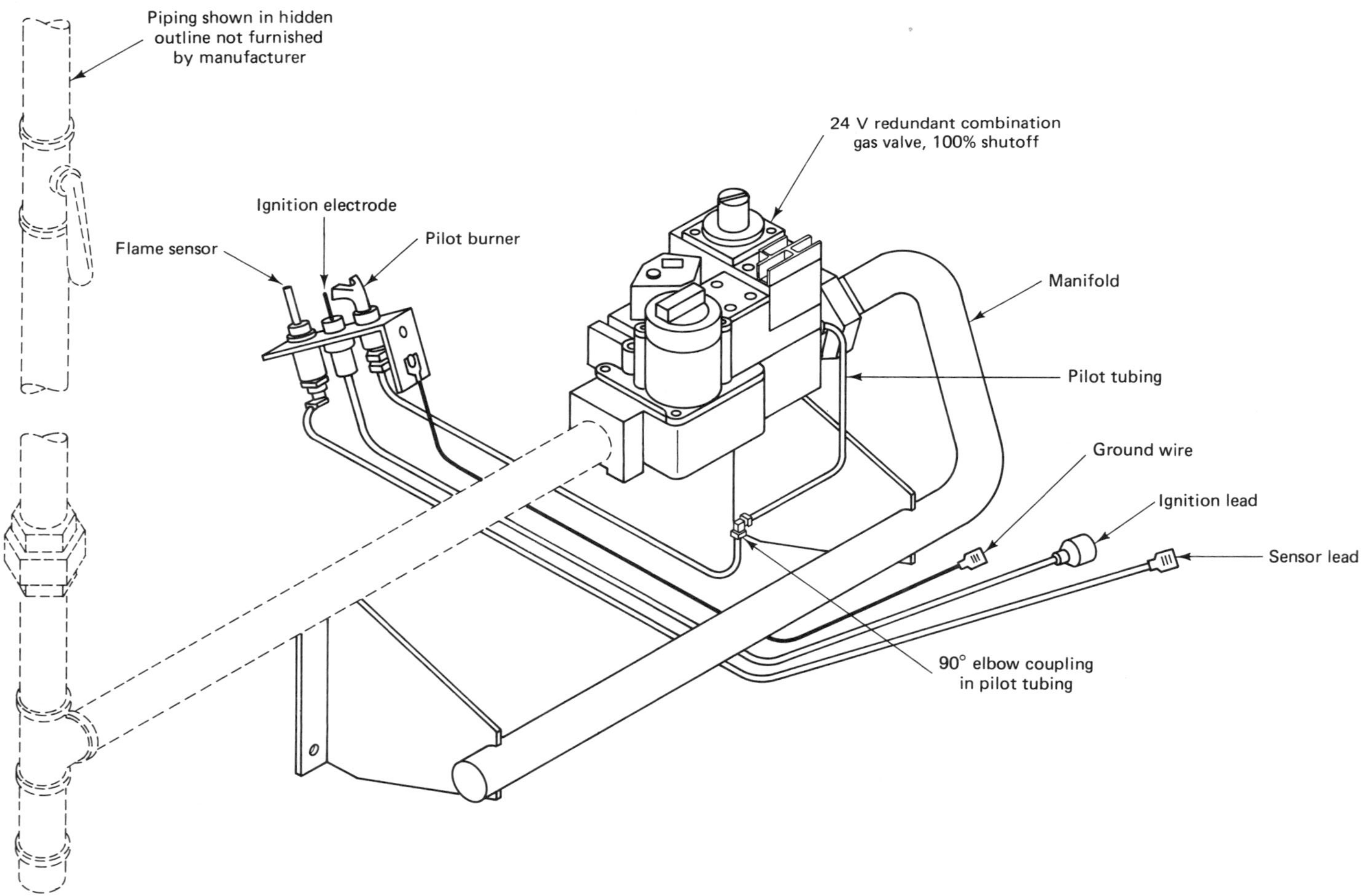

Figure 3.30 Pictorial of pilot and gas piping of 24-V electronic ignition gas control. (Courtesy of Burnham Corp.)

Figure 3.31 Air scoop and float vent assembly. This unit also features an automatic fill valve located between the air scoop and the expansion tank. (Courtesy of AMTROL Inc.)

Observation of the construction details of the air scoop in Figure 3.31 illustrates the placement of baffles that are engineered to direct trapped air bubbles to the top of the separation chamber within the air scoop (Figure 3.32). When sufficient air has collected at the top of the chamber, the float level drops, opening a valve at the top of the vent. Positive pressure within the air scoop forces the trapped air from the vent. As the air is eliminated, water replaces the air in the scoop chamber, raising the float, thereby closing the escape valve. Air separators should be installed horizontally on the main boiler feed lines as close as possible to the boiler.

Expansion Tanks

Most heating systems are "closed" in the sense that pressure built up within the heating system cannot back up and out of the system as the water is heated. Backflow preventers and water check valves are standard equipment on all automatic pressure reduction feed lines and are also required by almost all plumbing codes. During normal operation of the heating system, the water expands as it is heated and increases the system operating pressure. Provision must be made within the design of the system to accommodate increases in operating pressures to eliminate frequent discharge of the pressure relief value, which would cause fresh water to enter the system continually, which accelerates corrosion and can cause premature

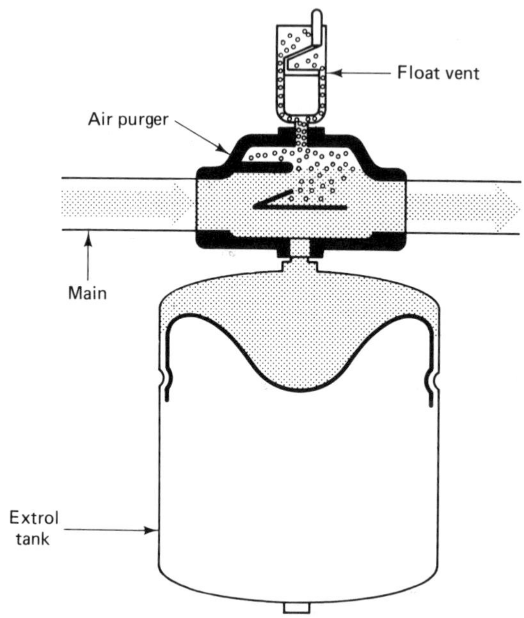

Figure 3.32 Continuous purging and venting of air from within the heating system. (Courtesy of AMTROL Inc.)

boiler failure. The device that handles this fluctuating pressure is called an *expansion tank*. There are two types of expansion tanks currently used for this purpose: closed air cushion tanks and pressurized diaphragm air tanks.

Air cushion tanks. A conventional air cushion tank is illustrated in Figure 3.33. When the heating system is initially filled with water, a pocket of air is trapped within the air cushion tank. Expansion of the heated water within the system compresses this air cushion. A rough rule of thumb for sizing expansion tanks is to allow 1 gal of tank capacity for each 5000 Btu of boiler heating capacity. It should be noted, however, that the actual tank capacity can vary greatly with total water content and operating temperatures designed for a particular heating system.

Pressurized diaphragm tanks. A prepressurized diaphragm expansion tank, illustrating the various stages of compression during normal heating system cycling, is shown in Figure 3.34. These tanks are generally charged to 12 psi at the factory. A flexible diaphragm separates the air within the tank from the water within the heating system. Assuming that both the expansion tank and the cold water fill pressure within the heating system are 12 psi, no water enters the tank (Figure 3.34a). When the heating system has risen to normal operating temperature, the increased volume of water due to expansion enters the expansion tank, flexing the diaphragm against the air cushion (Figure 3.34b and c). Since diaphragm tanks vary in design, manufacturers' instructions should be followed when sizing these units. Tables 3.2

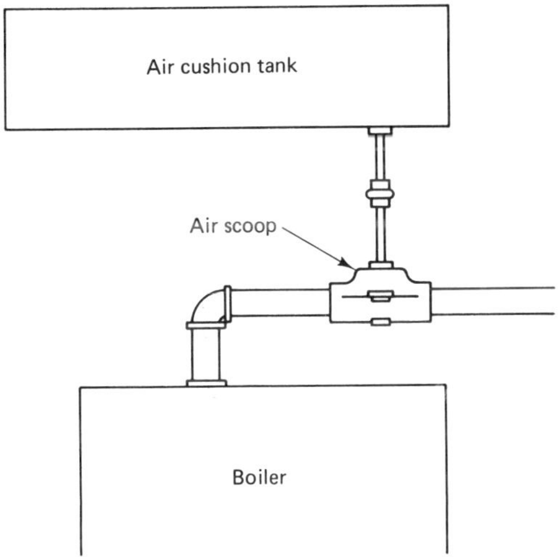

Figure 3.33 Location of conventional air cushion tank in hydronic heating system.

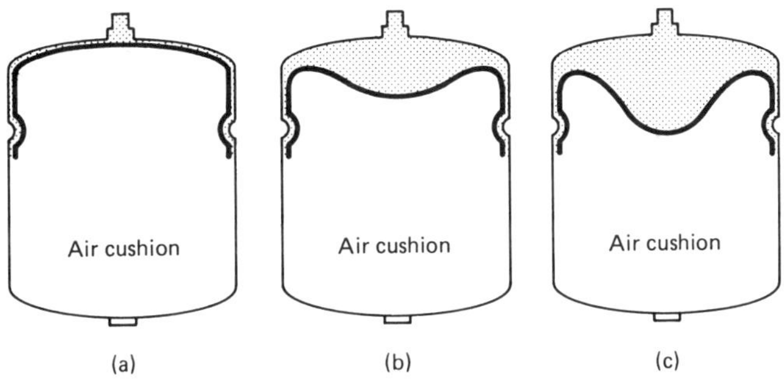

Figure 3.34 Operation of pressurized diaphragm expansion tank. (a) When the system is first filled with cold water, the charge pressure keeps the diaphragm flush against the tank's inner lining. (b) As the system comes up to temperature, the expanded water is received by the Extrol tank. (c) As the water temperature reaches its maximum, the Extrol diaphragm flexes against the air cushion to allow for the increased water expansion. (Courtesy of AMTROL Inc.)

and 3.3 illustrate typical sizing procedures and water capacities for pressurized diaphragm expansion tanks.

TRIDICATORS

Tridicators are sometimes referred to as *theraltimeters*. These gauges register three different boiler operating conditions: water pressure within the boiler, altitude of water as a function of boiler-water pressure, and boiler-water temperature. A typical

TABLE 3.2 SIZING TABLE FOR PRESSURIZED DIAPHRAGM EXPANSION TANKS BASED ON BOILER OUTPUT AND CLASSIFICATION OF HEAT DISTRIBUTION UNIT

Boiler net output (thousands of Btu/hr)	Type of radiation: Finned tube baseboard or radiant panel	Convectors or unit heaters	Cast iron radiators	Cast iron baseboard
25	15	15	15	15
50	15	15	30	30
75	30	30	30	60
100	30	30	60	60
125	30	60	60	90
150	30	60	90	90
175	60	60	SX-30	SX-30
200	60	60	SX-30	SX-30
250	60	90	SX-30	SX-40
300	90	SX-30	SX-30	SX-40
350	SX-30	SX-30	SX-40	SX-60
400	SX-30	SX-40	SX-40	SX-60

Source: AMTROL Inc.

TABLE 3.3 PRESSURIZED DIAPHRAGM EXPANSION TANK CAPACITIES AT VARIOUS SYSTEM OPERATING TEMPERATURES

Average system temp. (°F)	System content for Extrol model (gal): 15	30	60	90	SX-30
100	138	305	500	734	1102
110	101	232	380	557	835
120	80	183	300	440	660
130	65	149	244	358	537
140	54	125	204	298	447
150	47	106	174	256	384
160	40	92	150	220	330
170	34	78	128	187	280
180	30	68	111	163	245
190	26	60	98	144	216
200	23	52	86	126	189
210	21	48	78	114	171
220	19	43	71	103	155
230	17	39	64	94	141
240	16	35	58	86	129

Source: AMTROL Inc.

Note: Maximum working pressure, 75 psi; operating temperature, 40–240°F.

Figure 3.35 Boiler tridicator. (Courtesy of Triple S Specialties Corp.)

boiler tridicator is illustrated in Figure 3.35. The tridicator should be installed directly into a tapping provided on the boiler so that it is in direct contact with internal boiler water and exposed to the interior of the appliance without any obstructions.

PRESSURE AND TEMPERATURE RELIEF VALVES

Perhaps the most important safety device on the boiler and associated piping, pressure relief valves are designed to open and allow water and/or steam to escape should internal pressure reach the set point of the safety valve. Two types of valves are most commonly used for this purpose: pressure relief valves, used in both hot water and steam heating systems; and pressure/temperature relief valves, which respond to both temperature and pressure changes within the system.

The relief valves on residential hot water boilers are generally of the pressure relief type, set to discharge at 30 psi (Figure 3.36). The Btu/hr rating on the valve

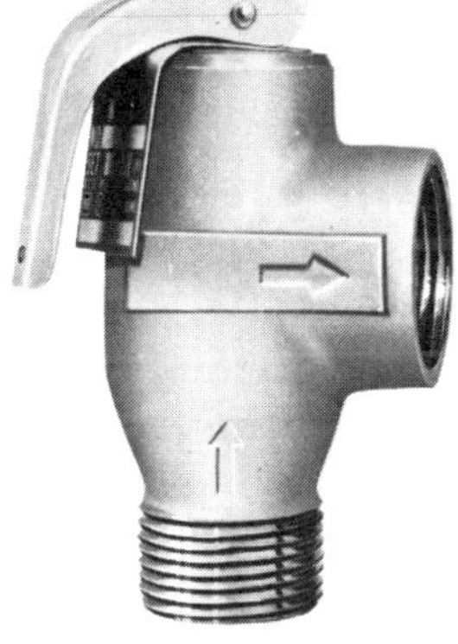

Figure 3.36 ASME-listed pressure relief valve used on residential hot water boilers. (Courtesy of Watts Regulator Co.)

should be matched to, or greater than, the gross Btu output of the boiler on which it is installed. This ensures that water or steam will be vented more quickly than the boiler can regenerate it in an emergency situation. These valves automatically reset when internal boiler pressure drops below the set point. Many of these valves also come equipped with a manual release lever. This feature is useful when either filling or draining the boiler by allowing the system to be vented through the relief valve when the manual release is engaged.

Pressure relief valves used on steam boilers are similar in most respects to their hot water counterparts with regard to operating characteristics. However, since most residential steam heating systems operate at pressures below those of most hydronic systems (3 to 8 lb is common), these valves are set to release at 15 psi rather than 30 psi residential hydronic pressure relief valves.

Combination pressure/temperature relief valves (PTRVs) are used in domestic hot water systems to maintain safe operating conditions. Although specific settings for these valves vary, most are set to discharge between 125 and 175 psi, and at temperatures greater than 210°F. Some valves are configured with adjustable settings for maximizing operation under a wide variety of system conditions. Pressure/temperature relief valves are usually equipped with a probe that is immersed in the tankless coil or domestic hot water heater for accurate temperature sensing along with a pressure sensor device. A typical pressure/temperature relief valve is illustrated in Figure 3.37.

When installing pressure/temperature relief valves on domestic hot water heaters, they should be inserted in a tapping provided by the water heater manufacturer that opens directly into the tank interior. These openings are clearly marked and should be used only for the appropriate relief valve specified for the water heater. When installing the relief valve on domestic hot water tankless coils, the valve should be installed in close proximity to the coil, with the sensing probe of the valve in direct contact with the heated domestic hot water.

All pressure relief and pressure/temperature relief valves should incorporate a discharge tube plumbed from the outlet of the valve, which directs any discharged water or steam to within 6 in. of the floor surface or to an appropriate floor drain. Under no circumstances should the diameter of the discharging piping be less than the outlet diameter of the valve, nor should any type of shutoff valve be placed in the discharge line.

AUTOMATIC FEEDWATER MAKEUP SYSTEMS

Provision should be made on all hot water and steam heating systems to replace water within the system that is lost through normal system operation. Feedwater makeup on steam systems is accomplished either manually with a separate valve that introduces feedwater into the bottom of the steam boiler, or automatically through a float-type device that maintains boiler water at a preset level. On both

Figure 3.37 Pressure/temperature relief valve used in domestic hot water applications. (Courtesy of Watts Regulator Co.)

manual and automatic systems, water level is checked on the gauge glass assembly. (Please consult Chapter 6 for more information on feedwater makeup in steam heating systems.) Hot water boilers usually employ automatic feedwater makeup to replace lost water within the system. Plumbing codes usually specify back-flow preventers on feedwater lines to prevent boiler water from flowing backward into the residential water system in case of water pump failure or power outages, which could result in low pressure within the domestic water supply system, leading to a reverse flow of water from the boiler into the supply system. Two different automatic feedwater systems are illustrated in Figures 3.38 and 3.39.

The feedwater line is usually located on the discharge side of the circulating pump or in a suitable boiler tapping. This location prevents water from being drawn into the heating system when the circulator is energized and a partial vacuum is

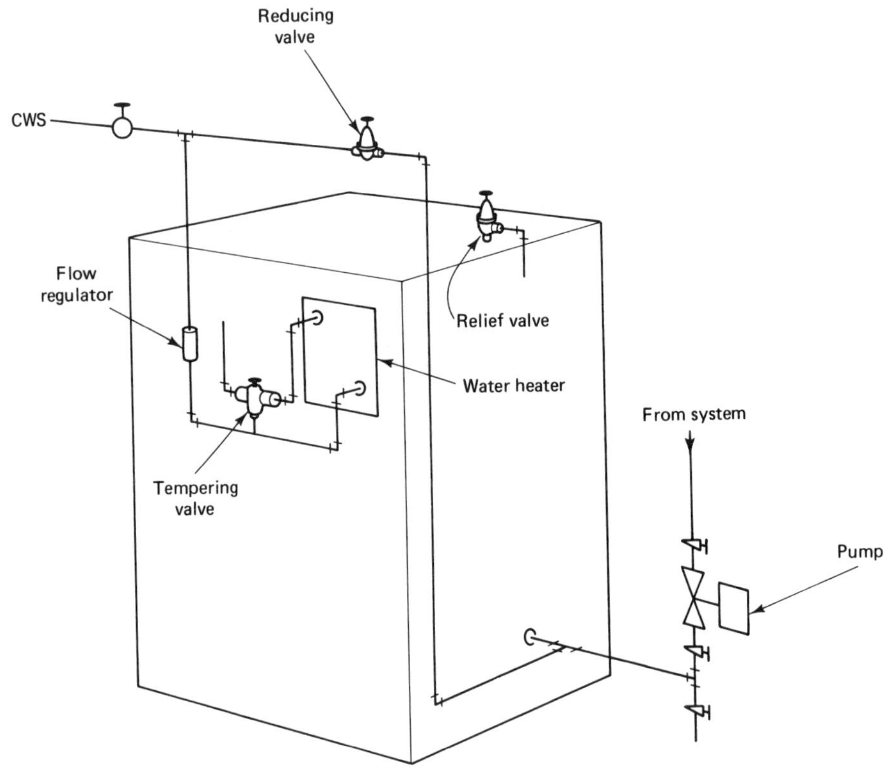

Figure 3.38 Automatic feedwater makeup and plumbing configuration of typical hot water boiler. (Courtesy of Taco, Inc.)

momentarily created by the circulator on the suction side of the pump. In Figure 3.39, the feedwater line is located at the base of the air scoop–expansion tank combination. This location is the point of no pressure change within the heating system and is relatively insensitive to pressure changes generated by the action of the circulating pump. Boiler feedwater valves reduce the pressure of the water to approximately 12 psi and are usually incorporated into an assembly that contains the pressure reduction valve, back-flow preventer, and pressure relief valve. This component assembly is illustrated in Figure 3.40.

Although feedwater systems require little maintenance over their operating life span, they should occasionally be inspected for leaking joints. Some units contain removable filter screens which should be inspected and cleaned on an annual basis. Also, the boiler tridicator should be checked periodically to ensure that boiler-water cold-start pressure is at approximately 12 psi. If the cold-start pressure varies significantly from this setting, consult manufacturers' instructions for pressure valve readjustment procedures.

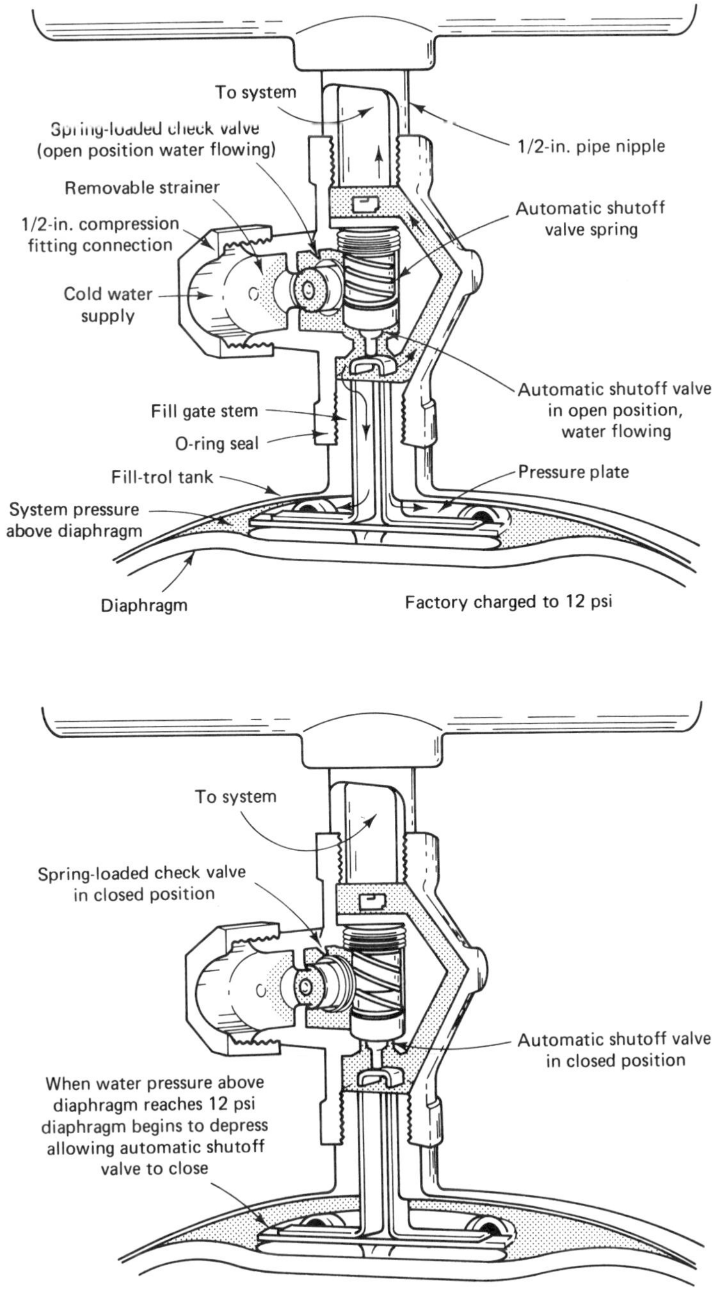

Figure 3.39 Automatic feedwater built into air scoop/expansion tank assembly. (Courtesy of AMTROL Inc.)

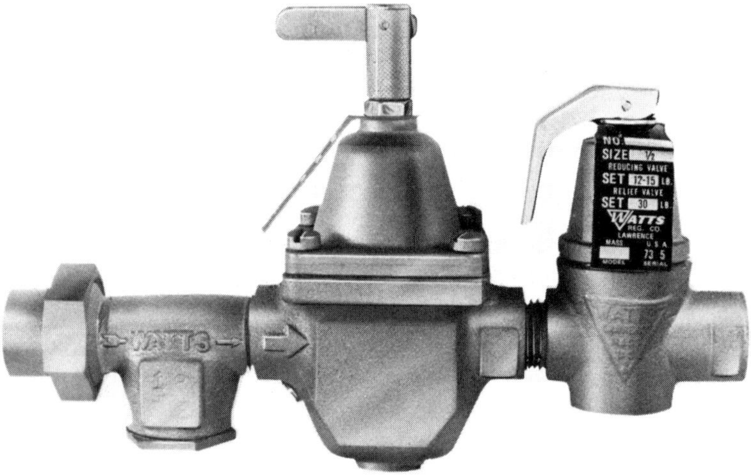

Figure 3.40 Combination feedwater pressure-reducing valve, back-flow preventer, and pressure relief valve component assembly. (Courtesy of Watts Regulator Co.)

FLOW-CHECK VALVES

The use of flow-check valves on forced-circulation systems prevents gravity circulation of system water when the circulator is not running and temperature must be maintained within the boiler. The most common instance of this would be in a heating system with a tankless coil which would always maintain the boiler at water temperatures sufficient for domestic hot water production.

A typical hydronic flow-check valve is illustrated in Figure 3.41. These valves generally employ either a preset spring tension or weighted valve seating system that opens up the water flow passage only upon discharge pressure from the circulating

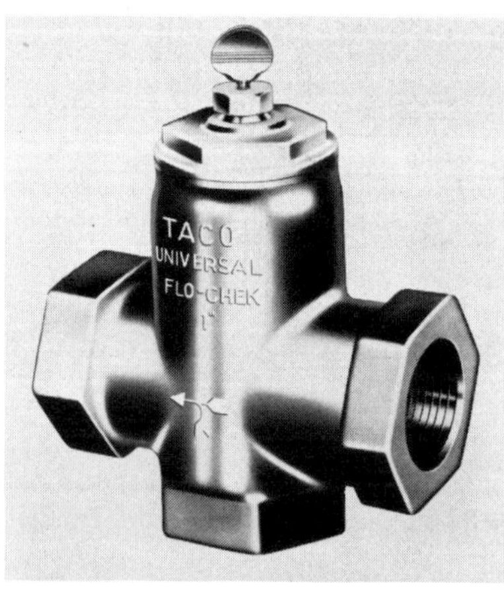

Figure 3.41 Flow-check valve. Incorporation of the flow-check valve prevents thermosiphoning of hot water through the baseboard radiation when system is in the off-cycle mode. (Courtesy of Taco, Inc.)

pump. These valves are recommended for zoned heating systems which employ circulators rather than zone valves for circuit control. Also, most flow checks incorporate a manual adjustment that permits the valve to be opened for gravity circulation. This bypass feature is used during power failures or other situations that result in loss of power circulation in the heating system.

FLOW INDICATORS

There are several devices available that give a visual indication of fluid circulation within the heating loop. One commonly used device, called a flow meter, contains a weighted metallic indicator ball that is placed in a clear plastic calibrated tube and installed within the piping circuit of the heating loop. Flow meters are available in a variety of sizes and capacities, depending on specific system applications, and are illustrated in Figure 3.42. The position of the metallic indicator within the calibrated flow meter tube is proportional to the fluid flow rate through the meter and reads in gallons or metric equivalents per minute of flow throughout the piping array. Another device that can be used to indicate whether fluid is circulating within the heating system is illustrated in Figure 3.43.

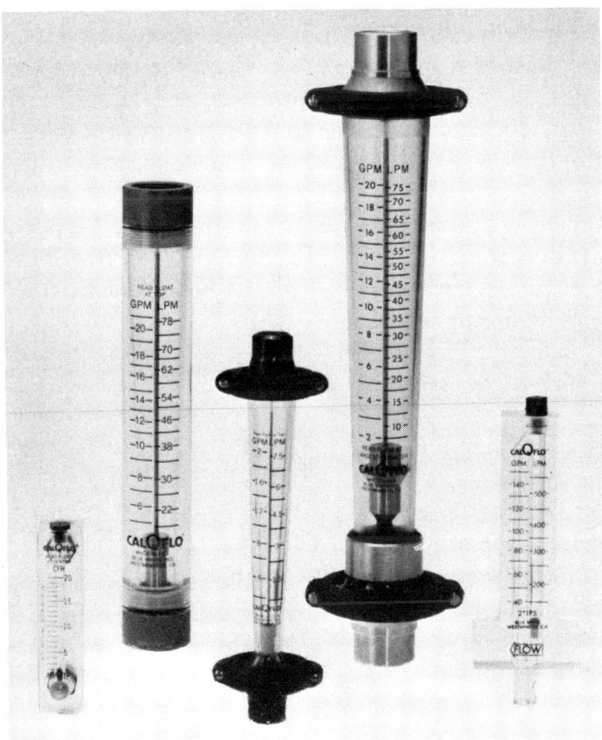

Figure 3.42 Flow meter. (Courtesy of Blue White Industries.)

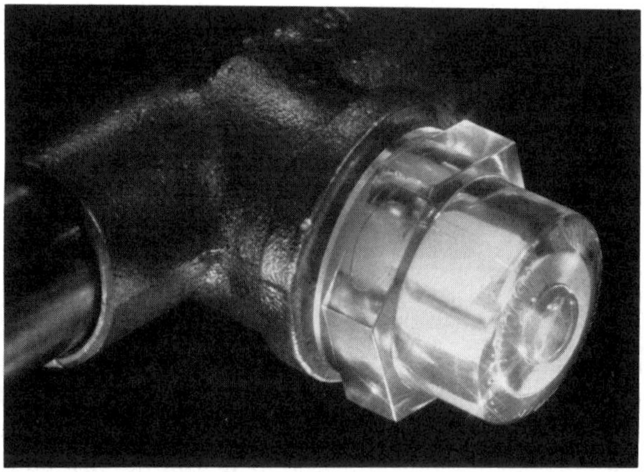

Figure 3.43 Spinning indicator flow meter. (Courtesy of Heliotrope General.)

SUMMER-WINTER DOMESTIC HOT WATER HEATING

Tankless Coils

Many homes with central hot water or steam boilers are equipped with a tankless coil to produce domestic hot water. Tankless coils are long coils of copper tubing that are wound into various configurations depending on the design of the boiler. Some coil configurations are illustrated in Figure 3.44. When the tankless coil is immersed in the hot water within the boiler, heat is transferred from the surrounding boiler water through the coil of copper tubing to the cold water within the coil. Hot water from the tankless coil is then piped throughout the residential service lines to individual appliances.

The location of the tankless coil assembly within a typical cast iron residential hot water boiler is illustrated in Figure 3.45. Note that the coil is inserted into the upper section of the boiler with convenient tappings on the boiler front panel for the hot and cold water service connections.

Conventional Domestic Hot Water Heating Systems

To understand how tankless coils function within the residential hot water system, an understanding of how the overall water system within a home functions is helpful. A typical domestic water system is illustrated in Figure 3.46. Cold water enters the building either from the local water mains or from a separate drilled well. Entering water pressures will vary between 20 and 60 psi in most locations. The entering cold water is channeled in two directions: One line enters the hot water heater or tankless coil, the other travels directly to the cold water outlets located on the sinks, showers, and commodes within the home. The system should be designed to enable all water lines to be completely drained so that the system can be totally shut down if necessary.

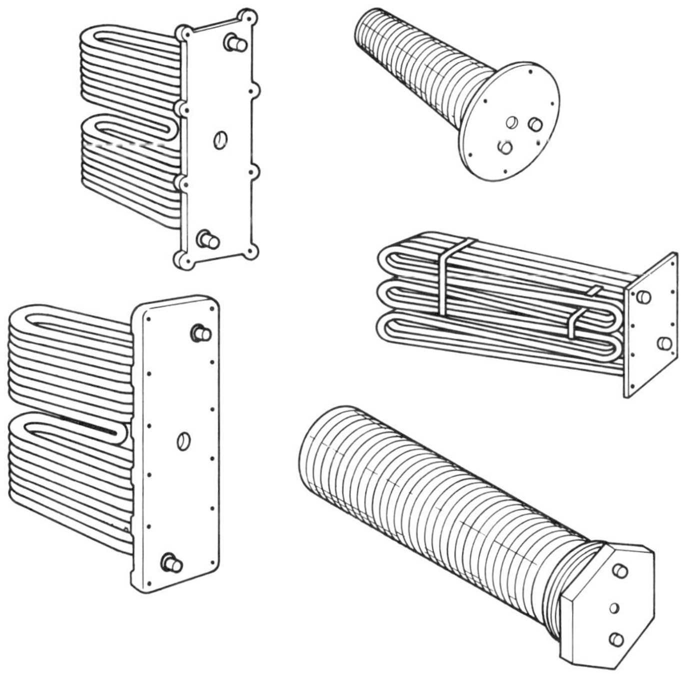

Figure 3.44 Tankless coil configurations. (Courtesy of AMTROL Inc.)

Fuel Consumption of Tankless Coil Systems

It is a commonly held belief that residential domestic hot water heating systems in which tankless coils are employed provide "free" hot water during the heating season as long as the heating boiler is providing space heating for the residence. This belief is a misconception. As is the case with so many modern appliances, the cost of operating the appliances compared to the benefits accrued are often subtle and require close examination.

One obvious advantage of the tankless coil domestic hot water heating system is that a separate hot water heater is not required. The major disadvantage of this type of system is that it is necessary to operate the heating boiler even during the summer months when space heating is not required in order to furnish the home with domestic hot water. For example, a family of four people consumes an average of 22 gal of fuel oil per month during the winter heating season to produce domestic hot water for the household (the assumed efficiency of the tankless coil in this instance is 55%). However, during the nonheating season, the same family of four requires over 80 gal of fuel oil for domestic hot water production. It can be seen that during the nonheating season, tankless coils turn out to be major consumers of fuel. Since domestic hot water is heated indirectly by the water in the boiler rather than directly as in an oil- or gas-fired water heater, the efficiency of the coil during

Figure 3.45 Tankless coil assembly within residential packaged hydronic boiler. (Courtesy of Slant/Fin Corp.)

the summer months drops to approximately 15 to 20%, as the boiler remains on standby solely for the production of domestic hot water.

This loss in system efficiency has assumed greater importance as the cost of fossil fuels has risen so dramatically. Thus, in certain parts of the United States, the use of tankless coils is discouraged in an effort to make more efficient use of available fossil fuels. Table 3.4 lists the approximate fuel consumption figures that have been computed for various fuels to produce domestic hot water using a variety of water heating methods. It can be seen from the table that regardless of the type of fuel used to produce it, domestic hot water represents a considerable expenditure of energy under most circumstances.

Temperature Control and Piping Configurations

The temperature of the hot water exiting the tankless coil is dependent on the internal boiler-water temperature. Assuming that the coil is fabricated from $\frac{3}{4}$-in copper tubing and is between 15 and 20 ft long, domestic hot water between 130 and

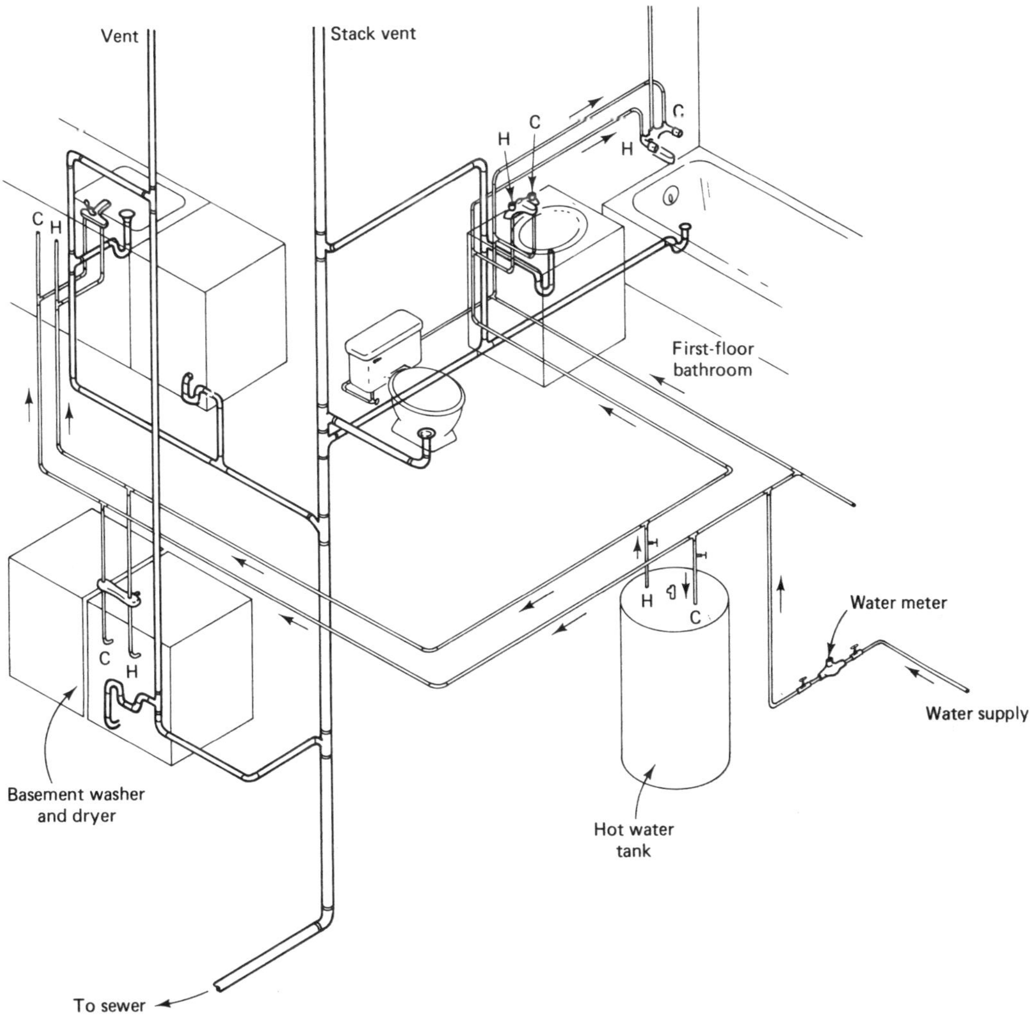

Figure 3.46 Typical residential water supply system. (Courtesy of Martin L. Greenwald and Thomas K. McHugh, *Practical Solar Energy Technology*, © 1985, p. 42. Reprinted by permission of Prentice-Hall, Inc., Englewood Cliffs, NJ.)

150°F will be produced when the boiler-water temperature is between 160 and 200°F. To ensure that hot water production is both uniform in temperature and is prevented from reaching excessively high temperatures, a hot water tempering valve should be installed in close proximity to the boiler. A temperature-sensitive compensating mechanism within the tempering valve automatically adjusts the amount of hot and cold water to produce a uniform flow of domestic hot water, depending on the setting of the valve. A typical hot water tempering valve, together with the instal-

TABLE 3.4 FUEL USAGE UTILIZING A VARIETY OF TECHNIQUES FOR PRODUCTION OF DOMESTIC HOT WATER

Number of people	Oil (gal/yr)		Gas: tank (therm/yr)	Electric: tank (kwh/yr)
	Tankless coil	Tank		
1	90	62.0	87.0	1,562
2	179	125	174	3,124
3	268	187	262	4,686
4	358	249	349	6,248
5	447	311	436	7,810
6	537	374	523	9,372
7	627	436	611	10,934
8	717	499	698	12,496

Source: Heliotherm, Inc.

lation configuration of the unit, is illustrated in Figures 3.47 and 3.48. (A tempering valve cannot, however, compensate for rapid pressure fluctuations. Therefore, where the system is subject to water pressure fluctuations, a pressure-equalizing valve should be installed.)

Note the installation of a check valve in the cold water feed line that enters the tankless coil in Figure 3.48. This check valve functions to prevent contamination of the home water supply system by the boiler water should the tankless coil rupture and a loss of cold water pressure occur. Also, the check valve will help to prevent thermosiphoning of hot water through the cold water feed lines as the coil brings the water up to service temperature. A spring-loaded check valve for this purpose is illustrated in Figure 3.49.

Sizing of tankless coils depends on two factors: water requirements within the

Figure 3.47 Hot water tempering valve. Incorporation of this valve into the hot water production system helps to prevent scalding and promotes even temperatures of hot tap water. (Courtesy of Taco, Inc.)

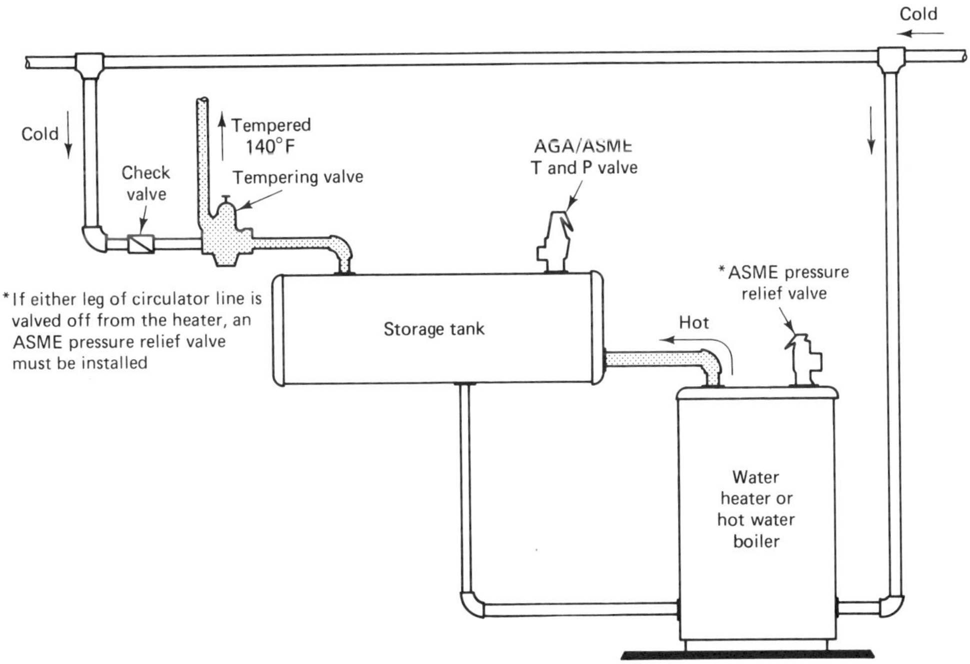

Figure 3.48 Plumbing configuration of hot water tempering valve installed in boiler with external domestic hot water heater. (Courtesy of Watts Regulator Co.)

home and the ability of the boiler to produce domestic hot water. To determine the tankless coil capacity in gallons per minute (gal/min) required for a specific household, consult the capacities listed in Table 3.5. Once the capacity of the tankless coil has been determined, the boiler manufacturers' literature should be consulted to determine which model can supply the required amount of domestic hot water.

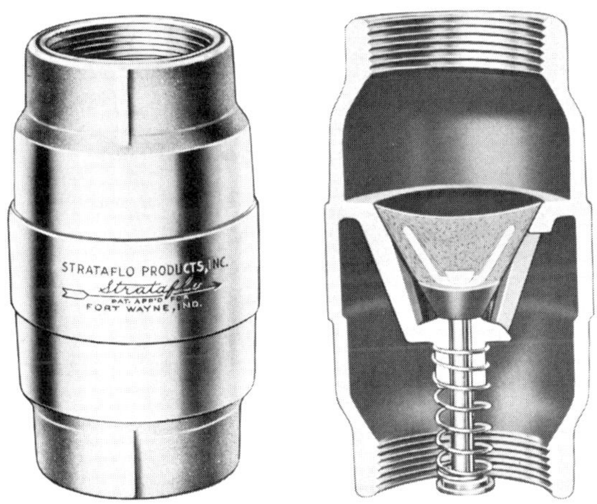

Figure 3.49 Spring-loaded check valve. (Courtesy of Strataflo Products, Inc.)

TABLE 3.5 RECOMMENDED CAPACITIES FOR TANKLESS WATER HEATERS

Number of bedrooms	Recommended heater size (gal/min)	
	Kitchen, laundry, 1 bath	Kitchen, 2 baths
1	$3\text{–}3\frac{1}{2}$	$3\frac{1}{2}\text{–}4$
2	$3\text{–}3\frac{1}{2}$	$3\frac{1}{2}\text{–}4$
3	$3\text{–}3\frac{1}{2}$	$3\frac{1}{2}\text{–}4$

Source: The Hydronics Institute.

Table 3.6 is typical of hot water production capacity available from one series of heating boilers.

It should be noted that the gal/min rating of the tankless coil in a specific boiler is *not* based on a continuous draw of domestic hot water from the unit but rather is rated on an intermittent draw of water. In other words, the hourly output of a coil rated at 3 gal/min by the manufacturer is *not* 3 gal/min × 60 min. On an intermittent draw, a 3-gal/min coil would have an hourly rating of approximately 42 gal. Also, the hourly capacity figures for tankless coil water production are based on a boiler-water temperature of 200°F.

Most boiler manufacturers design their units so that the coils can easily be replaced should this become necessary. Coil replacement usually involves little more than removing the hold-down bolts, taking out the old coil, and inserting a new gasket and coil assembly (Figure 3.50). The hold-down bolts of the coil should be torqued to manufacturers' specifications and checked periodically to ensure a watertight and pressure-tight seal.

TABLE 3.6 TANKLESS COIL HEATER CAPACITY AS A FUNCTION OF BOILER SIZE

Boiler number	Number	NPT connection	Rating[a]			
			Steam		Water	
			gal/min	Δp	gal/min	Δp
V-33	222	$\frac{1}{2}$	$4\frac{1}{4}$	36.0	$4\frac{3}{4}$	42.5
V-34	226	$\frac{1}{2}$	$5\frac{1}{4}$	17.5	$5\frac{3}{4}$	20.5
V-35	232	$\frac{1}{2}$	$6\frac{1}{4}$	22.5	7	27.5
V-36	445	$\frac{3}{4}$	$7\frac{1}{4}$	19.5	$8\frac{1}{4}$	23.5
V-37	445	$\frac{3}{4}$	$7\frac{1}{4}$	19.5	$8\frac{1}{4}$	23.5
V-38	445	$\frac{3}{4}$	$7\frac{1}{4}$	19.5	$8\frac{1}{4}$	23.5

Source: Burnham Corp.

[a]Gal/min is based on 200°F boiler-water temperature and 40–140°F rise; Δp is the pressure drop in psi.

Figure 3.50 Insertion of tankless coil assembly into boiler housing. (Courtesy of Slant/Fin Corp.)

One system to help overcome relatively low coil efficiency during the nonheating season is known as an *aqua booster.*

External Tankless Heaters

Aqua booster domestic hot water systems. An aqua booster is an adaptation of the tankless coil and is illustrated in Figure 3.51. In this arrangement, a separate storage tank with an internal heat exchanger is utilized. Such a system offers greater flexibility than the standard tankless coil in that it contains approximately 40 gal of water storage. Tankless coil capacity in common hot water boilers is limited to water within the coil, generally between 1 and 3 gal. Also, the buildup of lime and other mineral deposits associated with coil operation is eliminated, since the only water introduced into the heat exchanger in the storage tank is relatively oxygen-free water from the home heating system.

Side-arm heaters. Another type of external domestic hot water heat exchanger for hot water and steam boilers is the side-arm heater. Usually, these heaters are shell-and-tube heat exchangers. Water from the boiler circulates through the entire jacket of the heat exchanger, while the domestic hot water is run through a

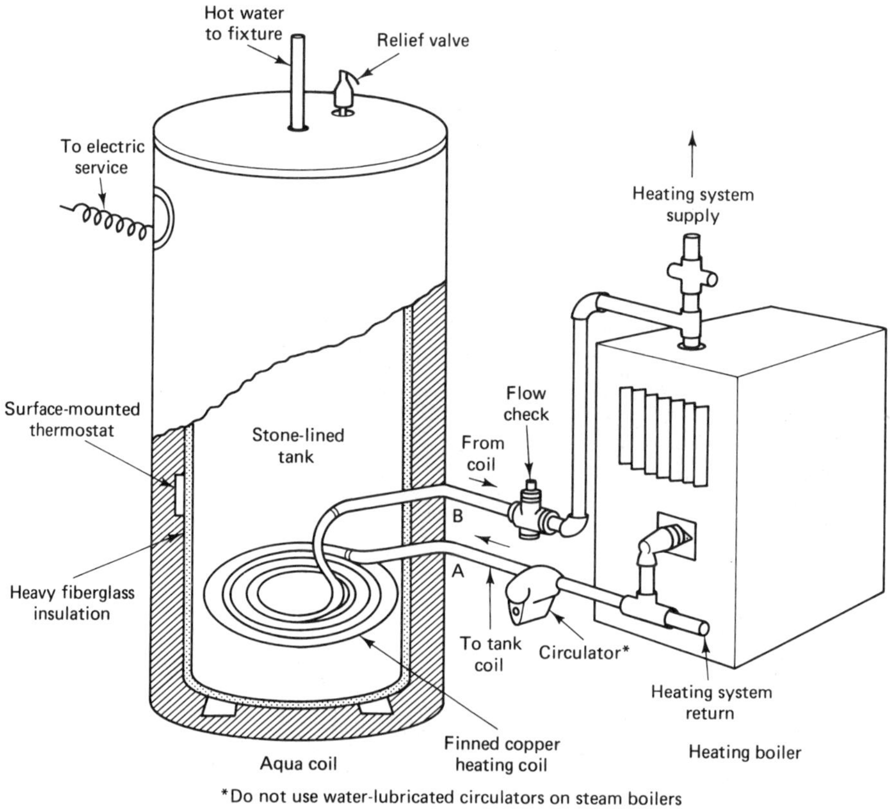

Figure 3.51 Aqua booster configuration installed in conjunction with a central heating boiler. Efficiency of domestic hot water production is greater than that of tankless coil alone. (Courtesy of Ford Products, Inc.)

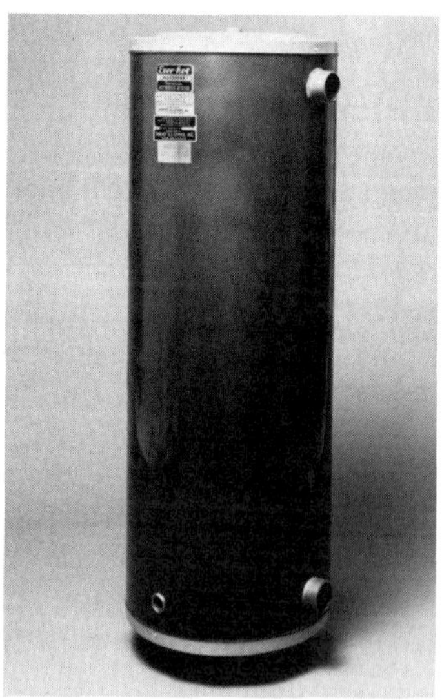

Figure 3.52 Side-arm heat exchanger. (Courtesy of Everhot All Copper, Inc.)

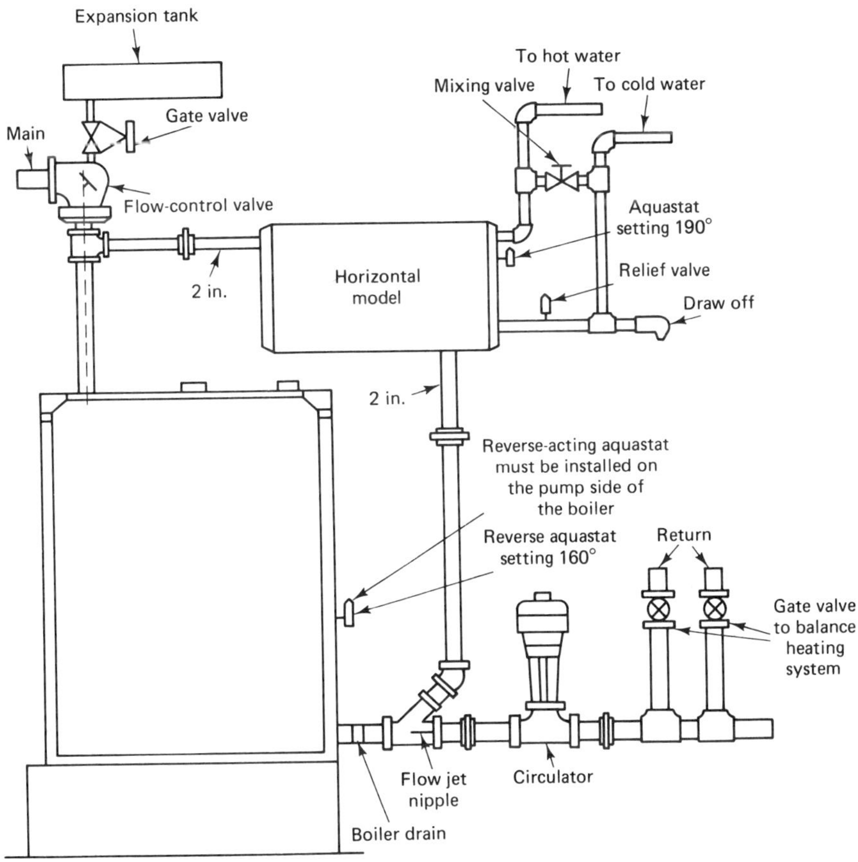

Figure 3.53 Details of side-arm installation. (Courtesy of Everhot All Copper, Inc.)

raised fin copper coil heat exchanger within the unit, picking up heat from the surrounding boiler water. A side-arm heat exchanger is illustrated in Figure 3.52. The name "side arm" is derived from the appearance that these units take on when they have been installed as part of the boiler trim. Figure 3.53 illustrates the piping details of a typical side-arm installation. Note that the piping arrangement should be designed to eliminate any possible short-circuiting of the boiler water as it travels from the boiler through the heat exchanger and back to the boiler return line.

4

Oil-Fired Heating Systems

CHARACTERISTICS OF FUEL OIL

Crude oil is rarely in usable form as it is pumped from the ground, and it must be refined. Crude oil is composed of approximately 84% carbon and 14% hydrogen, with the balance consisting of sulfur, oxygen, and other trace elements. The refining process, known as *fractional distillation,* separates various oil products into groups whose boiling points fall within relatively narrow limits of one another. Each of these groups is then further refined into specific products with which we are familiar: heavy heating oil, fuel oils, kerosene, gasoline, and various gas vapors.

Fractional distillation yields several different grades of crude oil. Their characteristics are listed in Table 4.1. Although both No. 1 and No. 2 fuel oils are used in residential heating systems, the use of No. 2 predominates. Note that fuel oils above grade 2 are used almost exclusively for industrial heating applications in which special equipment for proper combustion is required.

Combustion Characteristics

Efficient oil burner operation begins with knowledge of the combustion process. From Table 4.1 it can be seen that No. 2 fuel oil produces between 135,000 and 142,000 Btu/gal when completely burned. Of this total amount of energy produced, the combustion of carbon accounts for approximately 75,000 Btu and that of hydrogen the remaining 65,000 Btu. For complete combustion of each gallon of fuel

TABLE 4.1 FUEL OIL GRADES AND USES

Grade	Description	Heating value (Btu/gal)	Approximate gravity (°API)	Use of heating range	Approximate weight (lb/gal)	Flash point (minimum °F)	Pour point (maximum °F)
No. 1	Distillate oil intended for vaporizing in pot-type burners and in other burners requiring this grade	132,900–137,000	38–45	Vaporizing equipment	6.95	100 or legal	0
No. 2	Distillate oil for general-purpose domestic heating for use in burners not requiring No. 1	135,800–141,800	30–40	Domestic heating equipment	7.29	100 or legal	20
No. 4	Preheating is not usually required for handling or burning	143,100–148,100	20–28	Industrial heating power	7.78	130 or legal	20
No. 5	(Light) preheating may be required for handling depending on climate and equipment	146,800–150,000	17–22	Industrial heating power	7.94	130 or legal	
No. 5	(Heavy) preheating required for burning, and in cold climates heating may be required for handling	149,400–152,000	14–18	Industrial heating power	8.08	130 or legal	
No. 6	Preheating required for burning and handling	151,300–155,900	8–15	Industrial heating power	8.44	150	

Source: G. K. Hall & Co.

oil to take place, approximately 100 lb of air must be supplied, 79% of which is composed of inert nitrogen. The remaining 21%, oxygen, is responsible for the combustion process depicted in Figure 4.1 All the combustion air enters the fuel chamber at room temperature and exits the boiler at a temperature several hundred degrees higher. The proper amount of combustion air is critical for efficient burner operation. If too little oxygen is present, or if it is improperly mixed with the fuel oil, carbon monoxide (CO), smoke, and carbon dioxide (CO_2) will result. Carbon monoxide is smokeless, odorless, and highly toxic. Smoke wastes fuel by leaving a coating of soot on interior boiler surfaces, retarding efficient heat transfer: The by-products of incomplete combustion are not desirable from any standpoint.

To ensure that there is a sufficient quantity of oxygen available to unite with all the carbon and hydrogen atoms in the fuel oil, approximately 50 to 100% excess air should be present in the combustion chamber during fuel ignition. As the percentage of excess air increases, the temperature of the fire is lowered with an accompanying loss in combustion efficiency. However, this reduction is desirable because clean burning and complete combustion is possible only when sufficient quantities of excess air are available. Figure 4.2 illustrates the reduced efficiency resulting from 100% excess air (200 lb) in the combustion chamber.

Although the combustion process can be somewhat complicated from a chemical point of view, a variety of test instruments are available that make it relatively simple to determine the combustion efficiency of virtually any residential oil-fired appliance. Use of this equipment is discussed later in this chapter.

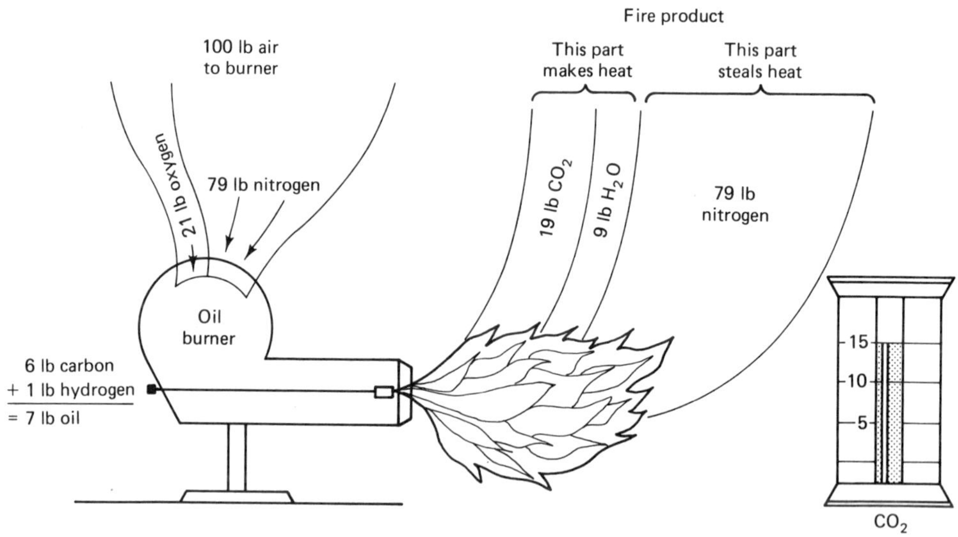

Figure 4.1 Combustion of fuel oil. (Courtesy of Dwyer Instruments, Inc.)

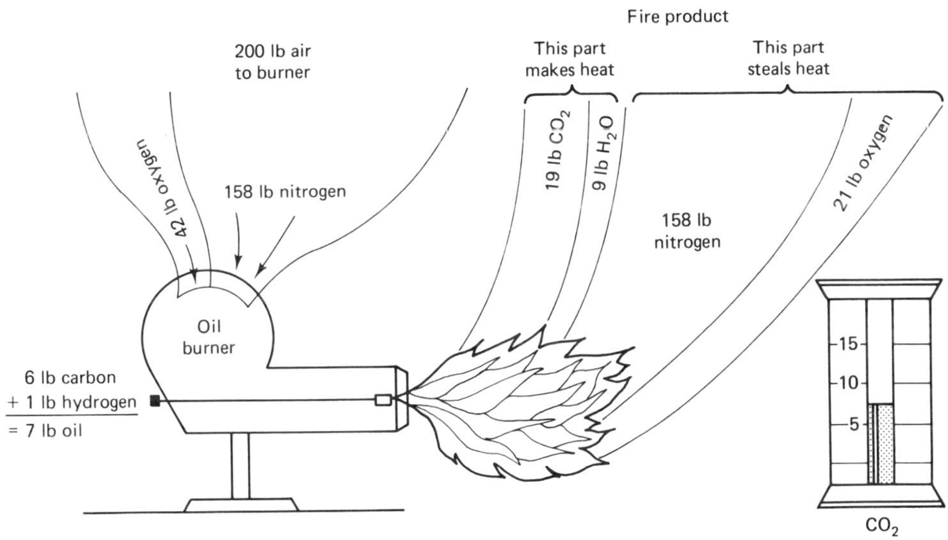

Figure 4.2 Approximate products of combustion of oil burners using 100% excess air. (Courtesy of Dwyer Instruments, Inc.)

STORAGE AND DELIVERY SYSTEMS

Oil storage tanks are available in a variety of sizes for residential storage purposes. The most commonly used material in tank construction is steel. These tanks may be installed in several different ways to match a variety of operating configurations.

Underground Tank Installation

Underground installation of the oil storage tank offers many advantages over aboveground installations. Underground tanks do not take up any space within or near the home; they are completely out of sight and safely removed from any potential fire or related safety hazard. A typical underground tank installation is illustrated in Figure 4.3. Underground tanks are usually supplied in capacities of either 550 or 1000 gal, although larger sizes are available. The tanks are manufactured from No. 7 gauge steel and come from the factory with a corrosion resistance coating. Note from Figure 4.3 that the tank is sloped to prevent sediment from accumulating near the feed and return lines. In addition to steel, underground storage tanks are also fabricated from heavy-gauge fiberglass.

Underground tanks require a two-pipe self-purging piping configuration for delivery of the fuel oil to the oil burner unless the bottom of the tank is located above the level of the oil burner. In the two-pipe arrangement, one of the pipes acts as the suction feed line, which ends at the input fitting on the fuel pump; the other

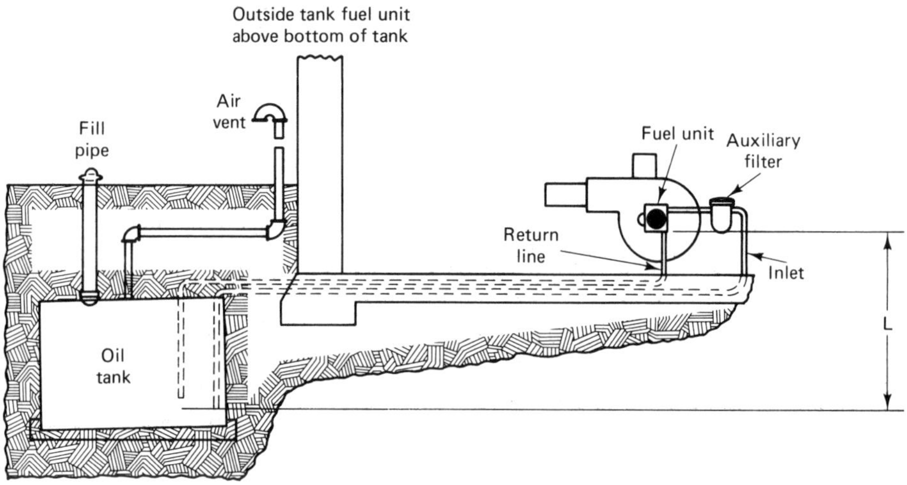

Two-Pipe Systems

Lift L (ft)	Length of Tubing[a] (ft)			
	Single-Stage Units		Two-Stage Units	
	3/8 in. OD	1/2 in. OD	3/8 in. OD	1/2 in. OD
0	53	100	68	100
2	45	100	63	100
4	37	100	58	100
6	29	100	53	100
8	21	83	48	100
10	13	52	42	100
12	Not recommended; use two-stage fuel unit		37	100
14			32	100
16			27	100
18			22	88

[a] Line lengths include *both* vertical and horizontal lengths

Figure 4.3 Underground oil tank installation. (Courtesy of Suntec Industries, Inc.)

pipe is a return line, which sends unused fuel oil back to the storage tank. Each of the fuel lines should terminate approximately 3 to 4 in. above the bottom of the tank and should be piped in $\frac{1}{2}$-in.-outside-diameter copper tubing. A minimum number of couplings and unions should be used in the piping runs to limit the possibility of air infiltration into the fuel lines or oil leaks from within. All lines should be buried or otherwise mechanically protected to prevent damage. Both feed and return lines should be equipped with shutoff valves located adjacent to the point at which the lines enter the foundation wall. Additional valves are located at the oil burner to facilitate service procedures.

Local building codes will usually specify the minimum allowable distances required between the oil storage tank and the exterior of the building foundation wall. These code requirements should be consulted prior to undertaking installation of the tank.

Aboveground Tank Installation

When oil tanks are installed above the ground, either indoors or out, the bottom of the tank must be no higher than 8 ft above the burner if a single-pipe feed system is to be used. A typical aboveground tank installation is shown in Figure 4.4.

At times, it may be desirable to connect two or more smaller oil tanks in tandem to increase the storage capacity of the system. There are several methods available to accomplish this, one of which is illustrated in Figure 4.5. Note that the tanks are piped in a series arrangement for filling and in a parallel arrangement for oil delivery to the burner. Each tank should be equipped with its own shutoff valve and fuel filter. Also, a third oil filter may be desirable at the fuel pump on the oil burner.

It should be noted that in certain localities, inside oil storage tanks are prohibited. In these instances, local codes may specify the minimum distances from residential foundation walls where oil storage tanks must be located. Prior to installation, local codes should be consulted to ensure compliance.

A shutoff valve with a fusible-link mechanism designed to turn off the oil flow in the event of excessive heat (usually 165°F) should be installed at the exit of the oil tank and next to the burner intake line. Shutoff valves of this type, together

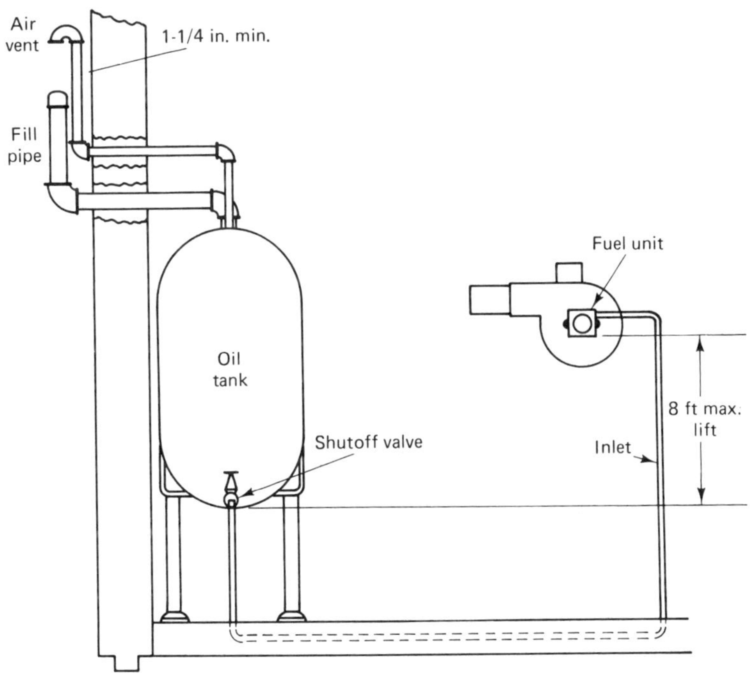

Figure 4.4 Aboveground oil storage tank installation.

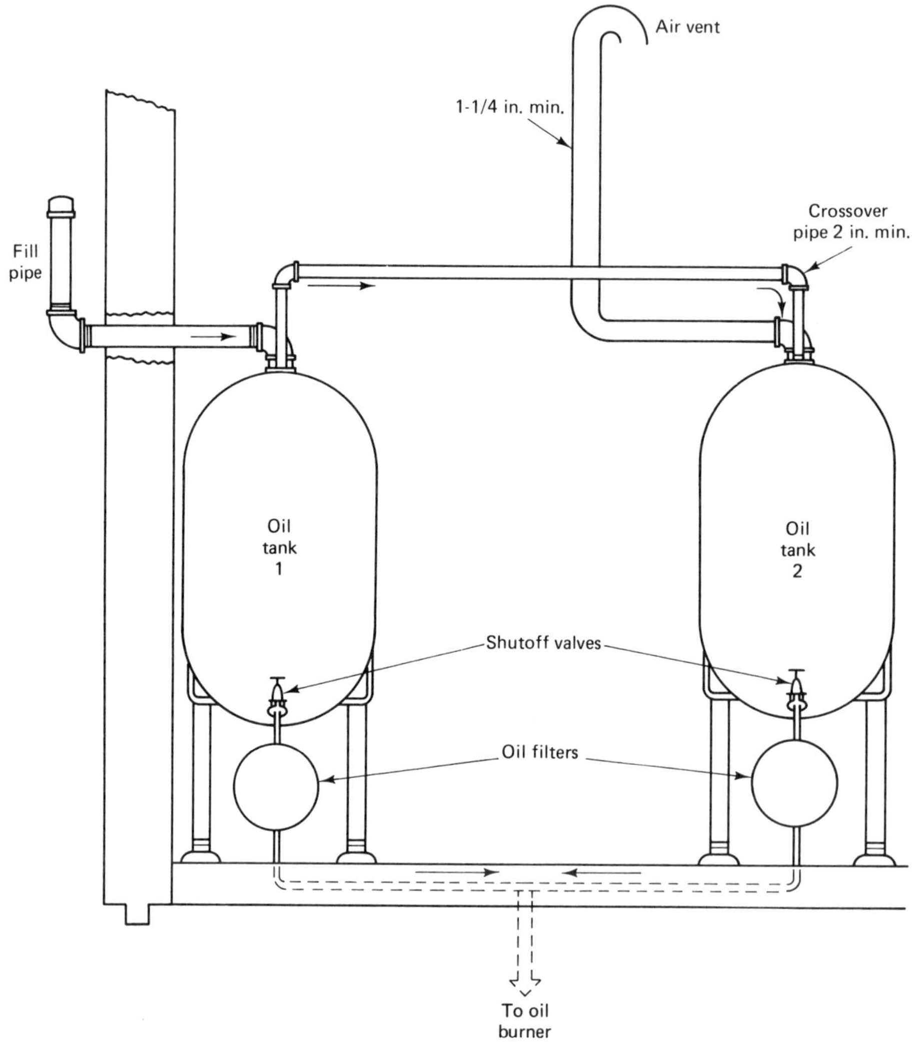

Figure 4.5 Tandem installation of two fuel oil tanks during normal fill procedures; oil first fills tank 1 and then fills tank 2 via the crossover pipe. Each tank is equipped with a separate fusible shutoff valve and filter assembly, terminating in a common feed line to the burner mechanism.

with their proper location, are illustrated in Figures 4.6 and 4.7. The tank should be provided with a drain plug for oil removal and tank cleaning if necessary. Piping from the tank to the burner should have a minimum inside diameter of $\frac{3}{8}$ in. and is usually made from seamless flexible copper tubing.

Minimum specified distances vary for indoor applications between the oil tank and the burner. In some instances fireproof separation walls must be provided be-

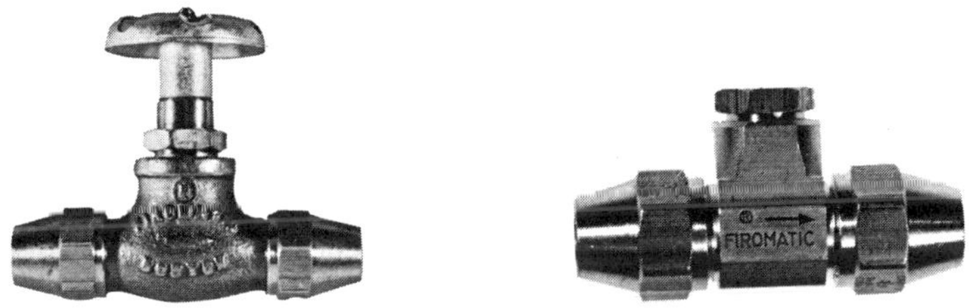

(a) (b)

Figure 4.6 Heat-sensitive oil shutoff valves: (a) cast bronze; (b) machined brass. These units are designed to stop the flow of fuel oil automatically if the temperature at the valve reaches 165°F. (Courtesy of Highfield Mfg. Co.)

tween the two appliances. Special code requirements should be researched carefully prior to tank installation.

Piping for Two Appliances

When two oil-fired appliances are fed from one storage tank, modifications must be made in the piping system to ensure proper fuel delivery to each. Figure 4.8 illustrates an installation in which an oil-fired water heater and heating boiler both draw fuel from one outdoor storage tank. In this configuration, the boiler operates on a two-line self-bleeding system. The water heater is fed from a supply T fitting in a single-pipe gravity-feed arrangement. If a conventional gravity-feed system is used rather than overhead piping as in Figure 4.8, both the boiler and water heater may be connected to one common feed line. When two appliances are being run from an underground storage tank requiring that the fuel be lifted for delivery to the appliances, the piping arrangement should be as illustrated in Figure 4.9.

Individual fuel lines should be used to supply each oil burner in this type of installation. The return lines can be manifolded together but should run to within 3 to 4 in. of the bottom of the storage tank. This results in a two-pipe system for the total installation in which air bleeding on each burner is automatic.

When individual oil lines cannot supply each burner, the piping arrangement in Figure 4.10 is suggested. In Figure 4.10, both the feed and return lines are installed on a manifold, with electrically operated solenoid valves on the inlet side of each burner provided to prevent a loss of prime in the fuel pumps when the burners are not operating. These valves should be wired in parallel with each burner motor to ensure that they both operate at the same time.

STORAGE TANK ACCESSORIES

Oil-Level Indicators

Indoor, outdoor, aboveground, and below-ground oil storage tanks are usually equipped with gauges that indicate both the level of the fuel inside the tank and can

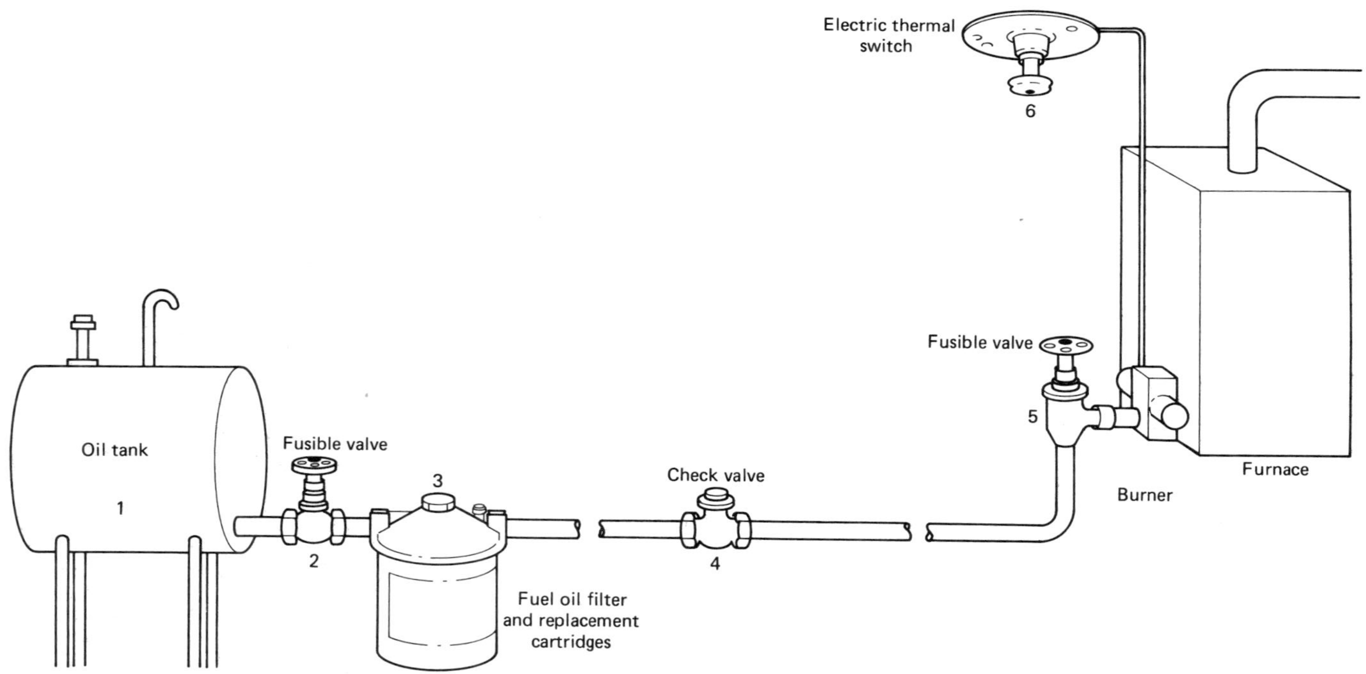

Figure 4.7 Recommended installation of fusible oil shutoff valves. (Courtesy of Highfield Mfg. Co.)

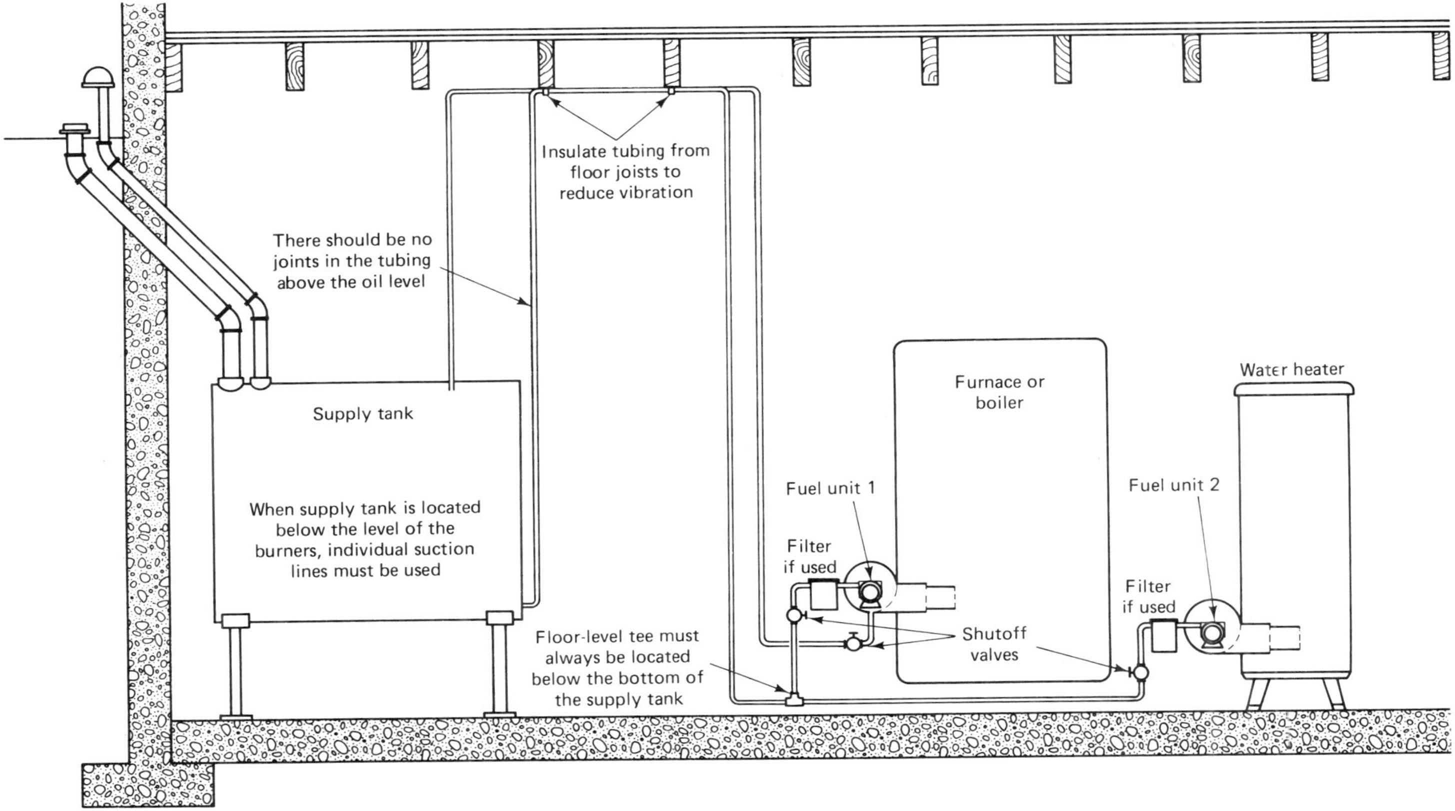

Figure 4.8 Two oil-fired appliances drawing fuel from single inside storage tank. (Courtesy of Suntec Industries, Inc.)

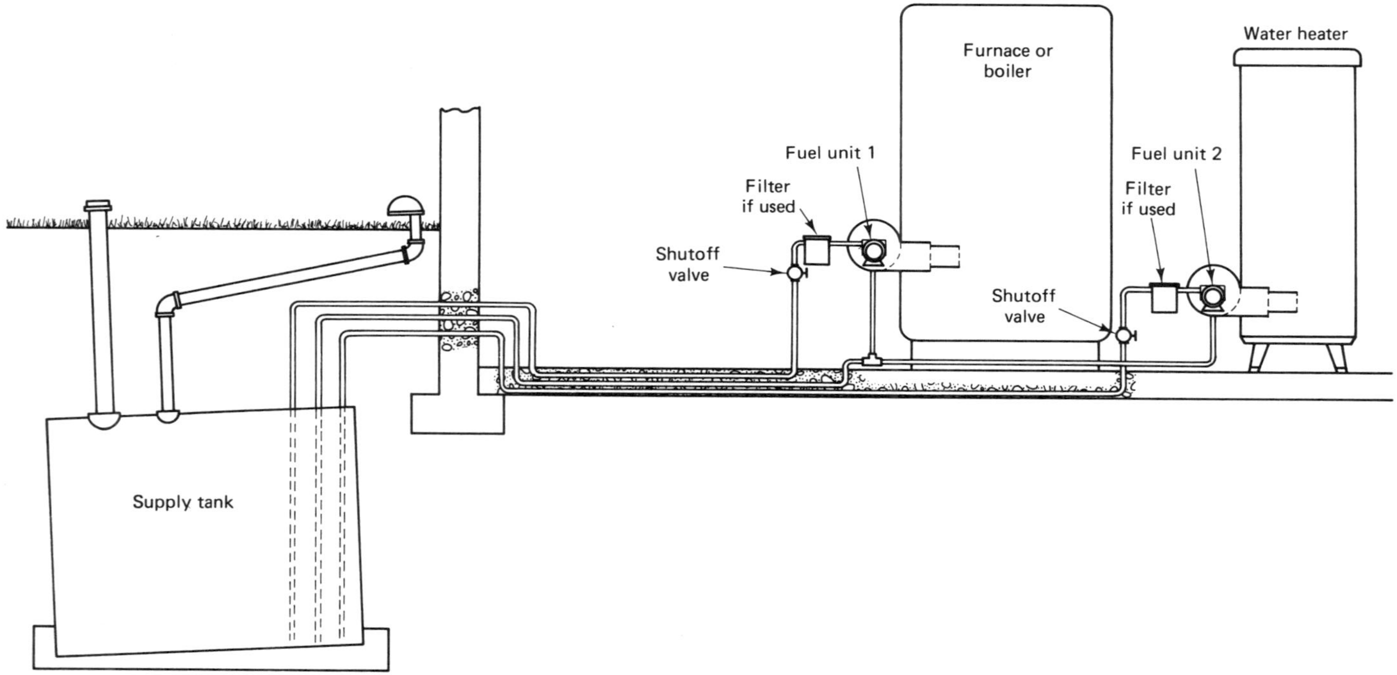

Figure 4.9 Two oil-fired appliances drawing fuel oil from single underground storage tank. (Courtesy of Suntec Industries, Inc.)

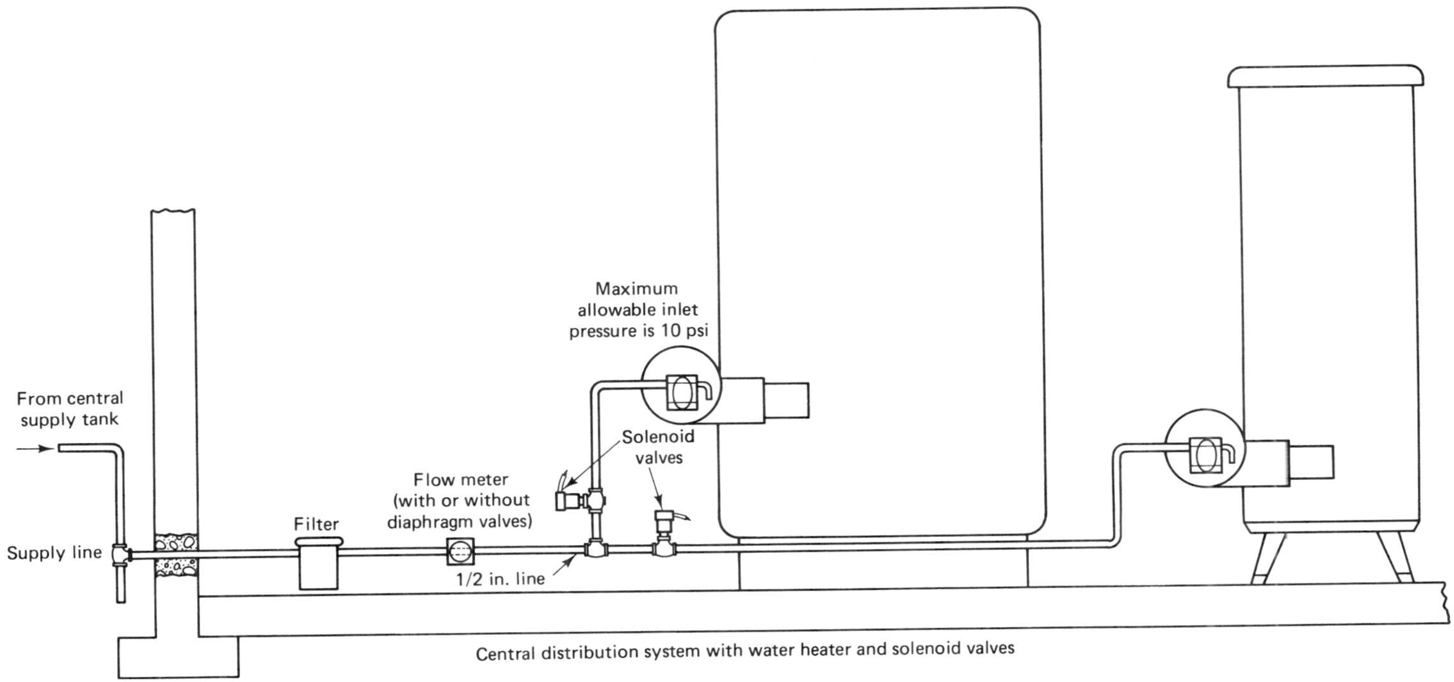

Figure 4.10 Using an existing underground fuel oil tank and piping system to supply two appliances. (Courtesy of Suntec Industries, Inc.)

sound an audible alarm during the filling procedure as the fuel oil approaches the top of the storage tank.

Aboveground oil-level gauges are activated by a lifting mechanism consisting of a cork attached to a level indicator via a lifting rod linkage. As the level within the tank rises, the cork lifts the indicator rod, which is located within a calibrated plastic sleeve.

Underground fuel oil storage tanks use barometric-type level gauges that read out the quantity of fuel in the tank on the height of a colored liquid in a vertical glass tube.

Audible alarms on the filling gauges are provided to help alert delivery personnel to the fact that the oil tank is approaching the full level. This avoids overfilling the tank, with resulting oil spills.

A typical combination oil-level gauge with audible alarm and barometric-type fuel-level indicating gauge is illustrated in Figure 4.11. If the tank is not equipped with an oil-level indicator, the volume of fuel in the tank can be determined by using a dipstick to determine fuel depth and then consulting Table 4.2.

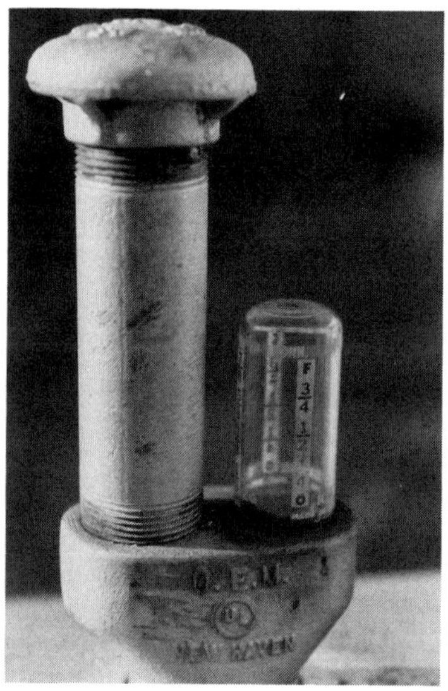

(a)

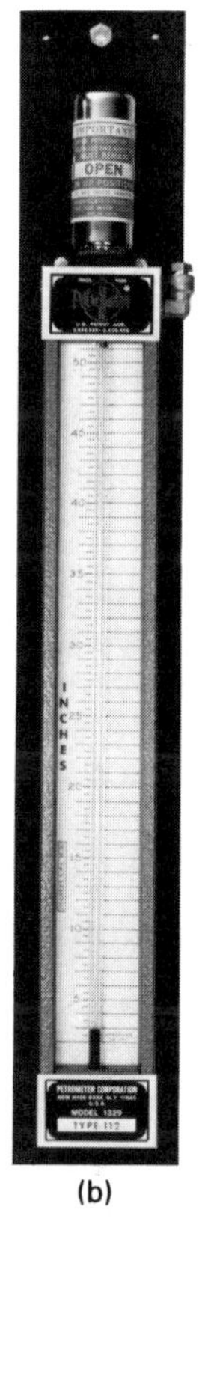

(b)

Figure 4.11 (a) Oil-level indicator gauge for aboveground storage tanks; (b) oil-level gauge for underground fuel oil storage tank. [(a) Courtesy of Oil Equipment Manufacturing Corp.; (b) courtesy of the Petrometer Corp.]

TABLE 4.2 TANK VOLUME CHART

Depth of fuel oil (in.)	Tank size		
	275 in. oval (upright)	550 in. round (48 in. diameter)	1000 in. round (48 in. diameter)
2	7	7	15
4	15–17	20	40
5	23	30	55
10	60	80	150
15	100	150	265
20	130	210–215	400
25	170	290	525
30	200	360	660
35	240	430	785
40	270	500	900
45	—	535	975

Note: Volumes given are approximate since individual tank configurations and dimensions will vary from one manufacturer to another.

Oil Filters

Fuel oil filters, similar in concept and principle to the oil filters found on automobiles, are installed on all oil-fired appliances.

When the oil tank is installed aboveground and in an accessible location, it is common for the oil filter assembly to be installed either after the tank shutoff valve, on the downstream side of the fuel pump, or in both locations. If the oil tank is installed below ground, the fuel filter assembly is generally installed either at the point where the fuel lines enter the house or just downstream of the inlet port on the fuel oil pump.

The filters used in these assemblies feature elements that remove virtually all the solid contaminants within the fuel and are suitable for both single-pipe gravity feed and two-line self-bleeding systems. A typical fuel oil filter housing and element assembly is illustrated in Figure 4.12.

Fill and Vent Piping

All oil storage tanks require a convenient and safe method for filling and venting the tank. Fill lines should be placed out of the way of pedestrian traffic. A weatherproof fill cap with a minimum diameter of 2 in. is commonly furnished. On underground storage tanks, the fill pipe rise should terminate just above ground level with a small clearing prepared to prevent contamination from the surrounding dirt when the tank cap is opened for filling.

Figure 4.12 Fuel oil filter assembly and filter elements. These elements should be changed annually for best results. (Courtesy of General Filters, Inc.)

On inside storage tanks, the fill lines generally penetrate the foundation of the building and extend partway up the exterior building wall. Often, these lines can be conveniently camouflaged with shrubbery.

Vent lines for both aboveground and below-ground storage tanks should be piped with a minimum diameter of $1\frac{1}{4}$-in. vent pipe extending above the highest snow line. A vent hood is placed on the line to prevent moisture or debris from entering the tank. The vent lines should be plumbed with a slope in all horizontal pipe runs. This procedure ensures that excess oil will drain from the vent back into the tank in case the tank is accidentally overfilled. Also, the possibility of pockets of trapped air forming in low spots or sags in the piping is minimized.

All fuel oil storage tanks, whether they are installed inside or outside the house, should be checked periodically to determine if any appreciable amounts of water have condensed and gathered at the bottom of the tank. This test is simple to perform and is done by applying a special water-finding paste to the end of a dipstick

Figure 4.13 Water-finding paste used to detect water accumulations within a fuel oil storage tank. (Courtesy of Oil Equipment Manufacturing Corp.)

that is inserted until it touches the bottom of the tank. The stick is then removed and the paste is inspected. The paste will change color if it comes into contact with water (Figure 4.13). If it is found that there is sufficient water to change the color of the paste, the tank should be drained or flushed until all traces of water have been removed.

OIL BURNER OPERATION

The modern oil burner is the result of sophisticated combustion technology that results in a highly efficient burner mechanism. Residential oil heating equipment falls within two general categories: conventional slow-speed and high-speed flame retention units. Each different burner design is examined below in more detail.

Conventional Slow-Speed Burners

In a conventional slow-speed oil burner, fuel oil is injected into a combustion chamber under pressure. A nozzle in the fuel line atomizes the fuel oil into tiny droplets which are mixed with combustion air and burned under controlled conditions. The term "slow speed" refers to the rotational speed of the burner motor, which is between 1725 and 1750 rpm. The flame pattern produced within the combustion chamber fired by a slow-speed burner is illustrated in Figure 4.14.

Note from Figure 4.14 that the flame base begins approximately 2 in. in front of the nozzle assembly. The flame is contained within the combustion chamber, which radiates heat from the burning fuel back into the flame area to help maintain a hot fire. The combustion chamber houses a target wall which also helps to promote heat recirculation. Also, recirculation aids in stabilizing the flame and promotes efficient mixing of the oil and combustion air.

High-Speed Flame Retention Burners

The high-speed flame retention burner is one of the latest improvements in oil burner technology. Flame retention burners rely on a specially designed combustion head and retention ring in which combustion air is rotated in a highly controlled manner directly in front of the burner head. Air is delivered into the combustion chamber by a high-speed 3450-rpm blower motor. This produces a high-intensity flame which is held to within $\frac{1}{2}$ in. of the burner head. A target wall is not needed to recirculate the flame pattern, which in a retention burner is highly stable. Efficient mixing of the combustion air and fuel oil can be accomplished without the use of a conventional combustion chamber. The flame pattern in a typical retention head burner is illustrated in Figure 4.15.

Thus retention burners yield higher operating efficiencies than do their con-

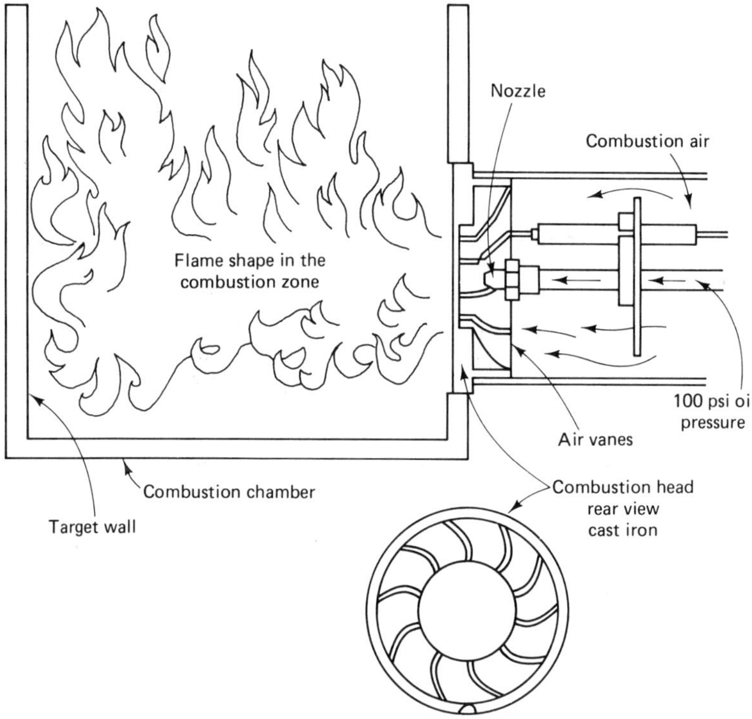

Figure 4.14 Conventional slow-speed oil burner flame and combustion configuration. (Courtesy of R.W. Beckett Corp.)

ventional slow-speed counterparts. Annual Department of Energy reports of combustion efficiencies of appliances equipped with flame retention burners show efficiencies as high as 85 to 88%. A high-speed flame retention burner is illustrated in Figure 4.16.

APPLIANCE DRAFT CONDITIONS*

All boilers and furnaces operate under one of three types of draft conditions: natural draft, induced draft, or forced draft. Each condition imposes certain restrictions on the burner and can in certain instances require special equipment to yield efficient combustion. Each of these draft conditions is examined below in greater detail.

*This section is from "High Speed Oil Burner Design as Related to Today's Applications," by Leonard A. Fisher, Director of Engineering, Carlin Company, Wethersfield, Conn. (no date). Used by permission.

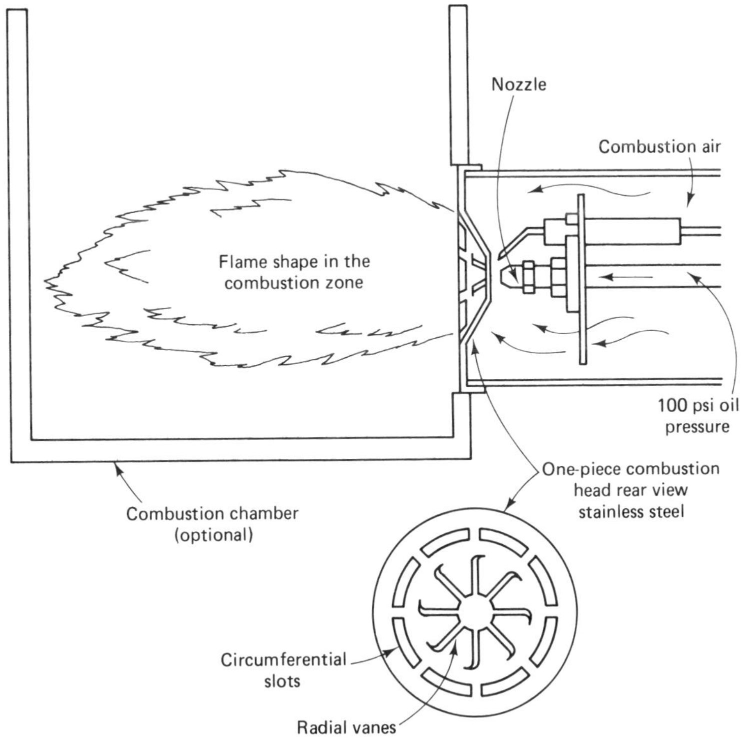

Figure 4.15 High-speed flame retention combustion pattern. (Courtesy of R.W. Beckett Corp.)

Figure 4.16 High-speed flame retention oil burner. (Courtesy of the Carlin Co.)

Natural Draft

Natural draft occurs due to the difference in the relative weights of the chimney air and the ambient outside air. As air within the chimney is heated its density is reduced and it rises. This action causes a negative or evacuative pressure within the chimney, which is referred to as *chimney draft.* (Chimney draft, and the specifics of chimney construction that affect draft, are examined in detail in Chapter 7.) A natural-draft system is illustrated in Figure 4.17.

Boilers that are designed to operate under natural-draft conditions have large-diameter flue passages and stack pipe connections that minimize frictional air losses within the system. Conventional slow-speed oil burners in conjunction with properly sized and shaped refractory combustion chambers operate with good performance in natural-draft installations. High-speed burners yield excellent results with or without refractory combustion chambers in natural-draft systems, although target

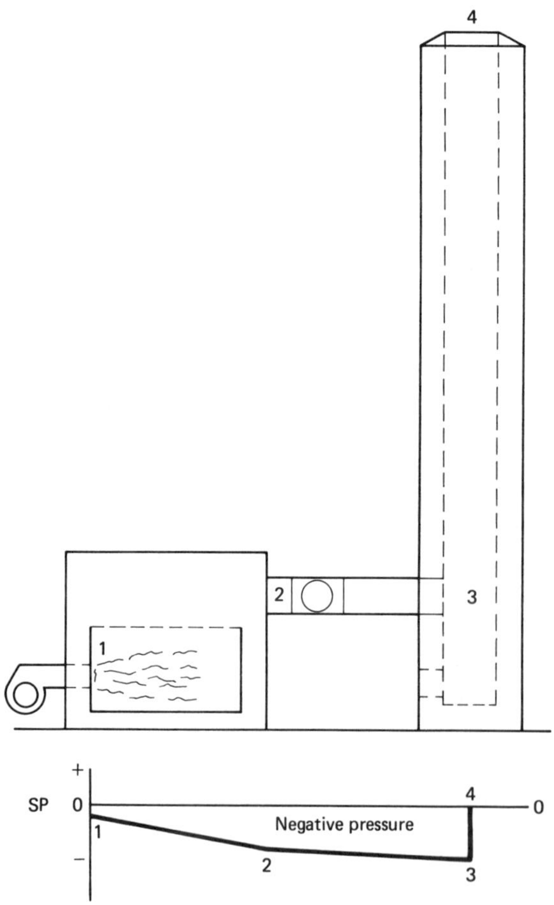

Figure 4.17 Natural-draft oil-fired combustion. (Courtesy of the Carlin Co.)

walls or combustion chamber "rugs" are generally used with low-range firing burners (capacities below 3 gal/hr).

Induced Draft

An induced-draft system uses a fan to provide the necessary draft for proper oil burner operation (Figure 4.18). Fan induction is usually necessary in this type of arrangement in order to overcome pressure losses in the system due to the following factors:

1. Restricted boiler or heat exchanger passages
2. Small chimney pipe breeching connections
3. Restrictions or aberrations in the chimney

Since the chimney and appliance draft can vary in this type of system, high-speed flame retention burners should be used exclusively, since they are not greatly af-

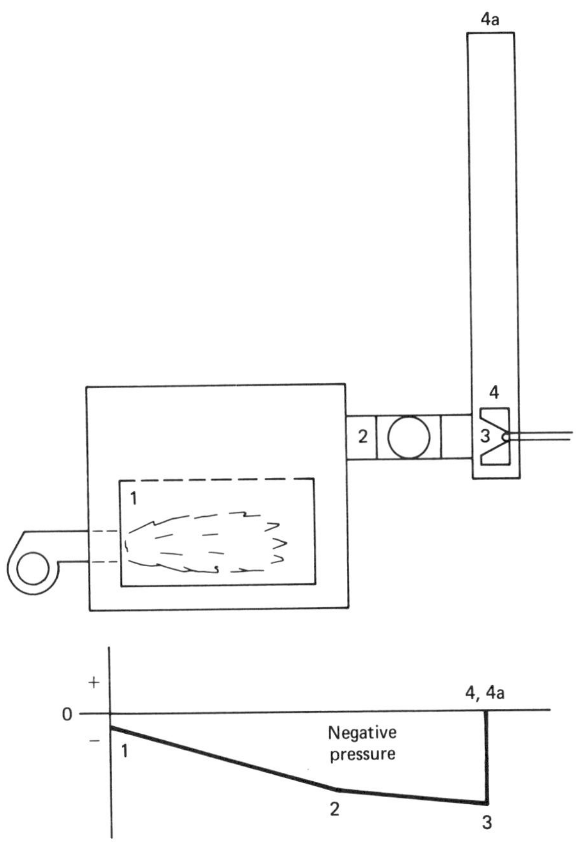

Figure 4.18 Induced-draft oil-fired combustion. (Courtesy of the Carlin Co.)

fected by changing pressures within the combustion chamber as are conventional slow-speed burners.

Forced Draft

In a forced-draft system, the by-products of the combustion process are propelled under pressure through the boiler flue passages and out the breeching connection without any assistance from natural draft within the chimney (Figure 4.19). Forced-

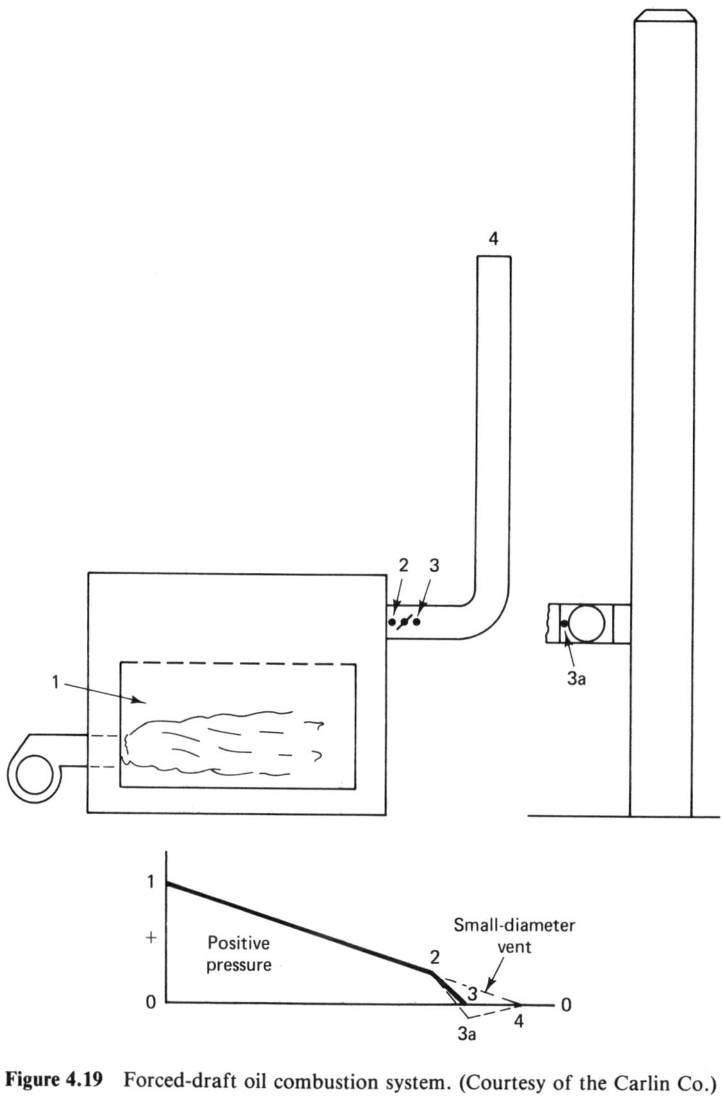

Figure 4.19 Forced-draft oil combustion system. (Courtesy of the Carlin Co.)

draft systems usually employ a blocking damper to set the combustion chamber draft pressure, together with a barometric draft damper that adjusts to changes in chimney draft as the boiler goes through normal cycling.

Forced-draft boilers require less horsepower than comparably sized induced-draft units. Forced-draft design features boiler and flue heat exchanger surfaces that are compact in size, since relatively high pressure drops are encountered during normal operation. These boilers are more common with high-firing-rate burners, usually found in commercial and industrial applications, with burner capacities in excess of 5 gal/hr.

OIL-FIRED BOILERS

Most residential oil-fired boilers are sold as prepackaged units and are manufactured from either cast iron or steel. A typical cast iron boiler is illustrated in Figure 4.20. Note that the wet-base design, in which water circulates beneath the combustion chamber as well as around it, together with the configuration of the interior boiler castings, yields the maximum amount of surface area to the flue gases to obtain highly efficient heat transfer. The tankless coil in the cast iron boiler is inserted into the side of the unit and is totally surrounded by heated water.

A variety of heating capacities in cast iron boilers are achieved by altering the number of boiler sections together with the capacity of the oil burner. The greater the number of sections comprising the boiler, the higher the firing range possible, hence the higher the heating capacity of the unit.

In the larger Btu sizes, cast iron boilers are not generally prepackaged but rather are sold as individual sections which are bolted together and sealed on the

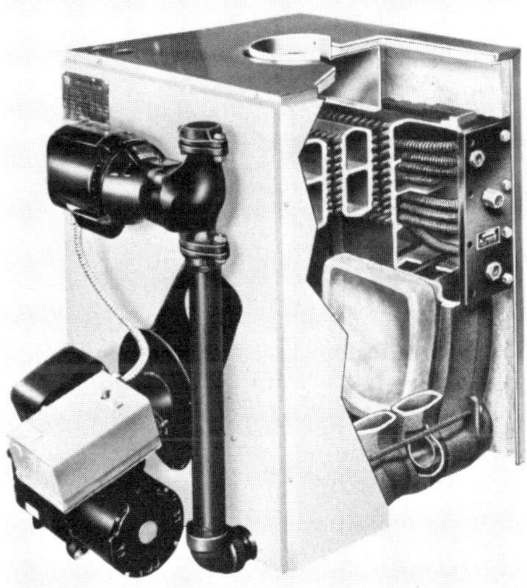

Figure 4.20 Packaged cast iron wet-base boiler. (Courtesy of Burnham Corp.)

job site. This technique allows for ease in handling otherwise inordinately heavy boilers that might be very difficult to move bolted together but in individual sections are easily moved about.

Variations in heating capacities, together with the number of cast iron segments comprising a typical cast iron residential boiler, are illustrated in Table 4.3.

A typical steel oil-fired packaged residential boiler is illustrated in Figure 4.21.

TABLE 4.3 HEATING CAPACITIES AND FIRING RATES OF A TYPICAL RESIDENTIAL CAST IRON BOILER SERIES

Boiler number[a]	Burner capacity (gal/hr)	DOE heating capacity (MBtu/hr)	IBR net ratings water	Net water rating (ft²)	DOE annual efficiency (%)
V-13A and RO/FO	0.75	89	77.4	516	83.8
V-14A and RO/FO	1.05	124	107.8	719	83.3
V-15A	1.35	159	138.3	922	82.8
V-16A	1.65	193	167.8	1119	82.2
V-17A	1.90	222	193.0	1287	82.2
V-18A	2.10	245	213.0	1420	82.2

Source: Burnham Corp.

[a]The second boiler number indicates the number of cast iron sections in the unit. For example, the V-14 is a four-section boiler.

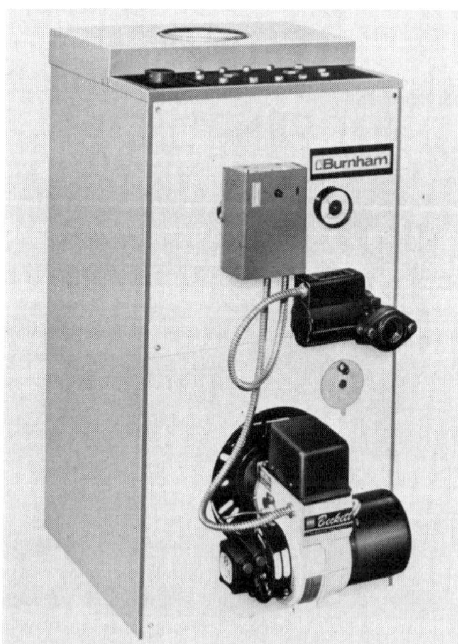

Figure 4.21 Packaged residential steel boiler. (Courtesy of Burnham Corp.)

Most steel boilers are not wet-base designs. Note the use of firetubes for flue gas removal as well as for maximizing heat transfer to the surrounding boiler water.

Since steel boilers are constructed as one-piece units rather than in segments as are their cast iron counterparts, variations in boiler heating capacity are determined by the overall physical size of the unit coupled with the appropriate firing rate of the burner.

Typical output ranges, together with physical dimensions and burner firing rates for a line of residential steel hydronic boilers, are given in Table 4.4.

TABLE 4.4 OUTPUT RANGES AND FIRING RATES FOR RESIDENTIAL STEEL OIL BOILERS

Boiler number	Burner capacity (gal/hr)	DOE heating capacity (MBtu/hr)	SBI net ratings water	DOE annual efficiency (%)	
				Less vent damper	With vent damper
RS-109	0.85	97	84.3	79.1	81.6
RS-110	0.90	104	90.4	79.1	81.5
RS-111	1.05	120	104.3	79.2	81.3
RS-112	1.25	142	123.5	79.2	81.1
RS-113	1.45	163	141.7	78.9	80.5
RS-114	1.70	190	165.2	78.2	79.9
RS-116	1.95	226	196.5	80.7	82.9
RS-117	2.40	268[a]	233.0	—	—
RS-118	2.85	318[a]	276.5	—	—

Source: Burnham Corp.

[a]SBI gross output, not rated for DOE capacity or efficiency.

COMBUSTION CHAMBERS

Many residential boilers have a combustion chamber in which fuel ignition takes place, whether the appliance is equipped with high-speed or conventional oil burner mechanisms. The combustion chamber serves a variety of functions within the boiler, the most common of which are:

1. The wall surfaces of the chamber reflect heat back into the oil flame, helping to stabilize it, and allow the boiler quickly to reach steady-state operating temperature.
2. The combustion chamber protects the interior boiler surfaces from the high heat of the flame. Also, it prevents the flame from directly impinging on the relatively cool interior boiler surfaces, which would cause soot and thus reduce the operating efficiency of the unit.
3. The combustion chamber helps to muffle the noise of ignition associated with oil-fired heating equipment.

High-speed burners have greatly reduced the need for combustion chambers in the larger burner sizes (3.5 gal/hr and higher) and minimized their need in the lower firing rates (3.5 gal/hr and lower). Many residential boilers, however, come equipped with some type of combustion chamber, target wall, or high-temperature rug to prevent direct flame contact with interior boiler surfaces, the result of which improves both flame and heat recirculation. This recirculation pattern is illustrated with both narrow- and wide-angle flame angles associated with typical residential oil burners, shown in Figures 4.22 and 4.23.

At the present time there are several types of combustion chambers in widespread use: precast, firebrick, molded high-temperature refractory materials, and vacuum-formed refractory fibers.

Precast Refractory Chambers

Precast combustion chambers come in a variety of sizes and shapes designed for specific boilers and firing rates. These chambers are composed of a number of individually shaped cast pieces which when bonded and fitted together form the combustion chamber of the unit, shown in Figure 4.24.

Refractory Firebrick

Refractory firebricks are made of lightweight, high-temperature material that possesses high insulating value (Figure 4.25). These bricks can be used in custom-designed installations by cutting or trimming them into specific shapes. The brick is usually available in a variety of temperature ratings, based on the firing rate of the boiler.

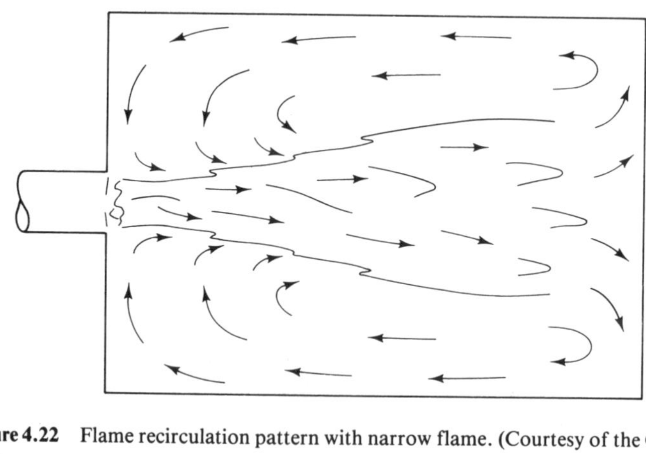

Figure 4.22 Flame recirculation pattern with narrow flame. (Courtesy of the Carlin Co.)

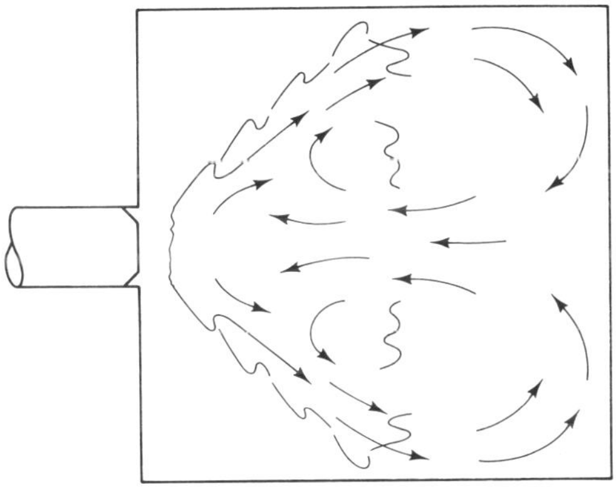

Figure 4.23 Flame recirculation pattern with wide flame. (Courtesy of the Carlin Co.)

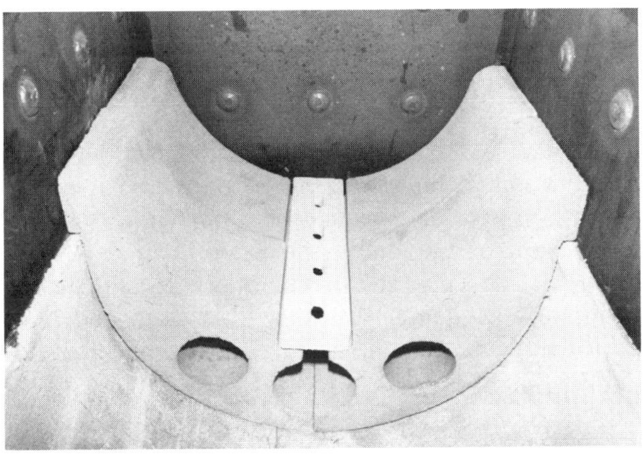

Figure 4.24 Precast refractory chamber. (Courtesy of Eshland Enterprises, Inc.)

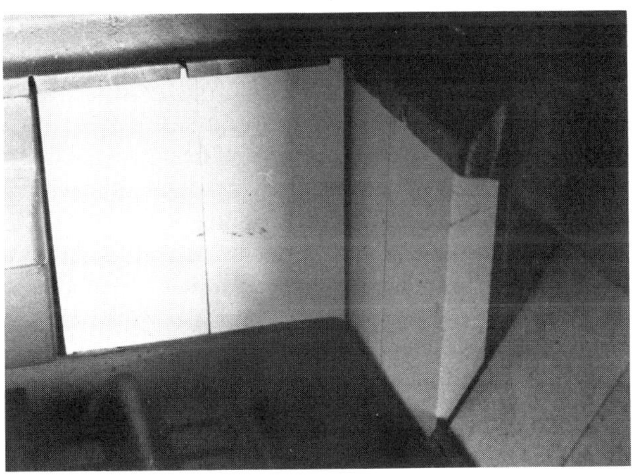

Figure 4.25 Refractory brick inserted into boiler combustion chamber. (Courtesy of Riteway-Dominion Mfg., Inc.)

Figure 4.26 Moldable refractory material. (Courtesy of R.W. Beckett Corp.)

Pliable Refractory Material

Sometimes referred to as a "wet pac," these materials come in a moldable form (Figure 4.26). To mold the combustion chamber, the pliable refractory material is inserted into the existing combustion area of the boiler and pressed securely into place to conform to the proper shape of the interior boiler surfaces. It will hold and maintain its molded configuration. This type of material is ideally suited to relining applications in older combustion chambers, as they work easily and assume almost any shape. They can usually be stored for up to one year in the factory-sealed vacuum container prior to use.

FUEL PUMPS

Fuel pumps designed to operate with No. 1 or No. 2 fuel oil incorporate a rotary gear mechanism that takes incoming low-pressure fuel oil and delivers it to the burner ignition assembly at approximately 100 psi. This high pressure is required for the fuel oil to be adequately broken up into tiny droplets, or atomized, by the nozzle assembly in the oil burner. Fuel pumps used in most residential heating applications accomplish this pressure increase in either one or two pumping stages, although the single-stage hydraulic pump is the most commonly used (Figure 4.27).

The operation of the pump takes place as follows. Fuel oil enters the pump from the oil storage tank, and first goes through a filter strainer to remove any trapped impurities and sediment. The oil then becomes trapped between the teeth in the pump gearing assembly; this increases the pressure of the fuel oil, which raises the nozzle cutoff valve. Oil is allowed to flow out of the exit port of the fuel pump and into the nozzle assembly, where it is atomized prior to ignition. Excess oil in

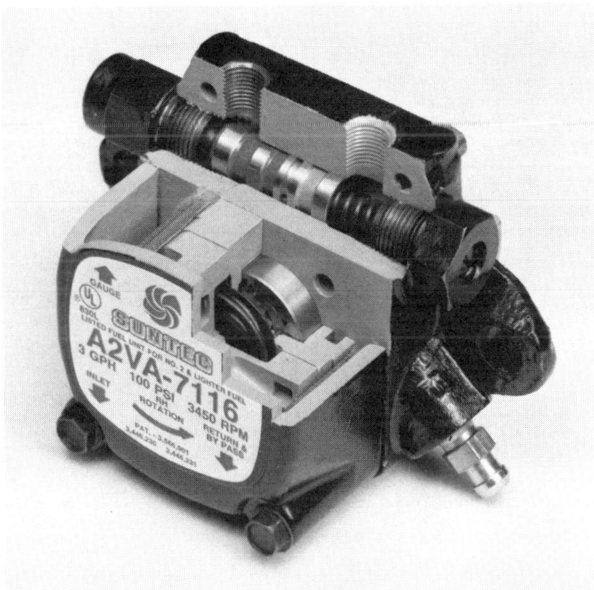

Figure 4.27 Single-stage fuel oil pump. (Courtesy of Suntec Industries, Inc.)

the pump is either returned to the oil tank in a two-pipe feed system or is returned to the strainer assembly in the fuel pump piped with a one-pipe gravity feed system. The circuit diagram for the single-stage pumping configuration is shown in Figure 4.28.

Two-stage fuel pumps are recommended for use in oil systems in which the oil either must be lifted higher than 8 ft or encounters long supply lines prior to entering the fuel pump. A typical two-stage pump is illustrated in Figure 4.29.

Routine Performance Checks

There are several conditions that can adversely affect the operation of the fuel pump. These can be classified as follows:

1. Incorrect pressure setting of the pump
2. Loss of pump vacuum
3. Loose supply line connections and associated fittings

Pressure settings. Most fuel pumps are equipped with a port into which a pressure gauge can be inserted to determine the pressure at which the pump is delivering fuel oil to the nozzle assembly (Figure 4.30). Normal operating pressure for residential fuel oil pumps should be 100 psi. The pressure setting of the pump should be checked every time the oil burner is serviced to ensure that that nozzle is receiving fuel at a pressure that is sufficient to atomize the fuel properly.

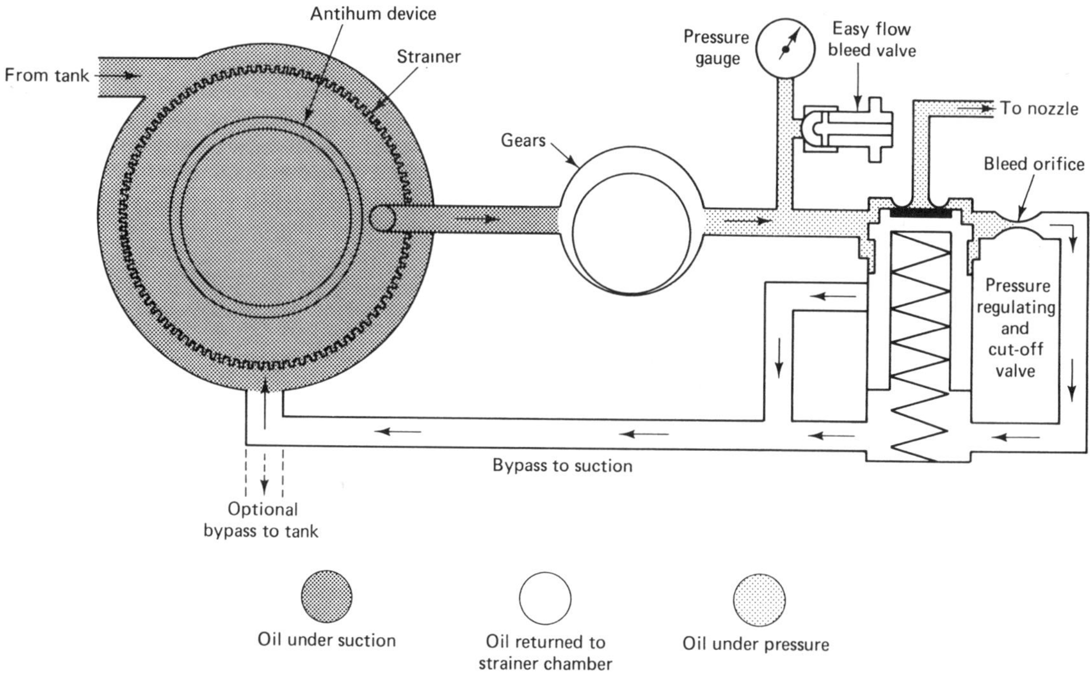

Figure 4.28 Circuit diagram for single-stage fuel pump. (Courtesy of Suntec Industries, Inc.)

Figure 4.29 Two-stage fuel oil pump. (Courtesy of Suntec Industries, Inc.)

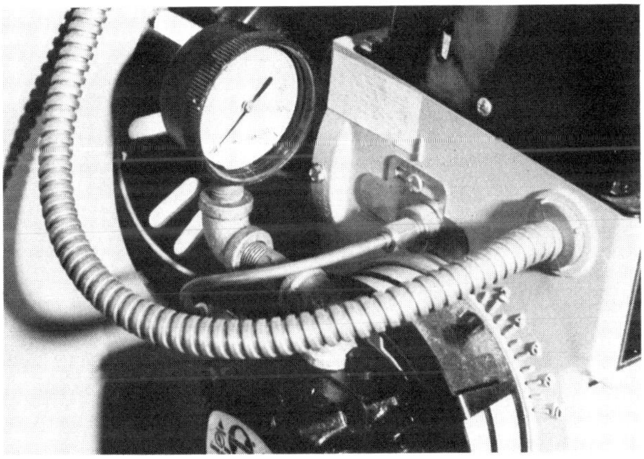

Figure 4.30 Pressure gauge inserted into gauge port of fuel pump.

Incorrect pump pressure can be caused by several conditions. Table 4.5 should prove helpful for troubleshooting some of the more common problems.

Improper nozzle cutoff. Improper flame cutoff at the nozzle when the burner shuts down is indicated by excessive smoke and after-fire within the combustion chamber. To determine if the fuel pump is at fault and needs to be replaced, a pressure gauge should be installed in the nozzle port of the pump (Figure 4.31). After turning off the burner, the pressure gauge should stabilize above 0 psi. If this does not occur, and instead the pressure reading drops to zero, the fuel pump should be replaced. Other causes that should be investigated relative to improper ignition cutoff are filter leaks, loose cover gaskets and screws on the pump case, air leaks in the intake line, or a partially clogged nozzle strainer.

Pump vacuum. Vacuum measurements are sometimes required and should be taken from an unused intake port on the fuel pump, similar to the procedures used for taking pressure measurements, using a good-quality vacuum gauge. Vac-

TABLE 4.5 FUEL PUMP PROBLEMS

Problem	Component at fault	Cause	Corrective action
Low oil pressure	Nozzle	Excess capacity	Replace with proper-size nozzle
	Pressure gauge	Defective	Replace
Pulsating pressure	Filter	Clogged	Replace
	Pump strainer	Dirty	Clean or replace
	Intake line	Air leaks	Tighten connections
	Pump cover	Air leaks	Tighten connections
	Fuel pump	Fuel leaks	Replace pump

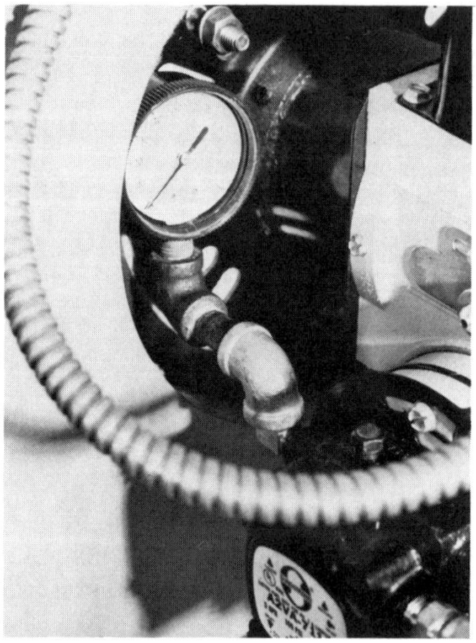

Figure 4.31 Pressure gauge directed into nozzle port on fuel pump.

uum readings are valuable to determine if the fuel supply system is operating properly. Under normal conditions, intake vacuum should be between 0 and 5 lb. Excessively high vacuum readings are indicative of either kinked fuel supply lines or improperly sized feed lines. Also, check the condition of the fuel filter to ensure that it is not clogged with dirt and sediment. Check all supply and return line fittings and manufacturers' specifications for proper line sizes if excessively high vacuum readings are encountered.

AIR TUBE ASSEMBLY

The burner air tube is designed to hold the nozzle assembly and ignition system components of the oil burner within a stream of high-velocity air supplied by the blower wheel and is illustrated in Figures 4.32 and 4.33. Note from Figure 4.32 that the air tube length is a function not only of the particular oil burner but the configuration of the combustion chamber of the appliance as well. The end of the blower tube/burner head combination should be $\frac{1}{4}$ in. back from the inside of the combustion chamber and should never extend into it, or premature failure of the combustion head assembly will result due to excess heat. There are a variety of air tube combinations available for most residential oil burners. When a new oil burner is being installed into an existing appliance, the manufacturer of both the burner and boiler should be consulted to determine the best air tube combination to be used.

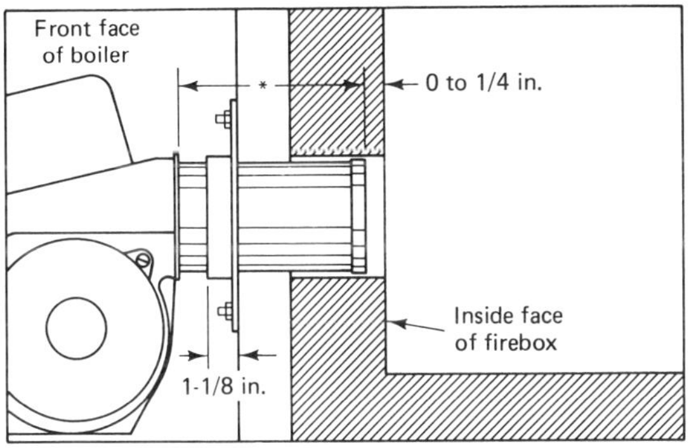

Stock air tube lengths

6-5/8 in.	9 in.	13 in.	16 in.

Figure 4.32 Oil burner air tube. The air tube length (*) is the distance from the front of the burner housing to the burner head and/or shield. (*Note:* Adjustable flange width—1 to $1\frac{1}{8}$in.) (Courtesy of R.W. Beckett Corp.)

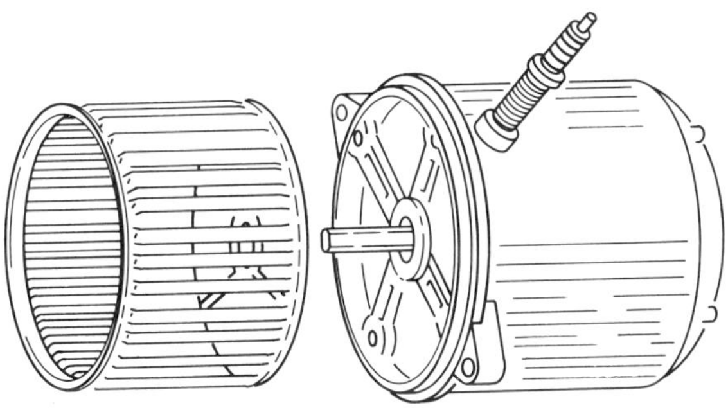

Figure 4.33 Oil burner blower wheel. (Courtesy of R.W. Beckett Corp.)

IGNITION SYSTEM

The fuel ignition system on modern residential oil-fired equipment is illustrated in Figure 4.34. In this ignition system, several processes are occurring simultaneously: An electrical transformer supplies a high-voltage current to the electrode rod extensions; this high current travels through the electrodes and jumps a preset gap, creating a spark at the tip of the electrode assembly; this spark lies in the direct path of the atomized fuel oil spray delivered by the nozzle assembly; combustion air travels from the blower through the air tube and air retention rings, where it combines with the fuel oil in the combustion chamber.

The nozzle and electrode assembly is illustrated in detail in Figure 4.35. The relative position of all components within this assembly is critical for the proper functioning of the burner. The electrodes should be positioned as illustrated in Fig-

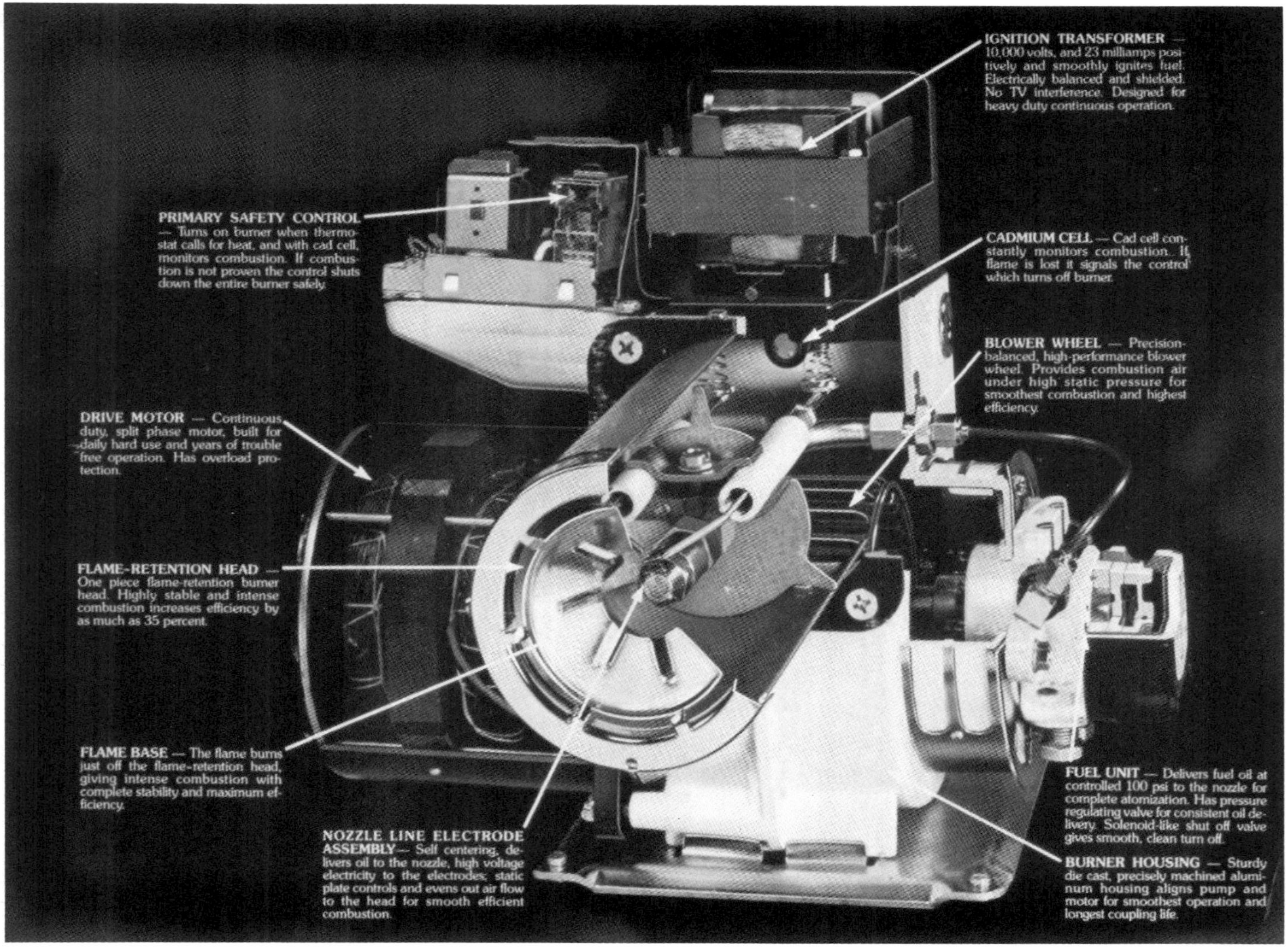

Figure 4.34 Fuel oil ignition system. (Courtesy of R.W. Beckett Corp.)

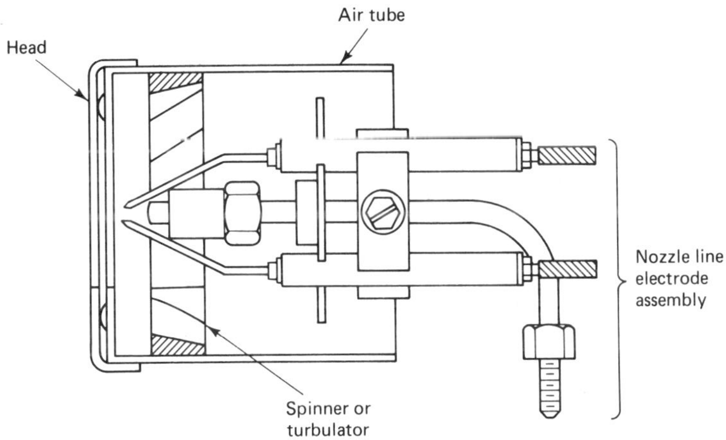

Figure 4.35 Nozzle and electrode assembly.

ure 4.36. Although the specific gap and position of the electrode assembly relative to the nozzle centerline may vary from one burner manufacturer to another, the basic configuration of the components is similar in all systems. Note the centering flange on the ignition assembly in Figure 4.34 to help ensure proper location of the unit within the blower tube.

Ignition Transformer

The ignition transformer (Figure 4.37) is specially designed for oil burner operation. Producing in excess of 10,000 V, these transformers are designed for heavy-duty continuous operation and are specially shielded to prevent radio and television in-

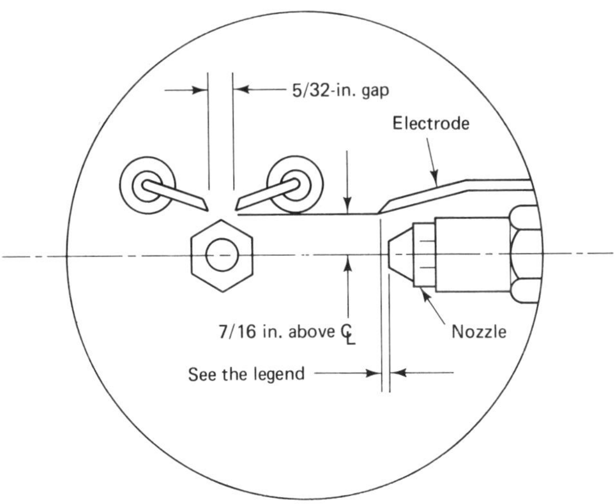

Figure 4.36 Electrode position and gapping. Note that the electrode position is ahead of the nozzle. F0 to F31 heads: 0 to $\frac{1}{16}$ in; F30 and F300 heads: $\frac{1}{8}$ to $\frac{5}{32}$ in.

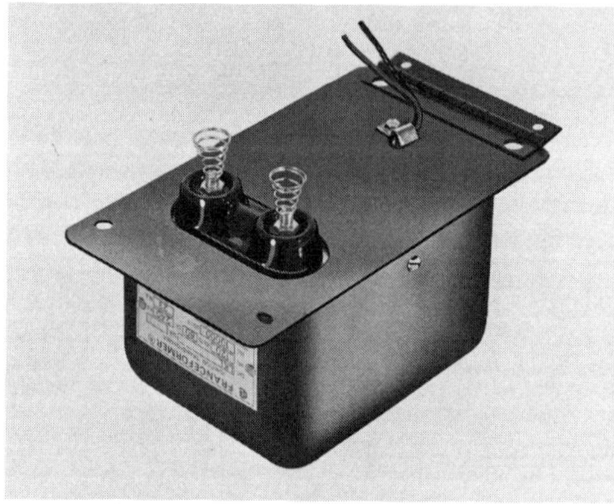

Figure 4.37 Oil burner ignition transformer. (Courtesy of France, a Scott Fetzer Co.)

terference. Two spring terminals extend from the bottom of the transformer to provide electrical contact with the electrode extension tubes.

Nozzle Assembly

The purpose of the nozzle assembly in the burner mechanism is to break up the fuel oil into tiny droplets. This atomization process greatly increases the surface area of the fuel oil, enabling the combustion process to take place at very high efficiencies.

Oil burner nozzles must be properly sized for the specific appliance in which they are to be installed. Nozzles used in most residential oil-fired installations are rated according to the nozzle capacity of No. 2 fuel oil delivered per hour (gal/hr) at 100 psi, the spray angle of the nozzle, and the spray pattern designation.

The capacity of the nozzle (gal/hr) must be matched to the Btu input for which the boiler is designed. Most residential oil boilers fire between 0.5 and 3.0 gal/hr. Although there is sometimes a small degree of variation that is allowable as far as overfiring or underfiring a particular boiler, the appliance manufacturer should be consulted prior to making any nozzle size changes.

In most instances, the appropriate information regarding nozzle sizing is stamped on the specification plate of the boiler. If the Btu input rating of the unit is available, the burner nozzle may be sized by using one of the following formulas for this purpose:*

1. If the gross Btu input of the boiler is available, the burner nozzle is determined by dividing the Btu input of the appliance by 140,000, the average amount of energy available from 1 gal of No. 2 fuel oil:

$$\text{nozzle size (gal/hr)} = \frac{\text{Btu input of boiler}}{140{,}000}$$

*Adapted from "A Total Look at Oil Burner Nozzles," published by Delavan, Inc., *Bulletin 884,* 1984.

2. If the boiler is rated in Btu output rather than input, the nozzle size may be determined by the following formula:*

$$\text{nozzle size (gal/hr)} = \frac{\text{Btu output of boiler}}{\text{boiler combustion efficiency} \times 140{,}000}$$

3. For steam heating systems, the nozzle size may be determined by using the total number of square feet of radiation used in the heating system, as follows:

$$\text{nozzle size (gal/hr)} = \frac{\text{total square feet of radiation} \times 240}{\text{boiler combustion efficiency} \times 140{,}000}$$

Modern oil burner nozzles are engineered to operate under the harsh combustion conditions within the appliance and yet deliver precisely metered fuel in a specific spray pattern and angle for extended periods. A residential fuel oil nozzle is illustrated in Figure 4.38.

For the nozzle to function properly, oil must be supplied to it at high pressure, usually 100 psi delivered from the fuel oil pump. The oil enters the nozzle through a built-in filter assembly and then into the swirl chamber through a series of slots that have been machined into the nozzle distributor at an angle. Oil entering the slots is forced into the swirl chamber, where centrifugal force throws the oil against the walls of the chamber, leaving an air gap in the center of the chamber (Figure 4.39).

Fuel oil exits the nozzle in the shape of a hollow tube or modified conical shape and breaks into thousands of tiny droplets. The shape of the spray is usually configured as a hollow, solid, or combination pattern. Each of these pattern shapes is illustrated in Figure 4.40.

Hollow cone nozzles are used in oil burners with a hollow air pattern and in

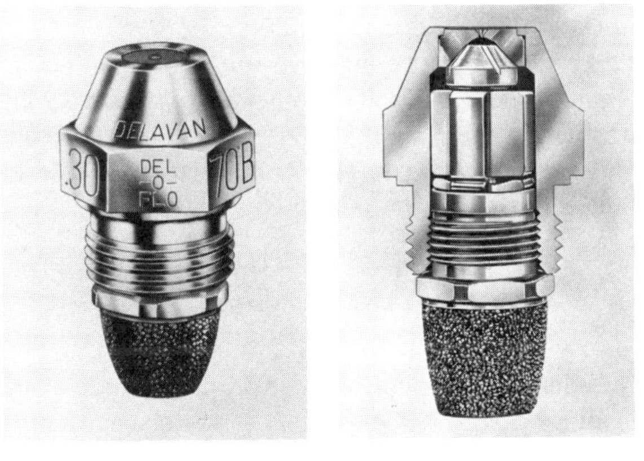

Figure 4.38 Fuel oil nozzle. Note the specifications on the nozzle that specify oil delivery rate and spray angle and pattern classification. (Courtesy of Delavan, Inc.)

*The figure that is appropriate for combustion efficiency will vary from one appliance to another. In general, for conventional low-speed oil burners, an efficiency figure of approximately 70% should be used. For high-speed flame retention burners, a combustion efficiency of 80 to 85% is appropriate.

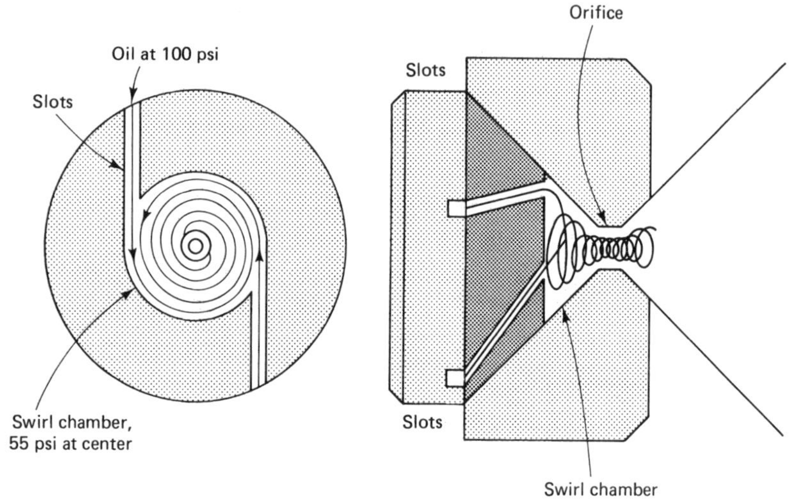

Figure 4.39 Operation of an oil burner nozzle. (Courtesy of Delavan, Inc.)

boilers with firing rates under 1 gal/hr. The spray pattern tends to be highly stable, a prerequisite for fractional gallonage firing rates where the droplet size can sometimes be irregular due to changes in fuel oil viscosity during operation of the unit.

Solid cone nozzles feature a spray in which oil is spread uniformly throughout the pattern. Solid cone nozzles are generally used in larger residential boilers that fire between 2 and 3 gal/hr and in units where long fire shapes are necessary.

Combination nozzles produce a spray which is neither totally hollow nor solid. These nozzles can be used effectively in applications from approximately 0.5 to 8 gal/hr. With these nozzles, the spray pattern at lower firing rates tends to be hollow; the higher firing rates produce a more solid spray configuration.

The spray angle of the nozzle should be matched to the design and configuration of the combustion chamber and will vary from one appliance to another. Sometimes, larger boiler sizes within the same manufactured series will require different spray angles on the nozzles. In each case, the manufacturer should be consulted for the appropriate angle of the nozzle to be used.

OFF-CYCLE AIR RESTRICTORS

Heat loss during the "off" cycles of the boiler can significantly increase fuel consumption throughout the heating season. Off-cycle heat losses occur when:

1. Air enters the air bands and air shutters on the burner housing. This air flows through the combustion chamber and boiler flue passages, where it picks up heat from the boiler water, and eventually flows out the chimney. This con-

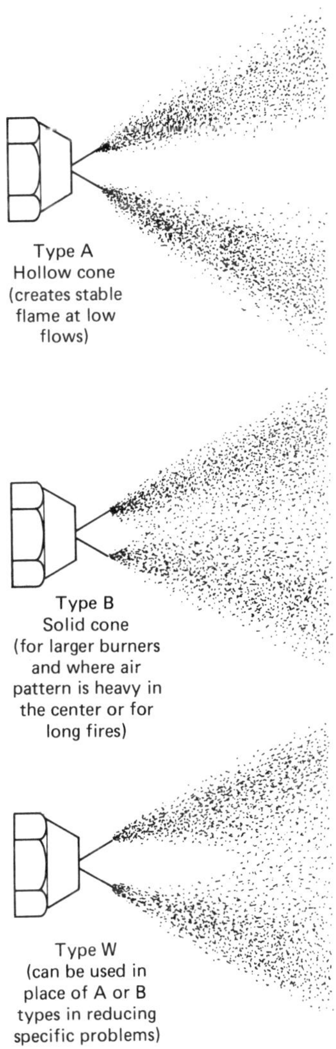

Figure 4.40 Nozzle spray patterns. (Courtesy of Delavan, Inc.)

dition lowers boiler water temperature on a continual basis, causing the unit to cycle more frequently due to continuing heat loss.

2. Air enters the barometric draft damper and travels up the chimney flue, cooling it in the process. This can both reduce chimney draft and retard heat transfer from the chimney into the interior of the home.

There are two popular methods for reducing off-cycle heat losses: installation of oil burner centrifugally operated intake dampers and installation of electrically operated stack dampers.

Figure 4.41 Centrifugal air intake damper. These dampers automatically close when the oil burner shuts off, restricting intake air. (Courtesy of R.W. Beckett Corp.)

Air Intake Dampers

An air intake damper is a device that seals off the air bands and air shutters on the oil burner when the burner motor is not operating. This damper replaces the standard coupling located between the burner motor and fuel pump (Figure 4.41). Operated by centrifugal force, the damper opens when the burner motor is running. When the motor stops, the damper closes by spring action, sealing the air intake shutters. This device is a simple, highly cost-effective method for reducing off-cycle air losses and should be standard equipment in all residential oil burner installations.

Electric Stack Damper

An electric stack damper is a device that is inserted into the vent pipe of the appliance and closes the pipe when the burner is not operating. This device is illustrated in Figure 4.42. The damper is connected to the oil burner relay in a series circuit configuration so that the burner can operate only when the damper has completely opened. When the burner shuts off, the damper automatically closes, stopping almost all airflow through the appliance by closing the appliance stack pipe.

Figure 4.42 Electrically operated stack damper for oil-fired boilers. (Courtesy of Honeywell, Inc.)

This damper will pay for itself many times over in fuel savings that result from significantly reducing off-cycle boiler heat loss.

COMBUSTION EFFICIENCY TESTING

To understand and interpret the procedures that are involved in combustion efficiency testing, a knowledge of the fundamentals of fuel oil combustion chemistry, together with an appropriate test kit, are required. A test kit suitable for residential fuel oil combustion efficiency testing is illustrated in Figure 4.43. We will now examine some of the chemical aspects of fuel oil combustion, together with specific procedures that are used in combustion efficiency analysis.

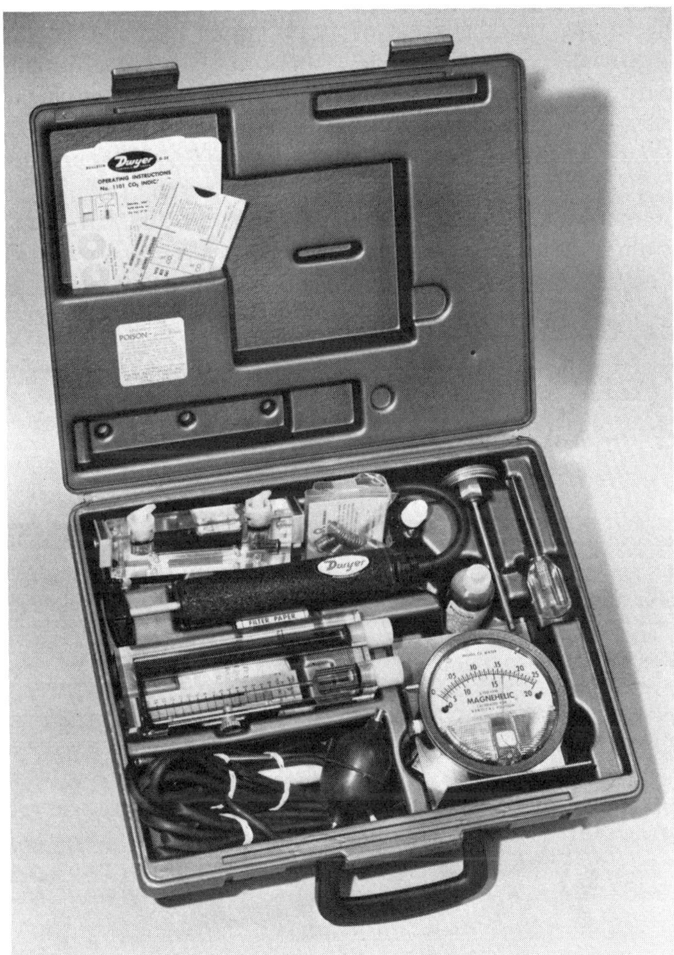

Figure 4.43 Combustion efficiency test kit. (Courtesy of Dwyer Instruments, Inc.)

Carbon Dioxide Test

The most critical adjustment to be made on the oil burner is the amount of excess air that is admitted into the combustion chamber. If pure carbon were burned with 100% efficiency, the major by-product of this process would be approximately 21% carbon dioxide (CO_2) by volume; the balance of the by-product would be nitrogen. This results from air being composed of 21% oxygen and 79% nitrogen. Nitrogen plays no part in this combustion process. Thus, if we were to make twice as much air available that is necessary for complete combustion of the carbon, CO_2 would account for approximately 10.5% of the combustion by-products. The level of CO_2 is an indication of the amount of excess air present when the fuel is burned. As the volume of excess air is increased, the percentage of CO_2 in the stack gases is reduced. Too much excess air results in low stack temperatures and excessive fuel consumption. To illustrate this point, if an oil burner is operating with 200% excess air, it will generally yield CO_2 levels of between 5 and 6% accompanied by excessively high fuel consumption. The test procedure used to determine the percentage of CO_2 present in the flue gases involves drilling a hole into the stack pipe in close proximity to the boiler through which a sample of the flue gases can be taken.

The flue gases are drawn into either a solid-state electronic tester or into a calibrated vial of fluid that absorbs carbon dioxide. A typical absorption fluid test instrument for determining carbon dioxide levels is shown in Figure 4.44.

When an absorption fluid tester is used, a small hand pump draws a sample

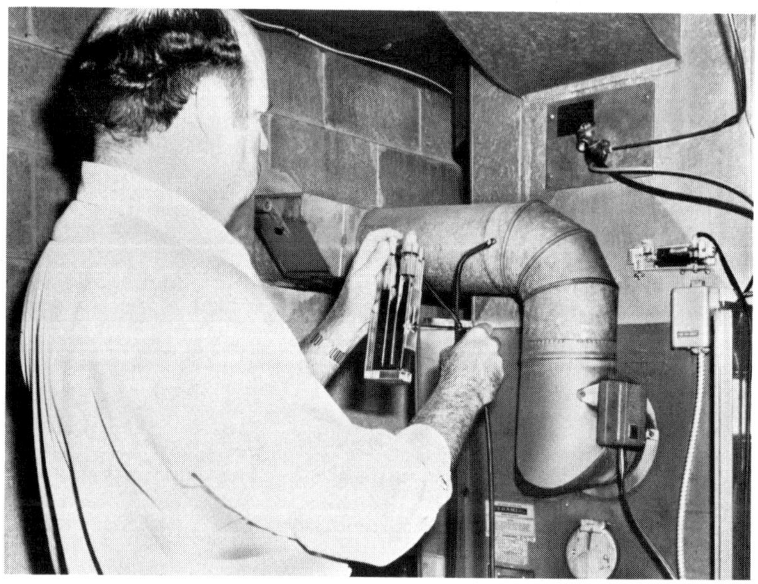

Figure 4.44 Sampling flue gases for a carbon dioxide test. (Courtesy of Dwyer Instruments, Inc.)

of the stack gases through the fluid in the test instrument. To ensure complete mixing, the fluid is carefully agitated. A reduction in the volume of the fluid in direct proportion to the amount of carbon dioxide absorbed within the test fluid can then be observed on a graduated scale (Figure 4.45). Be sure to observe manufacturers' instructions carefully when performing these tests to ensure the accuracy of the results obtained. In Figure 4.45, the fluid level indicates that 10% CO_2 has been absorbed during this test procedure.

The percentage of the CO_2 present in the flue gases indicates the quality of the combustion that is taking place within the boiler. High amounts of CO_2 (in excess of 10 to 12%) indicate relatively low amounts of excess air. Low excess air means that there is less air present to cool the fire. Thus a fire with little excess air will be hotter than one operating with high amounts of excess air.

There is a limit to the amount of CO_2 that can be achieved in residential oil-fired heating equipment under average firing conditions. If too much air is present, excessive smoke and resulting stack temperatures will be too cool to allow for efficient heating of the boiler. If too little excess air is present, smoke levels in the flue gases will rise to unacceptable levels, resulting in sooting of internal heat transfer surfaces within the boiler and reduced heating efficiency of the unit. Maximum smoke levels allowable for CO_2 readings are fixed at No. 1 smoke, a testing procedure that will be examined shortly.

To help the reader become familiar with problems associated with incorrect CO_2 stack gas levels, Table 4.6 presents some factors that influence the CO_2 levels.

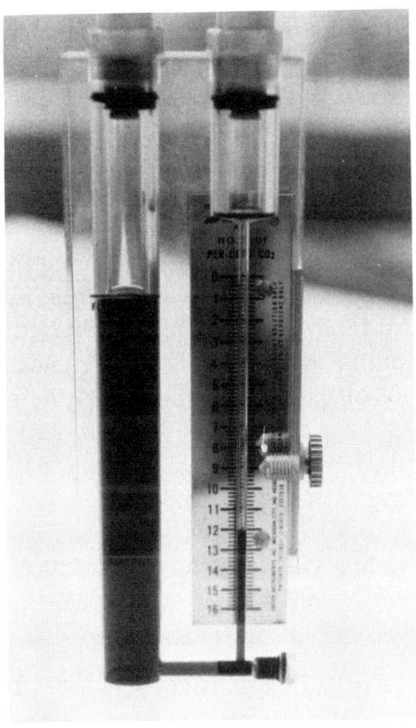

Figure 4.45 Carbon dioxide reading on a CO_2 sampling meter. (Courtesy of Dwyer Instruments, Inc.)

TABLE 4.6 TROUBLESHOOTING CO_2 LEVELS

Cause of low CO_2 levels	Cure
Insufficient fuel delivered to burner	Increase nozzle capacity
Improper mixing of combustion air	Clean air shutters, check condition of retention ring
Too much excess air	Reduce volume at burner
Plugged nozzle	Replace nozzle assembly
Damaged nozzle	
One-sided spray pattern Off-center firing	Replace nozzle, check for proper centering of assembly
Oversized combustion chamber	Reduce dimensions with new chamber lining
Poor insulation of combustion chamber	Replace with high-temperature brick or approved lining material
Unwanted air infiltration into appliance	Find and seal
	Install burner intake damper
Overfire draft too high	Recalibrate barometric damper

Smoke Test

All combustion efficiency test kits contain a smoke sampling device, which is illustrated in Figure 4.46. The level of smoke in the flue gases in oil-fired appliances is very important, since the point at which excess air begins to cause a fire to smoke is usually the point at which the highest CO_2 levels are achieved. Smoke points will vary from one burner to another, depending on the configurations of each specific installation. The smoke point in one boiler might be 6%; in another it might be 12%. Whatever the level for a particular oil burner, the objective of this procedure is always the same: to achieve the highest CO_2 levels possible with a clean-burning, low-smoke fire with minimum excess air.

Smoke levels are determined by inserting a sampling tube into the stack sampling hole and pumping a specific volume of gas through a small piece of white filter paper. The color of the filter paper is then compared to a standard smoke-level chart to determine the smoke level of the fire. This chart is illustrated in Figure 4.47. In almost all residential oil heating applications, the smoke level of the fire should never be above No. 1.

Stack Temperature Test

Stack temperature levels, together with CO_2 and smoke levels, are used to determine the combustion efficiency of the oil burner. A special thermometer with an expanded temperature range and corrosion-resistant probe is inserted into the sampling hole in the stack pipe to measure the flue gas temperature (Figure 4.48).

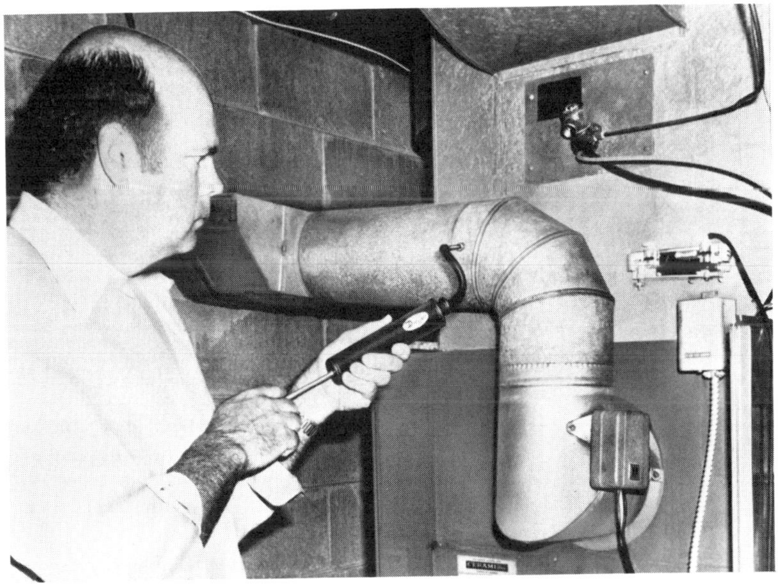

Figure 4.46 Drawing a smoke sample using a smoke sampling device. (Courtesy of Dwyer Instruments, Inc.)

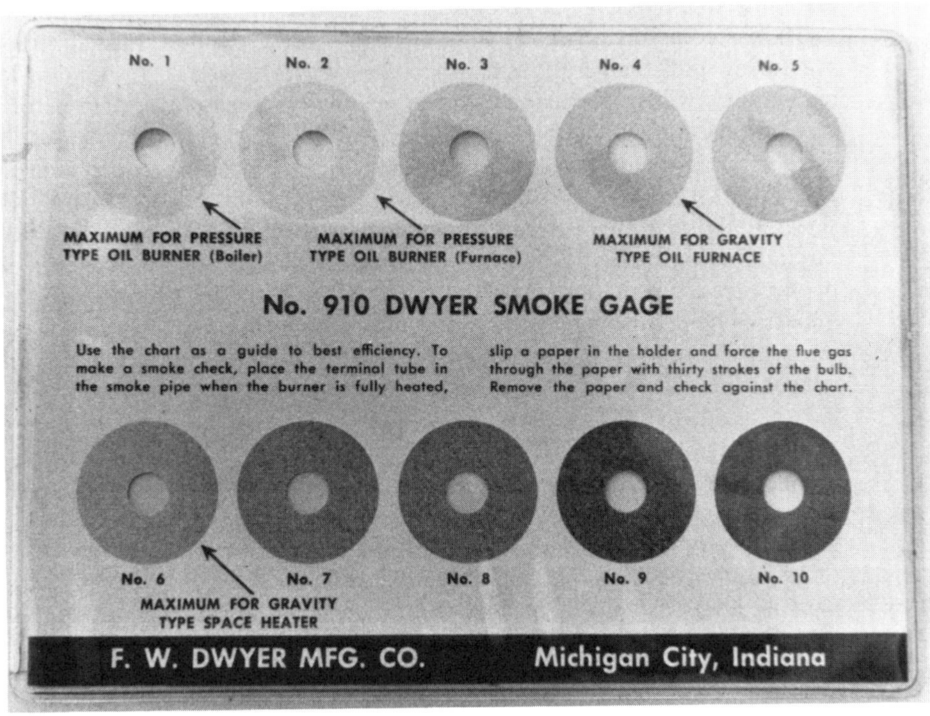

Figure 4.47 Smoke-level chart. A maximum smoke level of 1 is recommended for almost all residential oil-fired heating boilers. (Courtesy of Dwyer Instruments, Inc.)

Figure 4.48 Measuring flue gas stack temperature. Stack temperature readings should be taken only when the heating system reaches stable operating temperatures.

To obtain an accurate temperature reading, the thermometer should be left in place until the temperature stabilizes. This procedure should only be done after the boiler has come up to steady-state operation for accurate results to be obtained. Measuring stack temperatures from a cold-start boiler with corresponding low water temperature will result in temperature readings that are too low to be of value in correctly measuring boiler efficiency.

After the reading has stabilized on the thermometer, the room temperature of the boiler location should be subtracted from the stack gas temperature reading since combustion air was already heated to room temperature prior to passing through the oil burner. Stack temperatures are indicative of heat that is picked up by the excess air in the boiler; thus subtraction of boiler room temperature is an important step. Also, since the final combustion efficiency of the burner is a combination of both stack temperature and CO_2 levels, failure to subtract boiler room temperature will result in an efficiency figure lower that it truly is.

Since different appliances are designed to function most efficiently at different stack temperatures, it is often difficult to determine the one best reading for any given oil burner. For example, some burners operate most efficiently at temperatures in excess of 600°F. At these elevated temperatures the heat loss through the stack gases will rarely be less than 25%. Most oil-fired boilers operate with a high degree of efficiency when stack temperatures do not exceed 500 to 550°F, with accompanying levels of CO_2 at 10% or higher at No. 0 or 1 smoke levels. Also, ideal operating stack temperatures of low-speed old-style burners will often differ significantly from high-speed flame retention units.

Figure 4.49 graphs the stack losses of the appliance as a percentage of combustion efficiency, based on a variety of steady-state operating stack temperatures.

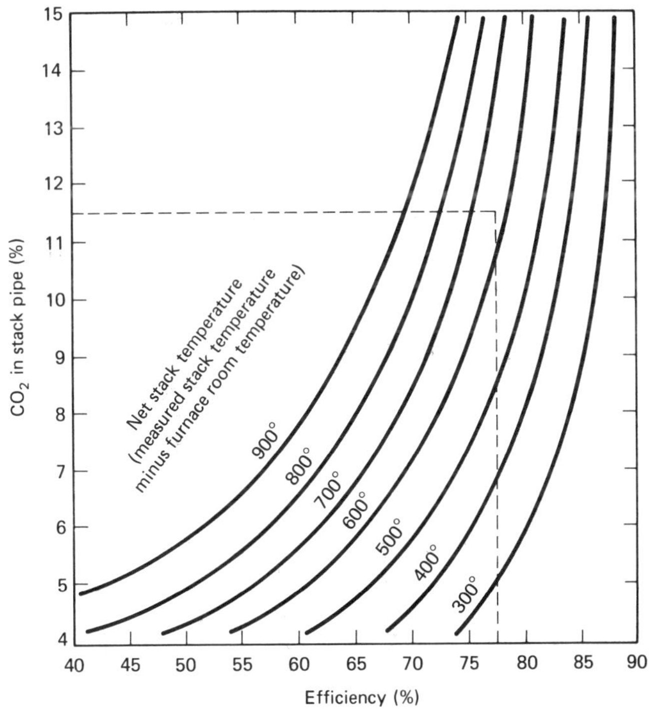

Figure 4.49 Stack losses as a percentage of combustion efficiency. (Courtesy of R.W. Beckett Corp.)

Draft Requirements

A draft gauge is standard issue in all combustion efficiency testing kits (Figure 4.50). *Draft* can be defined as the negative evacuation pressure on the combustion chamber caused by the chimney. Under ideal draft conditions, there will be sufficient negative pressure created within the chimney and venting system of the appliance to remove all the effluents from the combustion chamber without sucking excessive heat out of the boiler. Draft is measured in inches of water column with a U-shaped tube known as a *manometer* Figure 4.51).

Most residential oil burners require a chimney draft ranging between 0.02 and 0.03 in. of water column to remove safely and efficiently all combustion by-products. Since many chimneys produce draft in excess of that required for normal operation, a draft regulator should be installed within the system. A typical regulator for this purpose is the barometric draft damper, illustrated in Figure 4.52. The dampers must be installed in the proper location in order for the burner mechanism to

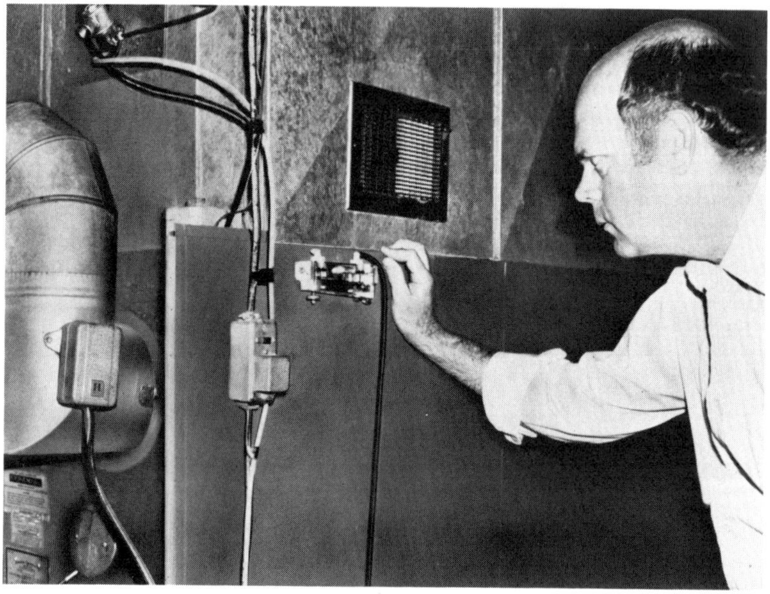

Figure 4.50 Checking draft with a draft gauge. (Courtesy of Dwyer Instruments, Inc.)

function correctly. Figure 4.53 illustrates both correct and incorrect locations for these dampers.

When adjusting draft dampers, it should be remembered that draft measurements should always be taken over the fire, not in the smoke pipe. Most boilers have a small test hole that opens into the combustion chamber for this purpose. If one is not present, a small hole can be drilled for this purpose and sealed when not

Figure 4.51 Manometer used for measuring chimney draft. (Courtesy of Dwyer Instruments, Inc.)

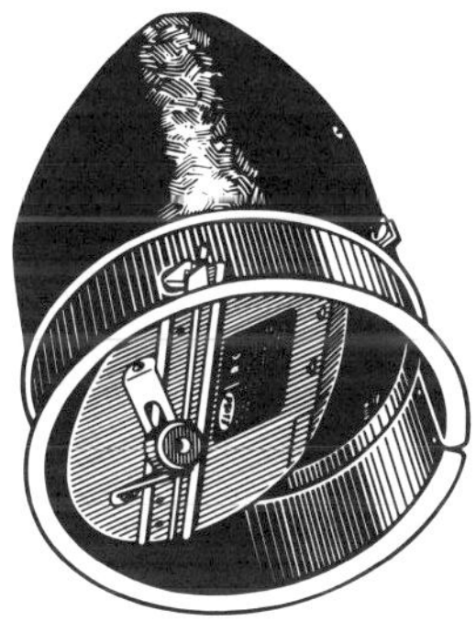

Figure 4.52 Barometric draft damper. These units, for use in oil and solid-fuel boilers and furnaces, automatically help to maintain proper over-fire draft under varying chimney draft conditions. (Courtesy of the Field Controls Company.)

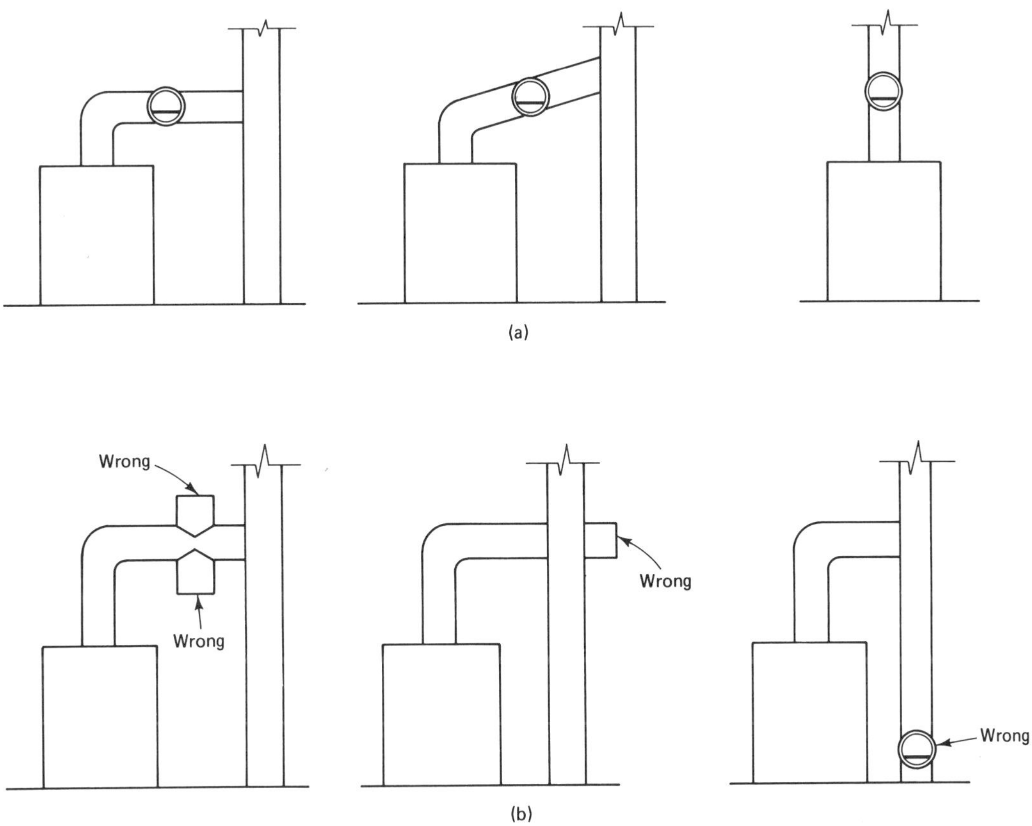

Figure 4.53 Proper (a) and improper (b) locations for barometric draft dampers.

being used. If a hole is to be drilled, it is imperative to locate the hole away from any water jacketing on the boiler. If there are any questions as to the best location for this sampling hole, the appliance manufacturer should be consulted.

Combustion Efficiency

After all appropriate measurements concerning boiler stack temperature, CO_2 levels, and draft measurements and adjustments have been completed, the efficiency of the appliance should be determined. All combustion efficiency kits contain a simple slide-rule calculating device into which all pertinent measurements from the efficiency tests are set. The combustion efficiency of the unit is then read out on a chart. A calculator of this type is illustrated in Figure 4.54.

With an understanding of how the combustion efficiency of oil burners is affected by excess air and how combustion variables can accurately be determined, proper adjustment of the particular appliance becomes the norm rather than the exception.

BOILER OPERATING SEQUENCES

Proper operation of an oil-fired hydronic heating system depends on the successful integration of three separate systems: fuel oil delivery and burner mechanism, electric control devices, and the integrity of the boiler itself. We examine aspects of all three systems to understand how their interaction results in successful heating system operation.

General Sequence of Operation

For purposes of illustrating a normal boiler operating sequence, let us examine a typical single-zone series-loop heating system with a tankless coil and boiler configuration, as illustrated in Figure 4.55. When the room thermostat calls for heat, a sequence of operations takes place whereby the thermostat energizes the aquastat relay to begin burner operation (Figure 4.56). Assuming that the boiler has been turned on for the first time and that the water in it is cold, no circulation of water within the heating loop will take place until it has been heated to the low-limit set point on the boiler aquastat, usually between 140 and 180°F. When the low-limit temperature setting on the aquastat has been reached, the circulator relay in the aquastat closes and heated water begins to move throughout the system. The circulator moves water from the boiler supply piping throughout the baseboard radiation loop and back into the bottom of the boiler, where it is reheated.

The oil burner will continue to fire until either the room thermostat setting has been satisfied or the boiler water reaches its high-limit set point. Assuming that the boiler reaches the high-limit set point and shuts off the burner while there is

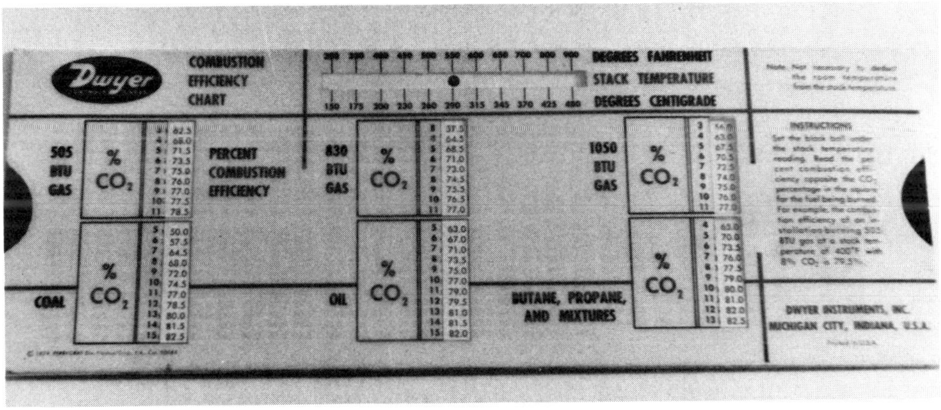

Figure 4.54 Combustion efficiency calculator. (Courtesy of Dwyer Instruments, Inc.)

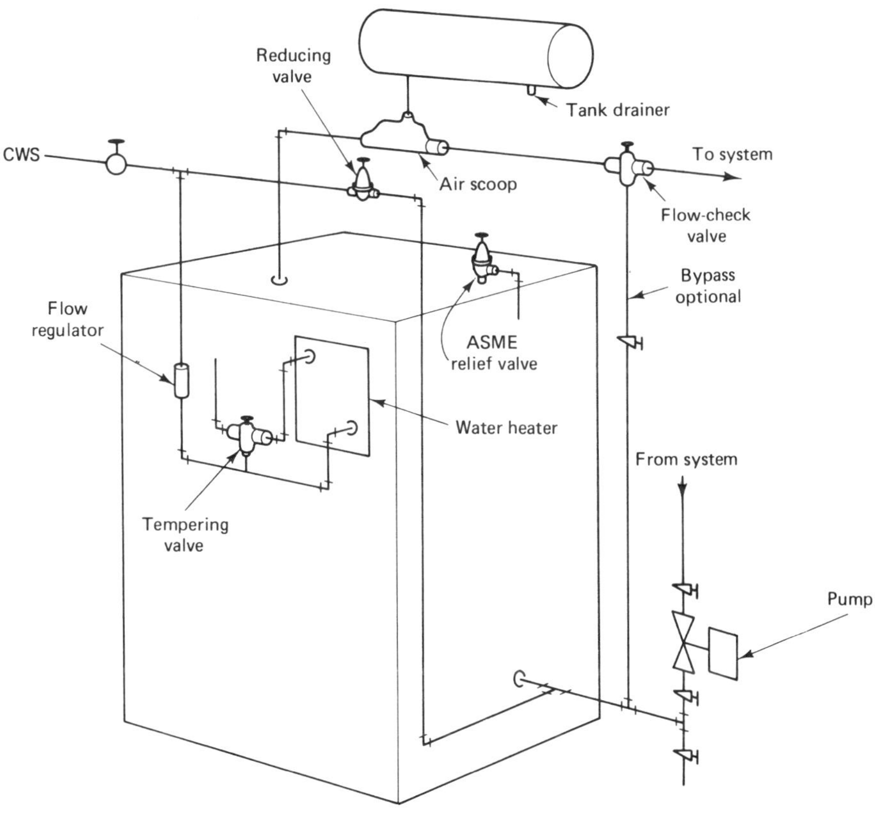

Figure 4.55 Boiler configuration for single-zone series-loop heating system. (Courtesy Taco, Inc.)

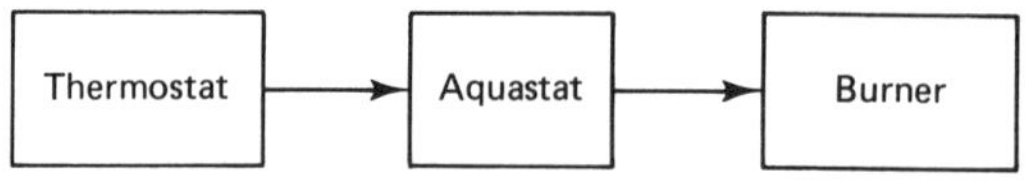

Figure 4.56 Ignition system sequence.

still a call for heat in the home, water continues to circulate throughout the baseboard radiation.

As the water circulates, it transfers heat to the interior of the home and returns to the boiler approximately 20°F cooler than when it left. This causes the burner to cycle in response to raising and lowering water temperatures throughout a typical heating cycle. When the thermostat setting has been reached, the burner and circulator will shut off.

In this particular illustration, we noted that the boiler contained a tankless coil for domestic hot water production. Thus, even during times when no heat is required within the home, the boiler will cycle on and off automatically to provide sufficient water temperature for the production of domestic hot water.

With aquastat settings of 160°F low limit and 180°F high limit and a 20°F differential, upon a drop in boiler-water temperature to 150°F (low-limit setting minus the 10°F built-in differential), the burner will ignite and continue to run until the water temperature reaches 170°F (cut-in point + adjustable differential of 20°F). The circulator in this instance will not operate as long as the water temperature is below 160°F, the low-limit circulator setting.

If no tankless coil is installed in the boiler, however, the operation of the system assumes a cold-start configuration in the aquastat that features only a high-limit setting. On a call for heat by the thermostat, the burner comes on until the boiler reaches the high-limit setting and then shuts off. Circulation of the water continues throughout the heating cycle. As the water temperature in the system drops below the set point, usually a fixed differential in the aquastat from the high-limit setting, the burner again comes on to bring the boiler water back up to temperature. When the room thermostat has been satisfied, the burner and circulator both shut off simultaneously. Since a cold-start configuration contains no low-limit burner control in the boiler aquastat, no cycling of the burner is required since no tankless coil is present, and the burner operates only upon a call for heat from the thermostat. This cycling arrangement is illustrated in Figure 4.57. Cold-start boilers sometimes incorporate a low-limit control that governs circulator operation to prevent cold water circulation in the heating system.

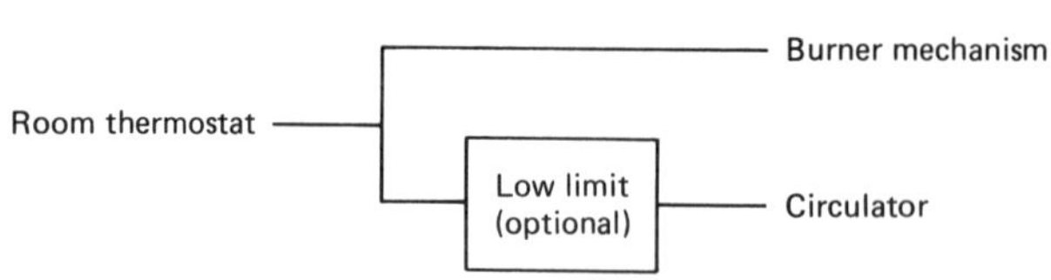

Figure 4.57 Control sequence of a cold-start boiler (no tankless coil).

Zoning Modifications

When more than one zone is incorporated into the heating system, modifications of the operating sequence described above are incorporated into the switching controls. These modifications are minor in nature and allow each zone to activate the circulator and burner independently of the other zones. This control arrangement is illustrated in Figure 4.58.

Most multizoned heating systems employ either zone valves or separate circulators for controlling hot water circulation in each separate baseboard circuit. When separate circulators are used for each heating zone, each is coupled to a separate switching relay. Each relay, in turn, is connected to the master aquastat controller. This electrical configuration is illustrated in Figure 4.59. Note that the zone 2 switching relays draw power from the master aquastat controller. This wiring arrangement incorporates low-limit circulator protection for the second heating zone by preventing the zone 2 circulator from coming on until the low-limit main aquastat temperature setting on the heating boiler has been reached. Each additional switching relay for added zones is wired through the low-limit circuit in the main aquastat to incorporate circulator protection if required. Further, each additional zone requires a separate field-added 24-V thermostat and R845 relay.

Each time a zone calls for heat, the burner is automatically activated. This is accomplished by wiring the zone switching relays to power the burner energizer terminal, ZR, in the main boiler aquastat whenever these zones are calling for heat (Figure 4.60).

When zone valves are used in multizone heating systems as opposed to separate circulators, the control configuration is similar to that illustrated in Figure 4.61. When a zone thermostat calls for heat, the zone valve is energized. This zone valve is an electrically operated water valve. When the valve is fully opened, a set of end switch contacts close, thereby closing the electrical circuit that enables the circulators to begin operating. Also, the burner will ignite when the circulators begin to operate. The boiler in this mode operates according to the dictates of the high and low limits of the aquastat in a tankless coil configuration, or on the demands of the high-limit control in the cold-start configuration. When the room thermostat in the zone has been satisfied, the electrical circuit deenergizes the zone valve, allowing it to close and terminate the burner ignition sequence. In zone valve configurations, one circulator is usually used for all the water circulation required in the heating system. As long as any zone is calling for heat, the circulator will remain in operation.

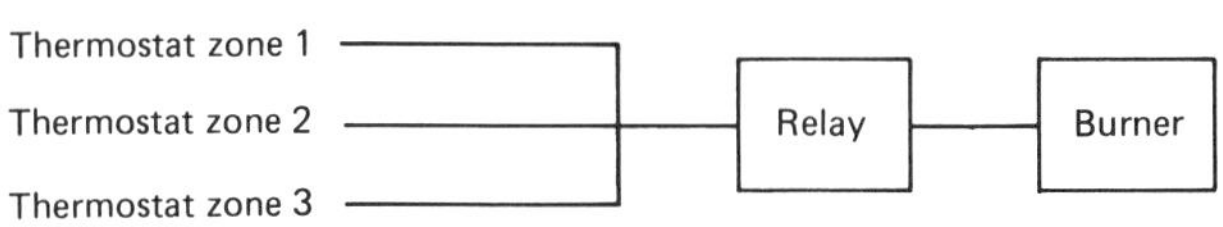

Figure 4.58 Multizone control sequence.

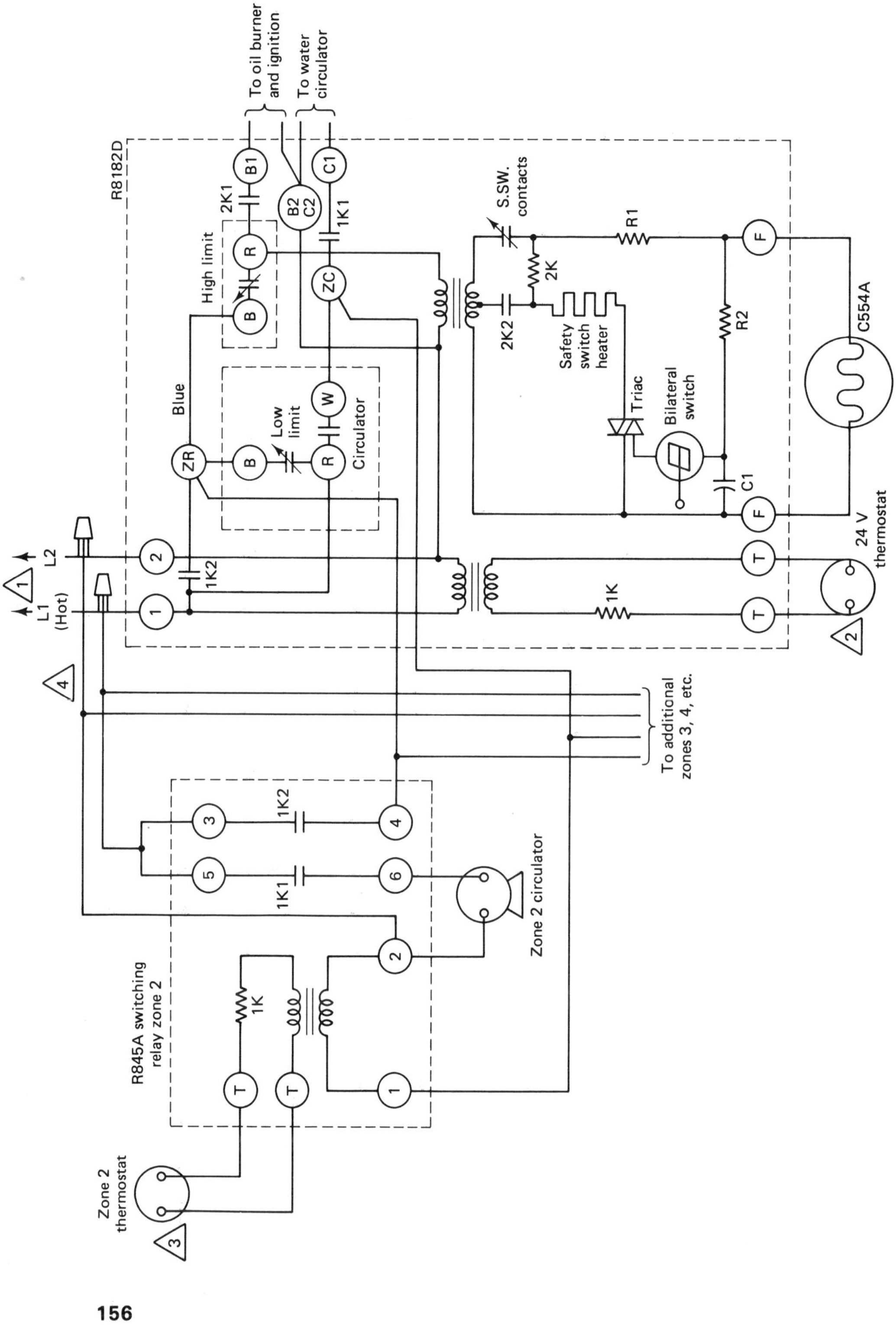
To oil burner and ignition
To water circulator
R8182D
B1
2K1
B2 C2
C1
1K1
High limit
B
R
ZC
Blue
ZR
B
Low limit
R
W
Circulator
S.SW. contacts
R1
2K
2K2
Safety switch heater
Triac
Bilateral switch
R2
C554A
C1
F
F
T
T
24 V thermostat
1K
2
1
1K2
L2
L1 (Hot)
1
2
4
To additional zones 3, 4, etc.
3
1K2
4
5
1K1
6
Zone 2 circulator
2
1K
T
T
1
R845A switching relay zone 2
Zone 2 thermostat
3

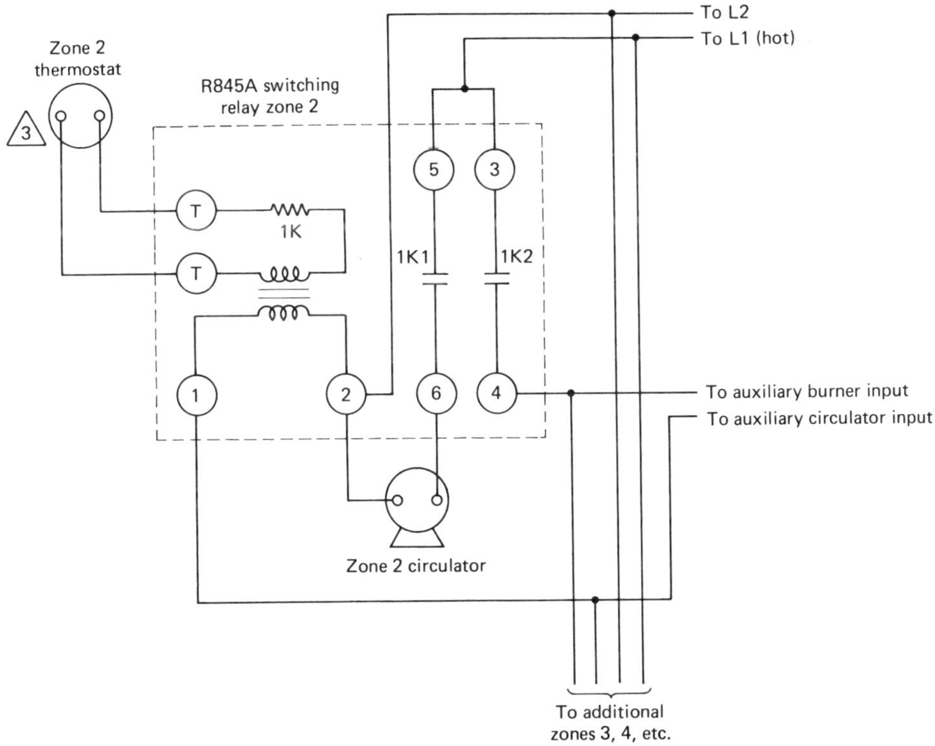

Figure 4.60 Wiring external switching relay to energize main boiler aquastat burner control. 1, 120-V ac power supply: provide disconnect means and overload protection as required; 2, thermostat heat anticipator setting, 0.2 A for R8182D; 3, thermostat heat anticipator setting, 0.4 A for R845A. (Courtesy of Honeywell, Inc.)

System Filling and Purging

One of the most critical procedures affecting successful boiler operation involves the initial filling of the heating system with water and subsequent purging of all air from within the boiler and associated piping. Filling and purging is not limited to startup procedures undertaken in new heating systems or boiler replacements. Often, service personnel must perform these procedures, which are necessitated by maintenance or replacement of various system components.

Properly engineered heating systems should have designed into them a suffi-

---

Figure 4.59 (*opposite page*) Electrical schematic of multizone heating system using separate circulators and switching relays. 1, 120-V ac power supply: provide disconnect means and overload protection as required; 2, thermostat heat anticipator setting, 0.2 A for R8182D; 3, thermostat heat anticipator setting, 0.4 A for R845A; 4, control case must be connected to earth ground (use grounding screw provided). (Courtesy of Honeywell, Inc.)

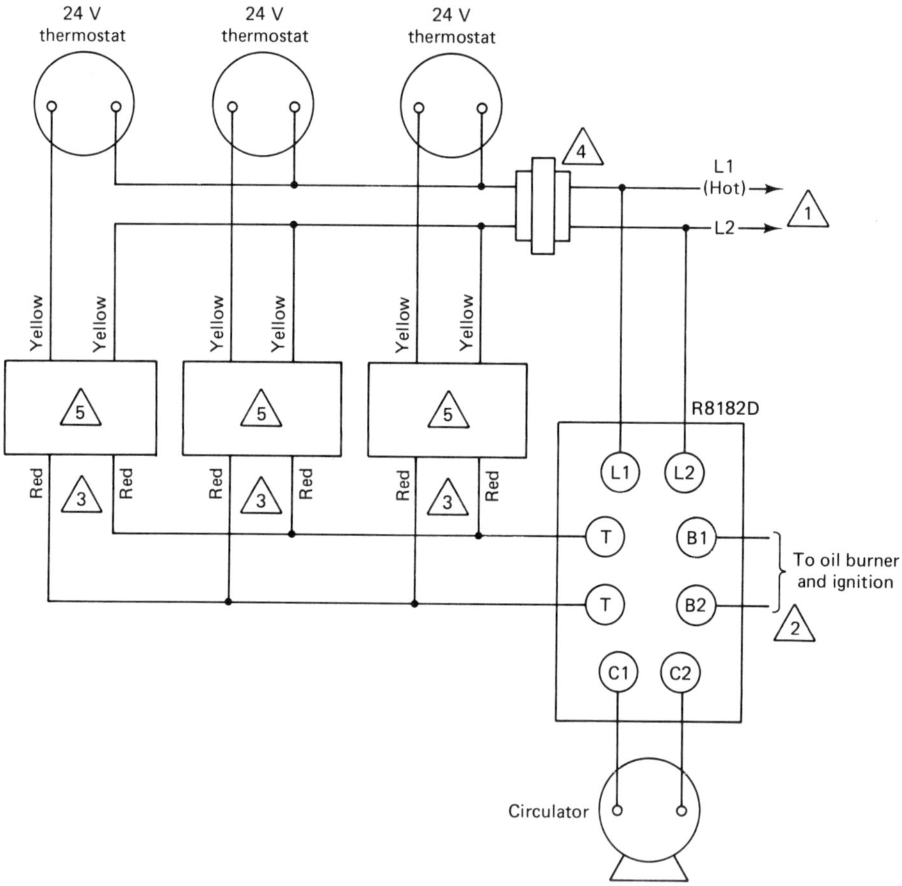

Figure 4.61 Multiple zoning with zone valves. 1, Power supply: provide disconnect means and overload protection as required; 2, R8182H has white and orange lead wires for oil burner and ignition connections; 3, red lead wires provided only on models with end switch; 4, choose AT72, AT87, or AT88 transformer to match maximum system load; 5, use V8043, V8044, or V8343 zone valves with auxiliary end switch only. (Courtesy of Honeywell, Inc.)

cient number of strategically located isolation valves, which enable the service and replacement of almost any system component without the need to drain the entire heating system. A typical valving arrangement of this type is illustrated in Figure 4.62.

The piping and valving arrangement in Figure 4.62 enables virtually any component within the vicinity of the boiler to be replaced or repaired without the necessity of draining either the boiler or the heating system. Complete isolation of critical components and zones is easily achieved for any service requirements, which

Figure 4.62 Boiler isolation valving. The placement of gate valves on the supply and return lines adjacent to the boiler allows for system servicing without the necessity of draining the heating system.

greatly simplifies filling and purging after all service operations have been performed.

Filling the system. Under most conditions, it is preferable to fill a heating system from the bottom of the boiler up through the piping array. In this manner, air is continually driven to the highest part of the system, where it can be purged either through the air scoop/vent assembly or through individually installed bleed valves on the high points of the radiation piping circuitry. An automatic fill valve with a manual override for fast filling of the system is generally used to perform this procedure. (See Chapter 3, "Automatic Feedwater Makeup Systems," and Figures 3.38, 3.39, and 3.40 for an in-depth discussion of fill valves.)

Using the override feature of the fill valve allows water to enter the heating system under normal house pressure in order to quickly fill the boiler and piping circuit. The manual override is used until the system is almost completely filled with water and the pressure reading on the boiler tridicator is below 5 psi. When this occurs, the override should be released. The boiler will now continue to fill automatically under reduced incoming pressure, which is normally limited to 12 psi. After the system has been completely filled, the pressure reading on the tridicator or pressure gauge on the boiler will stabilize at the set point on the feedwater control valve. If this pressure reading on a cold fill is over 20 lb or below 10 lb, it is likely that the pressure setting on the feedwater valve needs to be recalibrated or that the valve itself is defective and should be replaced. Under normal conditions, the cold-start water pressure in residential hot water boilers should be set between 12 and 15 psi.

To reset the feedwater valve if incorrect pressures are encountered, the manufacturer's specification sheet should be consulted. After the regulator has been readjusted, water and pressure should be vented from the boiler through the pressure relief valve or boiler drain, using a pail to catch the vented water. When this is being done, the cold water feed line shutoff valve should be closed to prevent makeup water from automatically being fed into the boiler as it is being drained. When venting the boiler, it is often convenient to attach a garden hose to the boiler drain valve for ease in channeling water flow into either a bucket or convenient floor drain. Allow the pressure in the boiler to drop approximately 5 to 10 psi, and then close the boiler drain valve. Next, open the feed line shutoff valve and observe the pressure on the boiler tridicator as the valve fills the boiler and stabilizes at regulator pressure. Since feedwater regulator adjustments are usually trial and error, this procedure might need to be repeated until the desired pressure setting is achieved. If, however, after repeated attempts, the cold fill pressure continues to remain at unacceptable levels, the valve should be replaced.

In an emergency, a boiler can be filled from the drain valve by using a garden hose or washing machine hose served by a normal water tap. This procedure should be done very carefully and only if an automatic feedwater mechanism is either inoperative or not installed. As water enters the boiler, the pressure gauge should be examined continuously and carefully, since water entering the heating system is at street pressure, which is considerably higher than the operating pressures for which hydronic heating systems are engineered. As pressure in the system slowly comes up to approximately 12 psi, the incoming water should be turned off. Failure to do this at pressures above 30 psi will cause the pressure relief valve to discharge.

Purging the system. The most common cause for loss of circulating water in a heating zone when the boiler and all associated controls appear to be functioning normally is due to air trapped somewhere within the piping circuit. To minimize the chance of this situation occurring, initial purging of the heating system is mandatory.

For this procedure to be accomplished effectively, the piping layout must contain a proper purging configuration (Figure 4.63). Return water from the baseboard heating circuit is moving toward the circulator under normal circumstances (all boiler circulators have a stamped arrow on the pump housing indicating direction of flow): The gate valves are open, and the boiler drain is closed. Also note from this illustration that the gate valves have been located so as to allow the circulator to be easily removed from the circuit without draining the heating loop. The purging should be done when the boiler and heating system is up to operating temperature with the circulators running.

To purge the heating system with this valving arrangement, the gate valve below the circulator should be opened, the one above should be closed. The boiler

Figure 4.63 Purge configuration on return line, consisting of a gate valve downstream of a boiler drain for air removal.

drain should be opened for short periods on a cyclical basis, usually between 10 and 20 seconds. This procedure allows for trapped air to be removed from the heating system by pushing air and water out of the boiler drain. The pressure gauge should be watched closely during this operation. When the boiler water pressure drops to 5 psi, the boiler drain should be closed to allow the feedwater system to make up the lost system pressure. The process of opening and closing the boiler drain is continued until warm water exits the boiler drain. When this occurs, the gate valve should be opened and the boiler drain closed. The heating loop is allowed to run for an additional 20 minutes. If the return line stays hot, a complete loop of water flow has been established through the heating circuit and the purging has been successful. If not, the process is repeated until warm water exits the return boiler drain. All air vent caps must be loose to allow vented air to escape.

In addition to the valving configuration illustrated in Figure 4.63, purging valves are commercially available that combine purge and flow balance functions into one housing, illustrated in Figure 4.64. This one-piece valve unit can incorporate a ball or butterfly valve, drain valve, and a check valve into one convenient assembly. The purge and balancing valve is inserted into the return loop of the heating system, directly above the circulator. If the circulator is located on the supply line of the heating system, the purge valve should be located on the return loop prior to its entry into the boiler return. To purge the heating loop, the valve assembly should be closed and the drain valve should be opened. The purging procedure is similar to that described for the gate valve/boiler drain arrangement. When the purging is complete, the ball valve or butterfly valve should be opened to allow water to circulate in the heating loop once again. After the system has been initially

TABLE 4.7 TROUBLESHOOTING PROCEDURES

Problem	Component at fault	Cause	Corrective action
Delayed ignition and puff-back	Electrode points	Incorrect gap	Check and reset
		Incorrect centering	Check and reset
		Dirt or soot	Clean and regap
	Transformer	Not delivering full voltage	Check for cracked insulators
			Replace transformer
	Nozzle	Clogged	Replace
		Incorrect capacity	Check manufacturer's specs, replace
	Air bands and shutter	Too much excess air	Check CO_2 and stack temperature
			Reduce air intake
Pulsation and rumbling	Nozzle	Solid instead of hollow	Switch to hollow cone
		Too-wide spray angle	Reduce spray angle in new nozzle
	Fuel pump	Pressure too low	Increase
	Oil burner	Low-speed unit	Replace with high-speed flame retention burner
	Combustion chamber	Incorrect size	Check specs and rebuild
	Inadequate combustion air	Insufficient air into boiler room	Install intake piping from outside air
		Incorrect air band adjustment	Readjust air intake on burner and test for CO_2 level
Burner will not turn on	Electrical wiring	Loose connections	Check all junction boxes and replace connectors
	Aquastat relay	Faulty relay	Replace relay
	Burner motor	Broken line connector	Replace line connection
		Broken coupling	Check coupling and replace
		Burned-out motor	Replace motor
Burner motor turns on but unit will not ignite	Fuel delivery system, fuel, fuel line, fuel filter	Empty tank	Check gauge, refill
		Kinked or loose connection	Replace faulty connections
		Filter clogged	Replace fuel filter
		Fuel line frozen	Thaw out
	Transformer	Open connection	Replace transformer
	Burner motor coupling	Coupling broken	Replace
Oil fumes and odors	Chimney	Too little draft due to blockage	Remove obstruction
		Incorrectly built	Replace

TABLE 4.7 *(continued)*

Problem	Component at fault	Cause	Corrective action
	Draft damper	Incorrectly set	Recalibrate with draft gauge
		Not functioning	Replace
		Loose	Tighten and level
	Line connections	Loose	Tighten or replace
Excessive fuel consumption	Nozzle	Too large	Replace with smaller size if approved by manufacturer
	Air bands	Too much excess air	Check CO_2 level, reset
		Off-cycle air infiltration	Install auto air damper
	Thermostat	Calibrated settings off	Recalibrate anticipator and thermostat or replace
	Aquastat	Limit settings too high	Lower
	Leaking hot water fixtures	Excess consumption due to cycling	Repair defective fixtures
	Chimney	Excessive off-cycle heat losses	Install electric damper
		Excessive draft	Calibrate barometric damper
	Fuel delivery system	Leaking	Check tank, lines, and connections, repair or replace defective parts
	Boiler heat transfer passages	Soot build-up retards heat transfer	Clean boiler according to manufacturer's recommendations
Burner ignites but does not maintain flame	Fuel filter	Clogged	Replace
	Cad cell	Defective or dirty	Clean or replace
	Nozzle	Clogged	Replace
	Fuel lines	Kinked or leaking	Locate and repair
	Air bands	Too much excess air	Check CO_2 level, reset air bands
Poor flame cut-off in burner	Air in oil line	Air expands when burner shuts off, pushing fuel out of nozzle	Bleed all fuel lines
	Cut-out solenoid	Solenoid not seating properly	Clean if possible, or replace
Poor tankless coil performance: demand outstrips supply	Inoperative or no flow restrictor	Water flows through coil in excess of coil rating	Install flow regulator with rating matched to unit
	Tempering valve	Incorrectly set or inoperative	Reset or replace
	Tankless coil	Clogged with sediment	Flush coil until it is clean

Figure 4.64 Combination purge and balancing valve. (Courtesy of Watts Regulator Co.)

purged, it is good practice to check the return lines for air within 24 to 48 hours to help ensure that no future problems develop.

Troubleshooting Procedures

The malfunction of an oil burner, as with any device that incorporates separate components, involves checking components on an individual basis to effectively isolate a problem area. Troubleshooting procedures are grouped into the following categories for ease in isolating specific problem areas: fuel delivery system, ignition system, and electrical controls. None of these categories is mutually exclusive; a problem can sometimes be caused by more than one factor. The general procedures offered in Table 4.7 are suggested as a starting point to help pinpoint problems and should not be used as a comprehensive service manual. These recommendations are based, however, on long-term experiences of a large number of service technicians.

5

Gas Heating Systems

The fuel gases that are used almost exclusively for space heating purposes are natural gas and liquid petroleum (LP) gas. The use of natural gas is common in urban and suburban areas where extensive underground pipeline and distribution facilities exist. LP gas is usually classified as a rural fuel, where tank trucks deliver the product to residential storage tanks. Gas-fired heating equipment is both efficient and reliable. Although similar in many ways, LP and natural gases differ significantly with respect to many properties that affect their combustion characteristics, which will now be examined.

CHARACTERISTICS OF NATURAL GAS

Natural gas, created over a period of many millions of years from the decay of plant and animal organic material, is a mixture of many different hydrocarbons. The primary ingredient of natural gas is methane (CH_4), which comprises approximately 90% of the natural gas mixture. The remaining 10% of the gas is made up of varying amounts of ethane (C_2H_6) and propane (C_3H_8). While natural gas deposits are sometimes discovered separately, most of the gas is usually found to accompany petroleum deposits and is most often trapped under great pressure.

In its raw state, natural gas is frequently found as a wet gas, with a methane content below 90%. Dry gas results from processing of the wet gas, increasing the percentage of methane to 90%.

The specific gravity of natural gas, that is, its weight compared to an equal volume of air under standard temperature and pressure conditions, is always lighter than air. For residential heating applications the specific gravity of natural gas is considered to be 0.60. Being significantly lighter than air, natural gas will disperse rapidly when released into the atmosphere, a desirable safety consideration in the use of the fuel.

Combustion Characteristics

In its unrefined state, natural gas has no odor. To enable leaks in the gas feed system to be detected, a chemical odorant such as ethyl mercaptan, the familiar "rotten egg" odor, is added to the gas.

For combustion of natural gas to take place, the relative percentages of natural gas and air must fall within narrowly defined limits. For example, in a mixture of natural gas and air, where the gas comprises less than 4% of the volume of the mixture, no combustion will take place. Also, should the percentage of the gas in the mixture exceed 14%, no combustion will result. These ranges, 4 to 14%, are the upper and lower limits of the combustion mixture.

For the air/gas mixture to burn, the gas flame must be able to spread from one gas molecule to the next while encountering a sufficient quantity of oxygen to support the combustion process. The degree to which this flame will spread is referred to as *flame propagation.* At the lower limits of the combustion mixture, below 4% gas, for example, too few gas particles are encountered to enable sufficient flame propagation for ongoing combustion. Conversely, at air/gas mixtures greater than 14% gas, too little oxygen is interspersed between the gas molecules to support the combustion process. The ignition temperature of natural gas, that is, the lowest temperature required to support flame propagation, is 1200°F.

While the heating value of 1 cubic foot (ft^3) of natural gas varies depending on the mixture of the specific gases, it is generally in the range of 900 to 1100 Btu/ft^3. The most commonly accepted heating value for a cubic foot of natural gas is 1000 Btu/ft^3. For complete combustion of 1 ft^3 of natural gas to take place, 10 ft^3 of air is utilized in the process. The by-products that result from the complete combustion of the gas are water vapor and carbon dioxide (CO_2). Should insufficient air be present during combustion, sooting of the combustion surfaces of the boiler will occur together with the production of various chemical compounds referred to as aldehydes, as well as carbon monoxide (CO). Aldehydes have objectionable odors associated with their formation and can be irritating to the skin. Carbon monoxide, an odorless, tasteless gas, is toxic in sufficient quantities. Thus, for safe operation of all natural gas appliances, complete combustion is essential.

Standard Flame Types

Natural gas is commonly burned under two sets of firing conditions: a yellow flame burner, or a blue flame, Bunsen-type burner. While there are gas appliances that

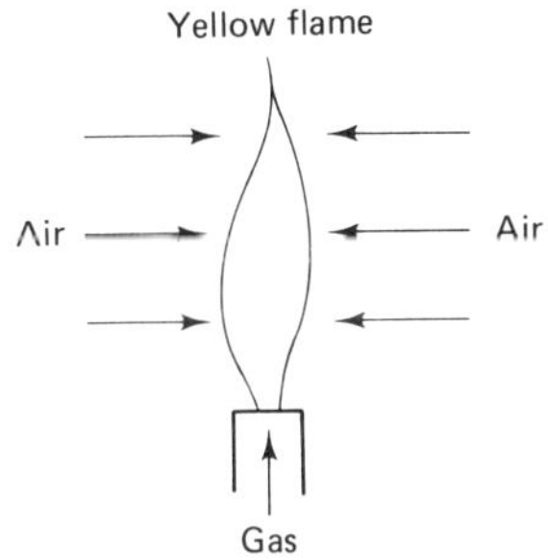

Figure 5.1 Yellow flame atmospheric burner. All combustion air is drawn from air surrounding the flame, giving it its characteristic color.

use modifications of these two types of atmospheric burners, the standard yellow and blue flame appliances are the most common.

Yellow flame atmospheric burner. A typical yellow flame burner is illustrated in Figure 5.1. In a yellow flame burner, no combustion air is premixed with the gas prior to burning. All combustion air is drawn from the air that surrounds the flame. This surrounding air is referred to as *secondary air.* As long as there is sufficient secondary air surrounding the flame, complete combustion can take place. It is important to design the flame position so that no contact takes place between the flame and any surface cooler than 1200°F, as incomplete combustion and sooting can result. The most common uses of yellow flame atmospheric burners are in low-Btu input pilot lights and illuminating lamps.

Blue flame burners. A blue flame atmospheric burner premixes combustion air with the fuel gas prior to its combustion at the gas outlet. A typical blue flame burner is illustrated in Figure 5.2. Blue flame burners are found on gas stoves, hot water heaters, and residential heating equipment.

Premixed air in the blue flame burner configuration is referred to as *primary air.* Generally, an air shutter arrangement is used to meter the amount of primary air that is admitted into the gas stream. If too much primary air is admitted to the

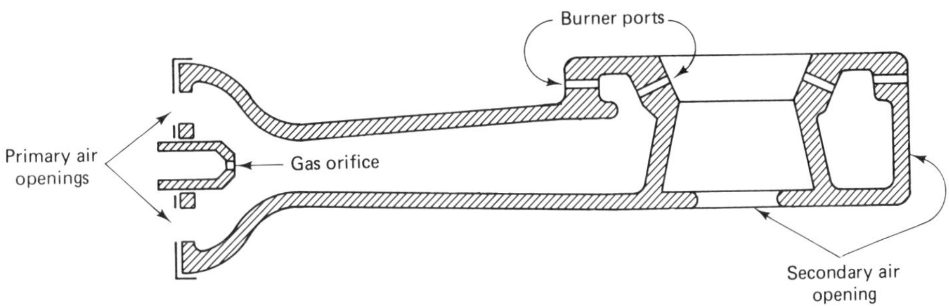

Figure 5.2 Blue flame atmospheric burner. Combustion air is premixed with the gas prior to burning.

burner, the flame will be noisy and unstable; too little primary air results in a yellow flame with excessive smoke and soot development. Properly metered primary air will yield a blue flame as illustrated in Figure 5.2. Occasional yellow flecks in the flame are due to dust particles and are of no consequence.

In blue flame burners, the outer zone of the flame can come in contact with surfaces that are cooler than 1200°F; the inner cone of the blue flame, however, must not contact these surfaces.

CHARACTERISTICS OF LIQUEFIED PETROLEUM GAS

Liquefied petroleum gas is generally referred to by its abbreviation, LP gas. LP gas is a mixture of propane and butane, propane being the dominant fuel gas for residential LP appliances. LP gas is available in six grades, which indicate the varying amounts of propane versus butane that make up the fuel mixtures. LP gas liquefies easily when pressurized and is usually delivered in tank trucks in the liquid state and piped into residential storage tanks adjacent to the home. In some instances, individual storage tanks are delivered to the residence instead of the bulk gas, depending on the quantity of gas used by the customer.

Propane vaporizes under normal pressure at temperatures above −44°F. In areas where winter temperatures are encountered for extended periods of time that fall below −44°F, some type of preheating device must be employed to allow the gas to vaporize in order for it to burn.

The specific gravity of propane varies depending on whether it is in the liquid or gaseous state. In liquid form, the specific gravity of propane usually varies from 0.5 to 0.6 when compared to water. In its gaseous state, propane has a specific gravity of approximately 1.5.

Due to its relatively high density, great care must be exercised in the handling of propane. When released into the atmosphere, propane, unlike natural gas, will not dissipate but rather tends to collect and settle at the lowest points in confined areas, for example, a boiler room or basement. For this reason, the safety shutoff mechanism for propane gas-fired heating equipment requires a complete 100% shutdown of both the pilot light and main gas lines whenever a loss of the pilot flame occurs.

Combustion Characteristics

LP gases, like their natural gas counterparts, are essentially odor free in their natural state and require the addition of chemical odorants for leak detection. The same chemical odors used for natural gases are used in LP gases.

The flammability limits for a gas/air mixture of LP propane gas fall between 2.4 and 10%. A mixture of 5% gas in the gas/air combination results in maximum flame propagation.

Complete combustion of 1 gal of commercial propane gas requires 24 ft^3 of air. The ignition temperature of the gas is approximately 950°F, somewhat lower than natural gas.

The heating value of commercial LP propane gas varies between 2200 and 2500 Btu/ft^3. Since the heating content of 1 ft^3 of LP propane is significantly higher than 1 ft^2 of natural gas, smaller burner orifices are used to achieve similar Btu inputs when converting an appliance from natural to LP gas.

The yellow and blue flame burners designed for natural gas appliances are essentially the same for LP appliances. In many instances when ordering a gas appliance equipped to burn natural gas when propane will be the specified fuel, a conversion kit is included which modifies the gas orifice and gas pressure regulator of the appliance; the basic burner is not modified. Since propane requires a significantly higher amount of combustion air than natural gas, primary air shutters are open further than in natural gas operating modes.

GAS PRESSURE MEASUREMENT AND REGULATION

Gas pressure measurements are taken in two ways: One method measures the gas pressures in natural gas feed mains and residential propane storage tanks in pounds per square inch (psi); the other method of measurement is taken on the output side of the appliance pressure regulator, which reduces entering gas pressure in psi down to what is referred to as pressure in inches of water column (w.c.). We will now examine in greater detail the methods and procedures for taking measurements in both natural and LP gas appliances using a manometer.

Manometer (Water Column) Pressure Measurements

Water column measurements are generally used to determine the gas appliance service pressures, that is, the pressure of the gas being fed into the gas orifices on the output side of the appliance regulator. A manometer is a device used to make these measurements. The U-shaped manometer in Figure 5.3 is a common type that is used for this purpose.

When using this device, a flexible hollow line is connected from the pressure tap on the appliance gas regulator control into the input port on the manometer. The pressure of the gas causes the water in the U-shaped column to be displaced a certain number of inches depending on the output pressure of the regulator. In Figure 5.3, the displacement is 4 in. of water column, the difference in height between the level of water in the two columns.

Natural Gas Pressures

Pressures within the natural gas feed lines vary somewhat depending on the size of the distribution system. Generally, delivery pressures range from 10 to 25 psi. Residential pressure regulators reduce the pressure to 7 in. w.c. within the residential

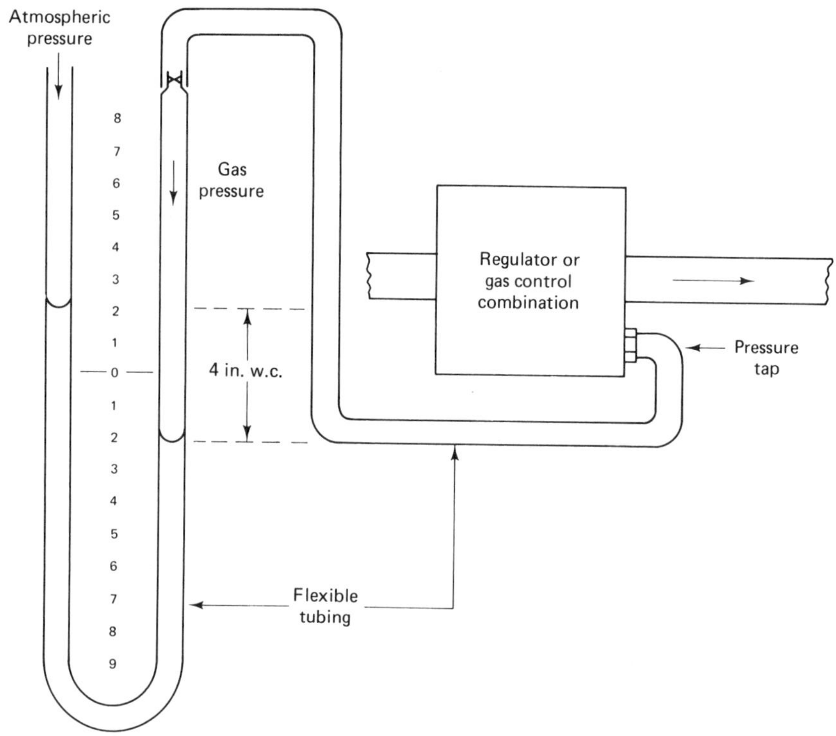

Figure 5.3 U-shaped manometer used to measure gas pressure. Gas pressure is measured as the difference in the height of water between the two glass columns. (Courtesy of Robertshaw Controls Company.)

service lines. Individual gas regulators on the appliances further reduce operating gas pressures to between 3 and 5 in. w.c. Most individual appliance regulators are adjustable and appliance instructions suggest that installing personnel check and adjust the regulators prior to placing a unit into service. A gas pressure regulator is illustrated in Figure 5.4. The pressure regulator is essentially a variable orifice, the opening of which is increased or decreased based on the outlet pressure sensed by the flexible diaphragm. For example, should the outlet pressure of the regulator decrease due to reduced gas flow or main line pressure, the diaphragm will open in response to the spring-adjusted setting. Decreased pressure drop through the regulator will tend to reestablish outlet pressure. An increase in outlet pressure will increase pressure against the diaphragm, moving it closer to the closed position to increase pressure drop through the regulator, thus dropping the outlet pressure.

LP Gas Pressures

The storage pressure in most residential LP tanks is in the 125 psi range, based on an ambient temperature of 80°F. Outlet regulators on the storage tank reduce the

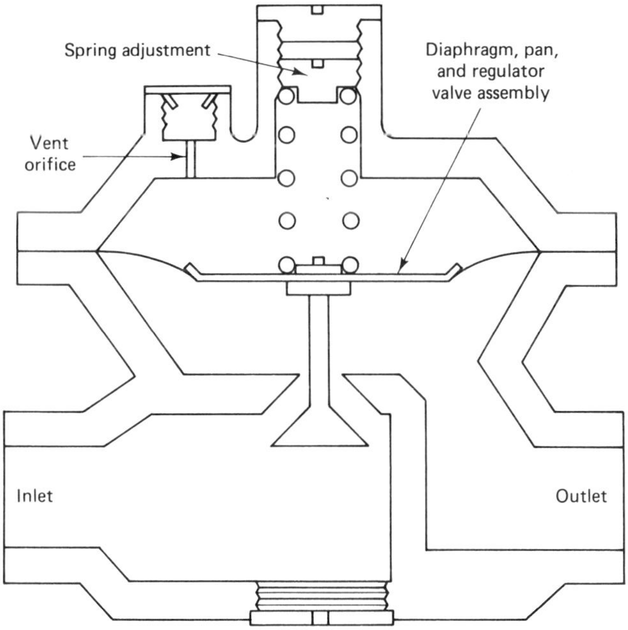

(a)

(b)

Figure 5.4 Simple gas pressure regulator. [(a) Courtesy of Robertshaw Controls Company; (b) courtesy of Honeywell, Inc.]

tank pressure to a service line pressure of approximately 11 in. w.c. for individual appliances. Some LP gas regulators employ a two-stage device: The first stage is located at the storage tank, the second stage at the beginning of the residential service line to accomplish the final pressure reduction.

Most LP gas appliances incorporate individual pressure regulators to hold operating gas pressure at 11 in. w.c. for the specific appliance. Note that the operating gas pressure for LP appliances is significantly higher than for natural gas units due to the higher Btu content per cubic foot as well as the higher specific gravity of LP gas versus natural gas. Also, smaller orifices are employed in LP units to obtain equivalent natural gas Btu input ratings. Higher orifice pressures increase primary air injected into the main burner. This assists not only in moving and mixing the heavier than air/propane mixture, but supplies the greater quantity of air needed for complete combustion of the LP gas.

CLASSIFICATION OF GAS COMBUSTION APPLIANCES

Although there are many manufacturers offering a large number of gas heating boilers available in a multitude of configurations and sizes, most units fall within one of four basic combustion design categories: conventional atmospheric, non-condensing induced-draft ignition, condensing induced-draft ignition, and pulsed ignition. The major design features of each of these units will be examined in greater detail.

Conventional Atmospheric Combustion Burners

Conventional atmospheric burners are the most prevalent type of combustion units used in gas-fired boilers. Typical of this type is the boiler illustrated in Figure 5.5. Conventional atmospheric burners are so labeled because they burn gas under standard atmospheric pressure with no pressurized mixing of the fuel gas and combustion air. Typical atmospheric burner configurations are illustrated in Figure 5.6. An atmospheric combustion boiler contains several individual burners, each of which taps off the main gas supply manifold located at the bottom of the unit (Figure 5.7). The specific number of burners within the unit depends on the Btu/hr input of the particular appliance. Almost all burners employed in this configuration are blue flame burners, premixing the primary combustion air and gas by means of the air-adjusting shutter illustrated in Figure 5.8. Note that the main burner flame has a clearly defined inner cone. Main burner flames should be checked periodically to ensure proper adjustment of the air shutter. Under most operating circumstances, the air shutter should be adjusted so that no yellow flame tipping appears. The inner cone should be well defined. Orange and yellow streaks that appear in the flame are caused by dust particles and should not be confused with yellow tipping of the flame, which is caused by insufficient primary air.

The combustion efficiency of atmospheric combustion boilers ranges between

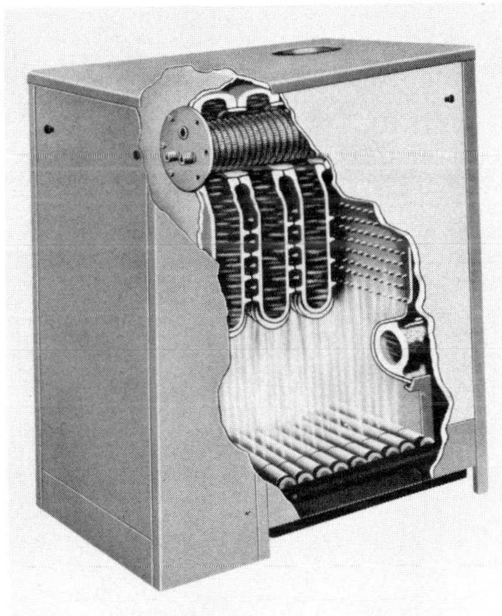

Figure 5.5 Atmospheric combustion hot water boiler. (Courtesy of Burnham Corp.)

60 and 75%. The seasonal operating efficiency of these units will be somewhat lower than the combustion efficiency when flue losses and jacket losses are taken into account. Flue gas temperatures in conventional atmospheric boilers vary between 350 and 550°F.

Burner units in these boilers can be fired by either a standing pilot or electronic intermittent ignition system. Each of these systems is examined in greater detail later in this chapter.

Induced-Draft Combustion Technology

Induced-draft combustion technology has resulted in the design and manufacture of gas-fired boilers capable of achieving very high combustion efficiencies, in the range 80 to 87%. Residential boilers that employ induced-draft technology are classified either as condensing or noncondensing units, depending on the degree to which exiting flue gas temperatures result in condensation of effluents within the exhaust stack. Both condensing and noncondensing boilers rely on a high-speed blower to push air into the boiler combustion chamber. The turbulence caused by this air injection ensures almost complete mixing of the air/gas combination, resulting in a highly efficient firing system within the unit. Exhaust gases are forced out of the appliance stack under the operating pressure induced by the blower unit. Figure 5.9 illustrates the flue gas travel in both conventional and induced-draft appliances. Note in Figure 5.9a that the flue gas travel in a conventional unit is up and through the sections of the boiler, out the top of the unit past the draft hood. This movement is accomplished solely by the natural draft created by the chimney and venting system of the appliance.

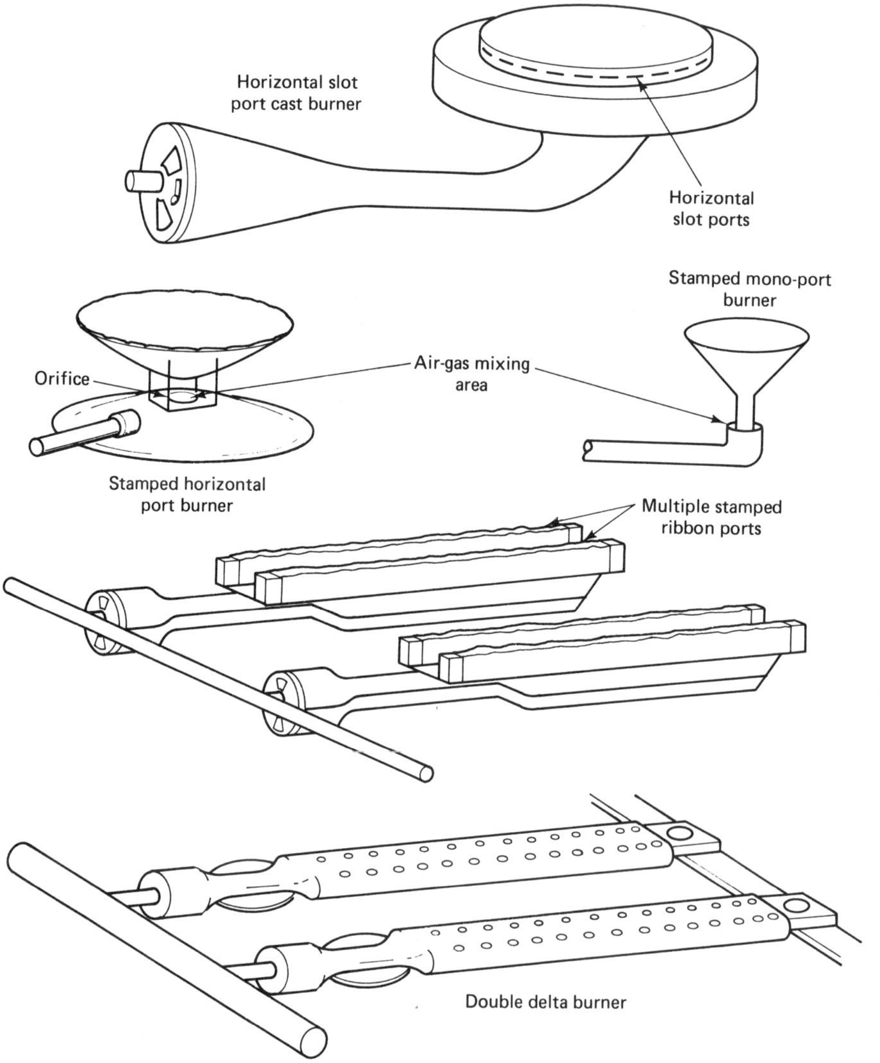

Figure 5.6 Typical configurations of atmospheric gas burners. (Courtesy of Robertshaw Controls Company.)

In an induced-draft unit (Figure 5.9b), a high-speed blower pushes the combustion gases and effluents throughout the boiler sectional assemblies and out into the venting system. No vent hood is used in the induced-draft configuration; all system draft is dependent on the pressure created by the blower unit.

Both condensing and noncondensing induced-draft appliances will now be examined in further detail.

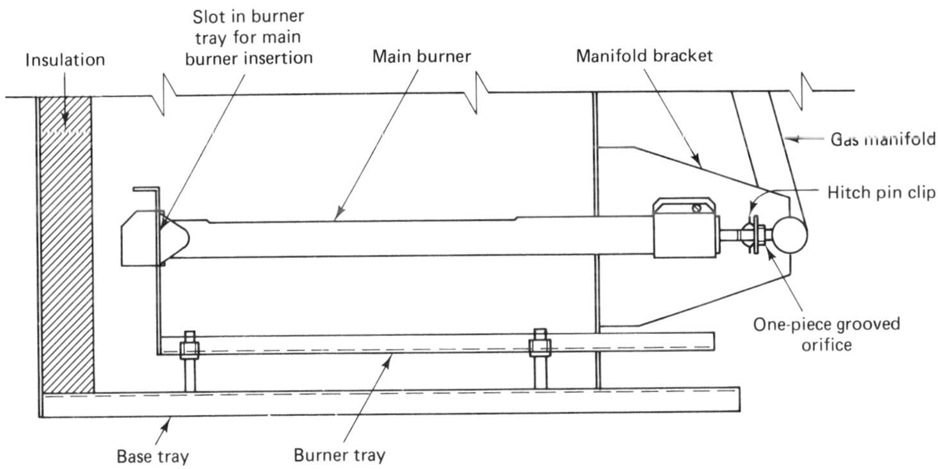

Figure 5.7 Burner manifold cross-section configuration. (Courtesy of Burnham Corp.)

Induced-draft condensing appliances. An induced-draft condensing boiler is illustrated in Figure 5.10. These boilers do not employ draft hoods in the venting system. In a typical condensing appliance, the flue gases are cooled to approach the condensation point of the vapors within the stack gases. In so doing, the latent heat of vaporization is recovered and transferred to the surrounding boiler water. The result of this process is an efficiency of combustion that can approach between 85 and 90%, with exiting stack gases in the range of 100°F.

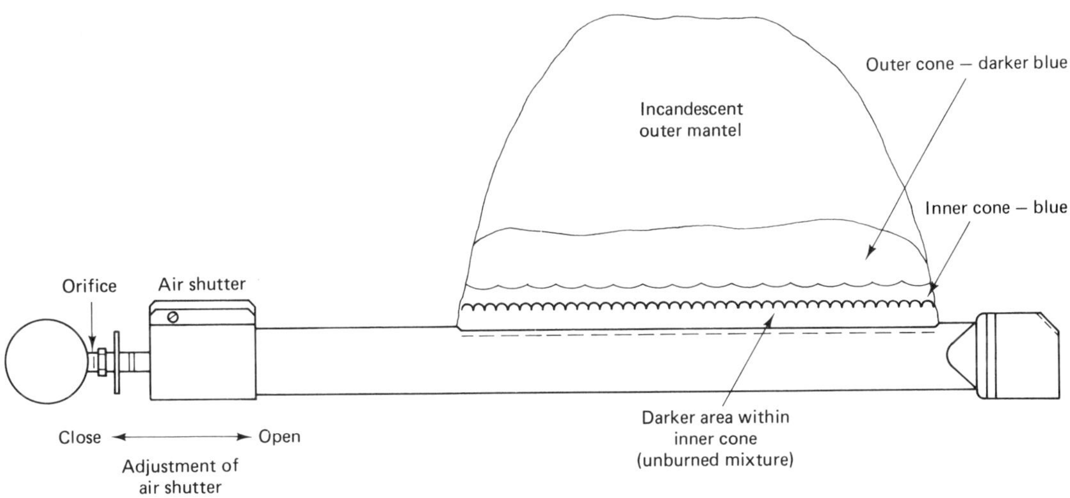

Figure 5.8 Main burner flame and air shutter adjustment. (Courtesy of Burnham Corp.)

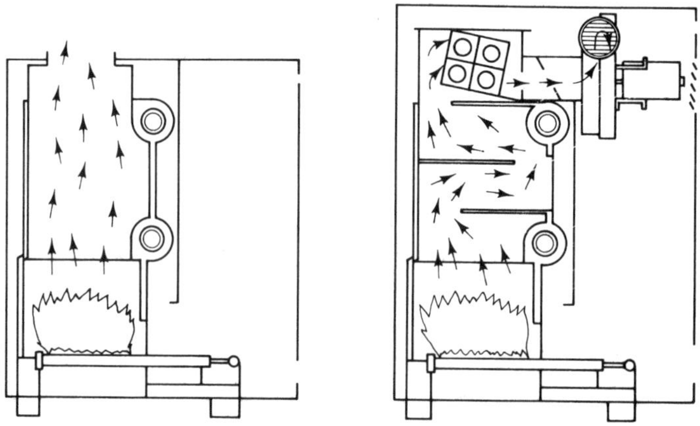

CGM (a) VHE (b)

Figure 5.9 Comparison of flue gas travel in (a) conventional and (b) induced draft boilers. (Courtesy of Weil-McLain, a Marley Co.)

Figure 5.10 Induced-draft condensing boiler. (Courtesy of Weil-McLain, a Marley Co.)

The condensate produced by the condensing boiler is channeled through one or more condensate removal tubes. Figure 5.11 illustrates the use of two condensate removal traps. One trap is connected to the exiting flue gas stack T, the other to the internal condensate trap within the boiler heat extractor. Since the condensate from the appliance has a relatively low pH (3.5 to 6.5 on the pH scale of 0 to 14), only stainless steel vent pipe recommended by the boiler manufacturer should be used in this type of installation.

The condensate removal tubes are usually run into an existing drain in the residence. While the chemical nature of the effluent has not revealed any adverse effects on the residential drainage or sewerage removal system, the effluent will deteriorate standard masonry chimneys. For this reason, these units *may not* be vented directly in standard masonry chimneys. Normally, the unit is vented through a wall or the roof using stainless steel vent pipe (Figure 5.12).

Greater detail on venting systems for condensing induced-draft boilers is given later in this chapter in a discussion of venting systems. The basic configuration of a condensing boiler is illustrated in Figure 5.13.

The main burner flames on these units should be inspected periodically to ensure proper operation of the unit. Overfired flames will tend to lick out from under the base of the boiler. Conversely, underfired flames will seem unusually small, with gaps occurring at or within the burner tubes. A properly adjusted flame of this type is illustrated in Figure 5.14.

In general, the primary air adjustment on these units is factory set to yield proper flame configuration. If the flame pattern is incorrectly set, adjustment should be made only by qualified service technicians experienced in the operation of these units.

Induced-draft noncondensing appliances. Noncondensing induced-draft boilers, like their condensing counterparts, operate with very high combustion ef-

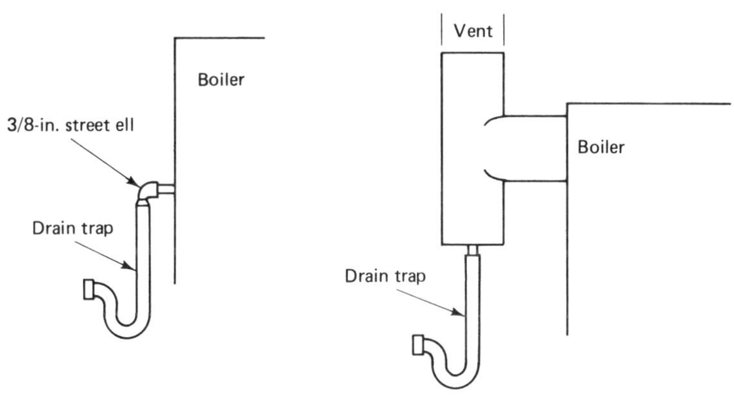

Figure 5.11 Condensate removal trap in condensing boiler. (Courtesy of Weil-McLain, a Marley Co.)

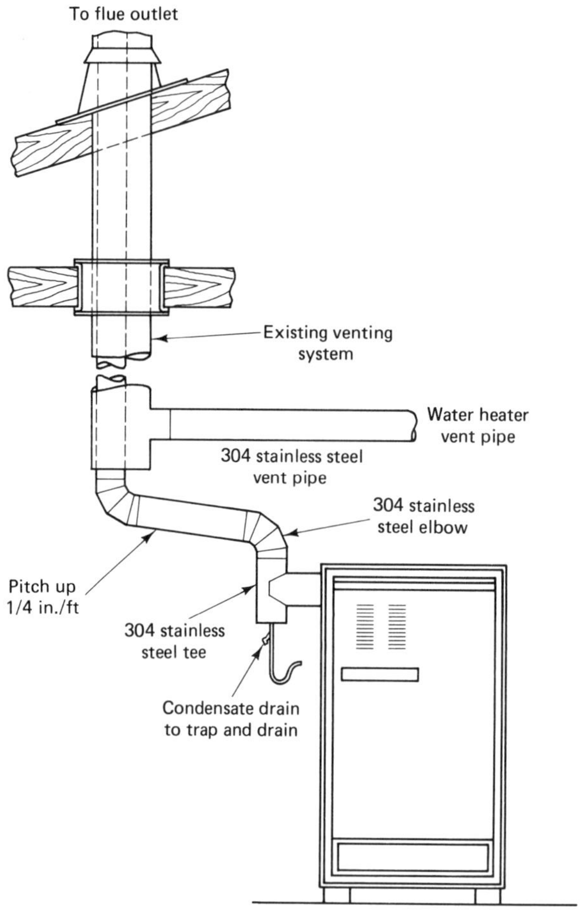

Figure 5.12 Venting a condensing hot water boiler. (Courtesy of Weil-McLain, a Marley Co.)

ficiencies, in the neighborhood of 85%. A noncondensing induced-draft boiler is illustrated in Figure 5.15.

Since no condensation of flue gases occurs in these units, venting of the appliance can be done through a standard masonry or AGA-approved gas venting chimney. A venting system for noncondensing boilers is illustrated in Figure 5.16. Also, the noncondensing mode of operation enables high-efficiency induced-draft technology to be applied to the construction of residential steam boilers in addition to the standard hot water baseboard units. A steam boiler using induced draft for residential installations is illustrated in Figure 5.17. Like their condensing counterparts, these noncondensing boilers use intermittent electronic ignition, which practically eliminates off-cycle heating losses.

The technology of the various ignition systems for burner operation is discussed later in this chapter.

Figure 5.13 Component arrangement of induced-draft condensing boiler. (Courtesy of Weil-McLain, a Marley Co.)

Domestic hot water heating with induced-draft appliances. Most gas-fired boilers do not contain built-in tankless coils for the production of domestic hot water. Since the standing-water temperature of the boiler must be kept at a minimum level sufficient for indirect domestic hot water heating, usually around 140°F, the efficiency of the boiler with the coil would be greatly reduced given the operational characteristics of induced-draft technology. There are, however, modifications that can be made to the boiler which will produce adequate residential hot water production without sacrificing boiler efficiency. While stand-alone domestic hot water heaters are readily available, there are add-on exterior heat exchangers, sometimes referred to as "side-arm" heaters, that can be used for domestic

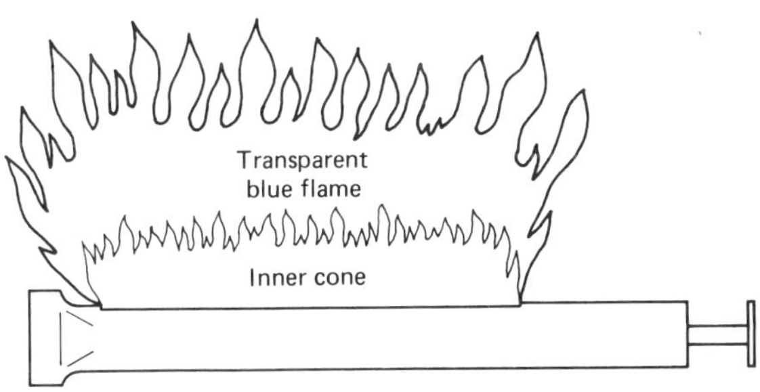

Figure 5.14 Properly adjusted flame pattern for condensing boiler. (Courtesy of Weil-McLain, a Marley Co.)

Figure 5.15 Noncondensing induced-draft boiler. (Courtesy of Burnham Corp.)

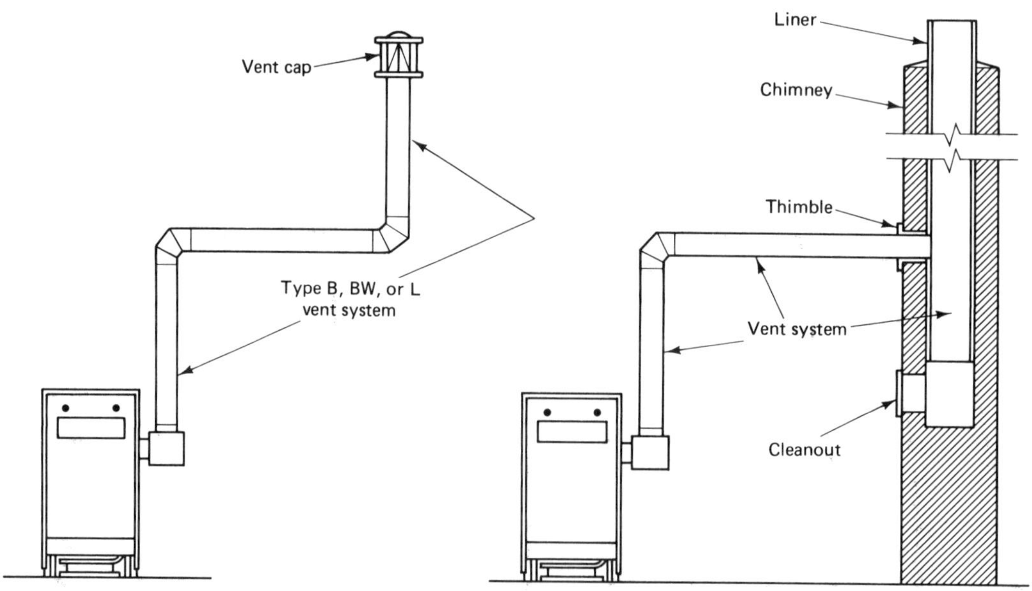

Figure 5.16 Venting configuration of a noncondensing induced-draft boiler. Note that no draft hood is used in venting an induced-draft unit. Chimneys shown are inside the heated space. (Courtesy of Burnham Corp.)

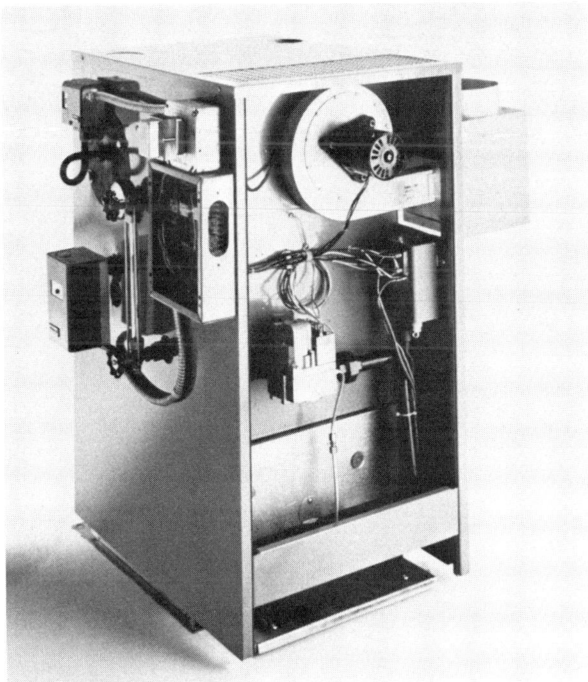

Figure 5.17 Induced-draft steam boiler. (Courtesy of Burnham Corp.)

hot water production. A unit of this type is illustrated in Figure 5.18. In this type of installation, the exterior heat exchanger is treated as an additional heating zone. When the water in the heat exchanger tank drops below a preset level, hot water from the boiler is piped through the heat exchanger coils, heating up the water in the heat exchanger tank until the set temperature has been reached.

External-type domestic hot water heaters are available from several different manufacturers, offering a variety of operating configurations. In Figure 5.19, an arrangement is illustrated in which water from the boiler runs through an internal raised-fin copper heat exchanger located in the bottom of a water storage heater. Note that a separate zoning configuration is required to bring the water in the storage tank up to a preset temperature, operating the boiler just as any heating zone would do.

Pulsed Combustion Condensing Systems

Pulsed combustion hot water boiler technology differs markedly in design from conventional atmospheric and induced-draft burners. Instead, it is an offshoot of an internal combustion engine design. A pulsed combustion boiler series is illustrated in Figure 5.20. Examination of the operating characteristics of these units reveals the cross-sectional design illustrated in Figure 5.21.

Figure 5.18 External side-arm heat exchanger for domestic hot water production. (Courtesy of AMTROL Inc.)

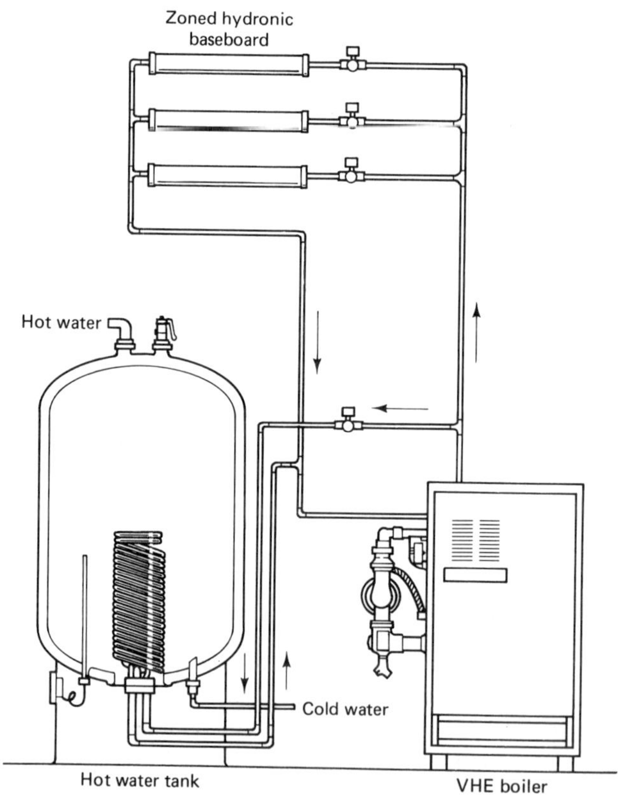

Figure 5.19 External domestic hot water production from a hot water boiler. The hot water tank is treated as separate heating zone. (Courtesy of Weil-McLain, a Marley Co.)

Figure 5.20 Pulsed combustion boiler series. (Courtesy of Hydrotherm, Inc.)

In pulsed combustion units, a heating cycle occurs as follows: Upon a call for heat from a standing cold start, a small blower forces outside air into a sealed combustion chamber. Gas is fed into the combustion chamber and a spark plug is energized, igniting the air/gas mixture (Figure 5.22).

When ignition occurs in the combustion chamber, the air/gas intake disks are closed, which forces the exhaust gases through the heat exchanger passages, illus-

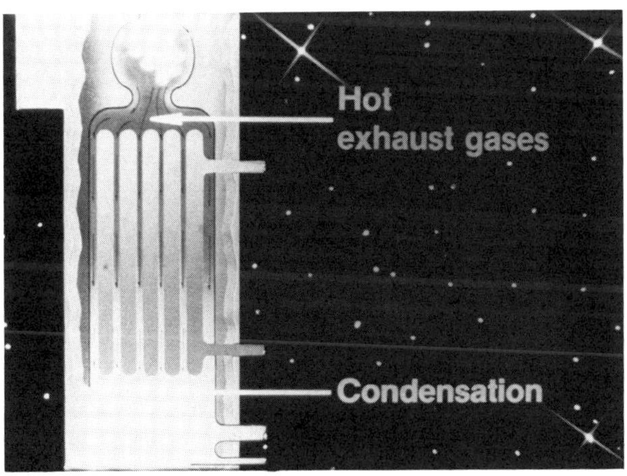

Figure 5.21 Cross section of pulsed combustion boiler. (Courtesy of Hydrotherm, Inc.)

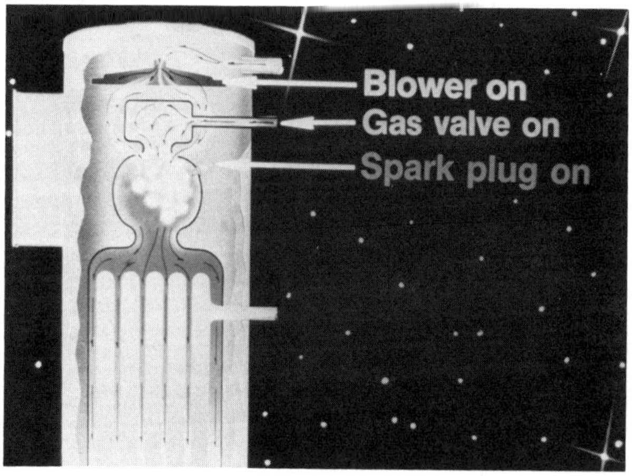

Figure 5.22 Ignition of air and gas in pulsed combustion boiler. (Courtesy of Hydrotherm, Inc.)

trated in Figure 5.23. After the first ignition cycle has taken place, the spark plug and blower are deenergized. Each subsequent air/gas mixture is ignited by residual heat and/or flame from the previous combustion cycle. The vacuum caused by the exiting exhaust gases draws in a fresh air/gas mixture. This ignition cycle continues at the rate of approximately 60 to 70 cycles per second until either the room thermostat settings have been satisfied or the high-limit water temperature of the boiler has been reached.

Pulsed ignition boilers generally operate in thc condensing mode, with resulting temperatures of the ignition exhaust gases below their dew point. Condensation of the flue gases takes place, recovering the latent heat of vaporization of the flue gases. This latent heat, amounting to approximately 10% of the potential energy of the fuel input, is transferred to the boiler water.

Pulsed combustion units are unique in that they require neither a pilot ignition

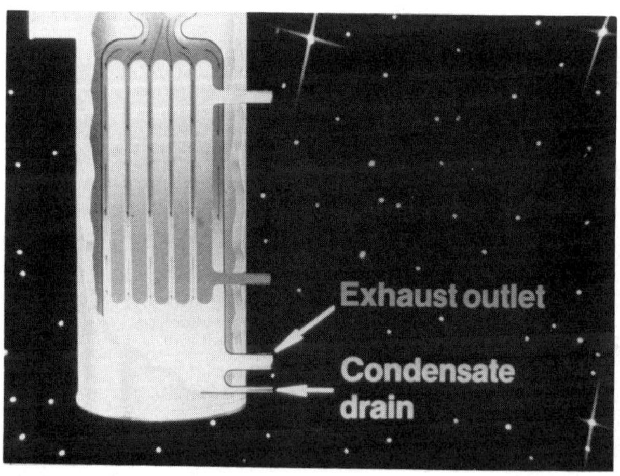

Figure 5.23 Exhaust gas travel through heat exchanger passages. (Courtesy of Hydrotherm, Inc.)

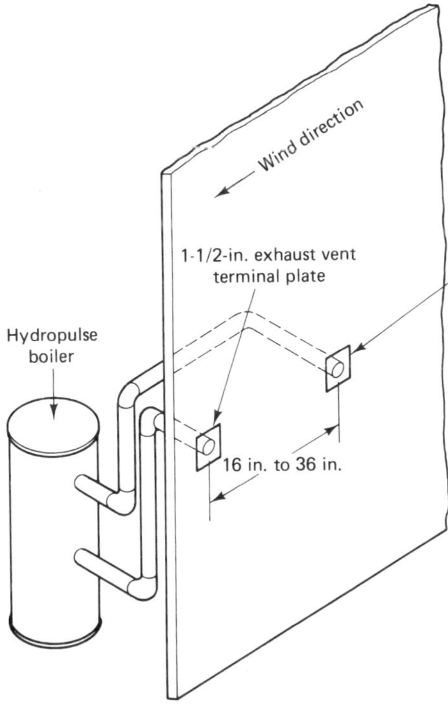

Figure 5.24 Intake and exhaust system arrangement in pulsed combustion boiler. (Courtesy of Hydrotherm, Inc.)

system nor a standard chimney venting system. Rather, two ducts are connected to the boiler: One supplies outside combustion air; the other removes the exhaust products from the unit (Figure 5.24). All other connections to the boiler water supply and return piping and circulator configuration are undertaken in a conventional manner.

We now move on to a discussion of the techniques used in the installation of the various types of gas-fired boilers that we have examined. The techniques discussed include gas piping and delivery systems, factors to consider in the location of gas-fired boilers, and venting system procedures applicable to specific types of gas-fired boilers.

BOILER INSTALLATION PROCEDURES

Proper installation of the appliance is mandatory for both safe and efficient operation of the heating system. The general procedures that govern the installation of gas-fired heating equipment fall under the auspices of four organizations: the American Gas Association (AGA), the National Fire Protection Association (NFPA), the American National Standards Institute (ANSI), and the American Society of Mechanical Engineers (ASME). Although these organizations neither inspect nor approve the use of specific materials and procedures utilized in any individual installation, most local authorities who have jurisdiction over installations such as local building and electrical inspectors require evidence of compliance with the procedures outlined in the applicable installation and code guides that are published by these organizations. While the ASME and NFPA guidelines address themselves primarily to specifications relating to appliance manufacture and electrical wiring procedures, the ANSI code deals with regulations covering appliance location, combustion air requirements, gas piping procedures, and a variety of venting situations. This ANSI publication is referred to as the National Fuel Gas Code, ANSI Standard Z223.1 (NFPA Document No. 54), and is cosponsored by the American Gas Association (AGA) and the NFPA. Copies of this code book are available from the American Gas Association, 1515 Wilson Boulevard, Arlington, VA 22209.

We will now move on to a discussion of the specific aspects of most residential gas-fired boiler installations.

Locating the Appliance

All gas boilers must be located so that they satisfy several basic safety and operating requirements. Adequate clearance between the boiler jacket and any combustible material must be maintained. Also, clearances between the boiler vent pipe and walls or ceilings are critical to prevent overheating of combustible surfaces. For noncloset-type installations, the minimum clearances to be observed are listed in Table 5.1. Note the following:

1. Although the clearances listed in Table 5.1 are the minimum for safety purposes, a minimum of 24 in. from the left, right, and front are recommended for ease in servicing the unit.
2. For noncloset installations in which rooms are listed as being large in com-

TABLE 5.1 MINIMUM CLEARANCES BETWEEN BOILER COMPONENTS AND COMBUSTIBLE SURFACES

Component	Clearance (in.)
Top of boiler	59
Front of boiler	18
Flue	6
Back of boiler	6
Right side	1
Left side	9

Source: Burnham Corp.

Note: For listed and approved units only.

parison with the size of the heating equipment, a large room is defined as having a volume equal to at least 16 times the total volume of the boiler.

For closet installations, the clearances above may be lessened to those distances specified by the specific appliance manufacturer. Figure 5.25 illustrates safety clearances suggested by one manufacturer for both closet and noncloset installations. While the specifications will apply to most ASME-listed gas-fired boilers of the atmospheric and induced-draft combustion types for use in the United States or Canada, specific criteria for each application and each specific locality should be researched prior to undertaking any installation. Also, local building and energy codes may vary from the specifications listed in Table 5.1 and Figure 5.25. Be sure to consult local authorities for specific details.

Combustion Air Requirements

Gas-fired boilers will ordinarily be installed in buildings featuring either very tight construction or conventional frame and masonry construction. When the unit is installed in an unconfined space, such as a large basement in a building of conventional construction, air leaks through the structure will ordinarily provide sufficient combustion air for the appliance and thus no special procedures need to be taken to provide for combustion air.

When the appliance is installed in an unconfined space in a building featuring tight construction techniques, a single opening to an outside wall with a screened metal rodent trap must be provided. The opening in the outside wall must be sized to provide 1 $in.^2$ of surface area for each 5000 Btu of appliance input.

When the boiler is to be installed in a confined space within a building of conventional construction techniques, one of the following procedures must be followed to ensure adequate air for both combustion and appliance ventilation.

1. When installed in an interior confined room, two openings in a wall that adjoins another interior area which has adequate outside air supply must be provided. In this instance, each opening should have an area of not less than 1 $in.^2$ for each 4000 Btu of appliance input. One of these openings should be located within 12 in. of the boiler room ceiling, the other within 12 in. of the boiler room floor (Figure 5.26).
2. When the unit is installed in an interior confined space and all combustion air is to be drawn from the outside, two openings as described above in procedure 1 must be provided. However, these ducts must open into or be ducted into areas that draw in outside air freely, such as an attic or crawl space. In this instance two guidelines must be followed:
   a. Where vertical ducting is to be used, or ducting through outside walls is to be provided, 1 $in.^2$ for each 4000 Btu of appliance input must be provided (Figure 5.27).
   b. Where horizontal ducting connects the boiler room to the outdoors, not

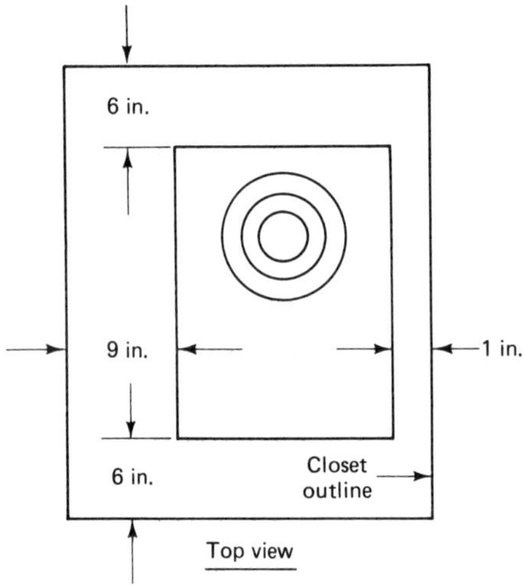

Boiler size	Dimension "W"
203	14-7/16 in.
204	17-11/16 in.
205	20-15/16 in.
206	24-3/16 in.
207	27-7/16 in.
208	30-11/16 in.
209	33-15/16 in.
210	37-3/16 in.

Note: Area of each opening – 1 in.^{2} for each 1000 Btu/hr input – with a minimum of 100 in.^{2} height of opening should be 1/2 width

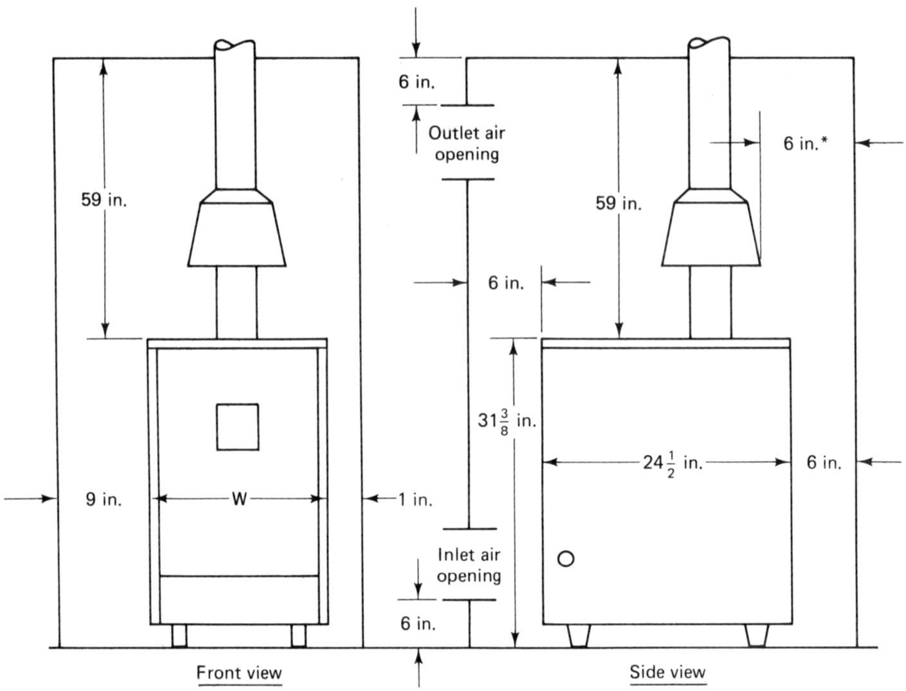

*Minimum clearance around draft hood and breeching

Figure 5.25 Minimum clearances for both closet and noncloset installations of hot water boilers to combustible surfaces. (Courtesy of Burnham Corp.)

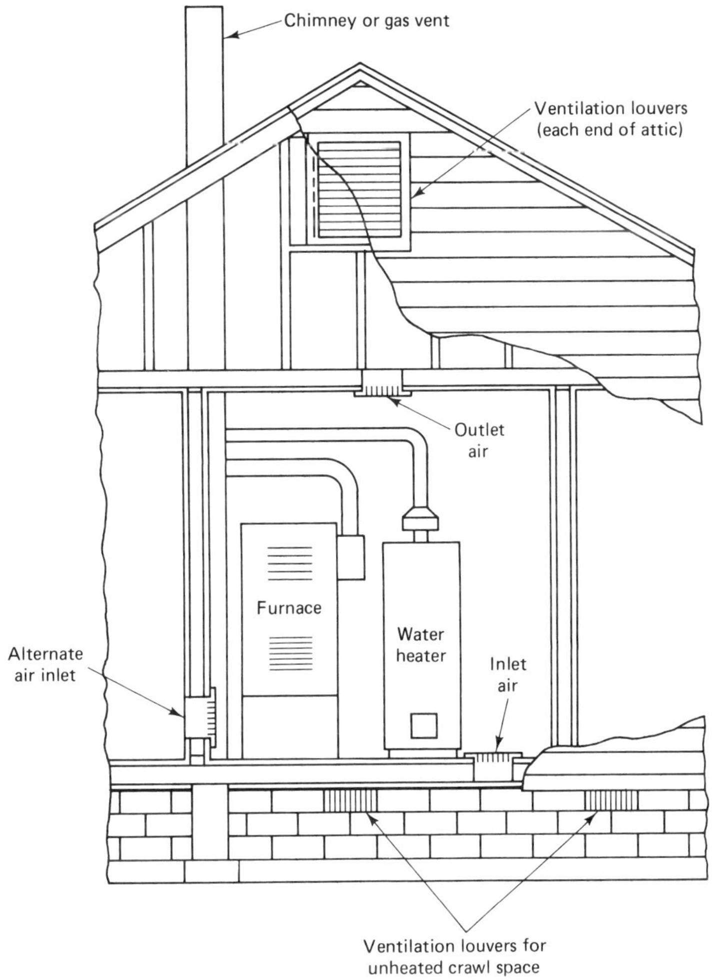

Figure 5.26 Boiler installed in interior confined room. (Copyright 1984 by the American Gas Association and the National Fire Protection Association. Used by permission of the copyright holder.)

less than 1 in.^{2} for each 2000 Btu of boiler input should be provided (Figure 5.28).

3. In the third instance, where the boiler room will be vented to an adjoining room and the adjoining room will be vented to the outdoors, two openings in the boiler room must be provided as in procedure 1. These two openings should be sized to provide 1 in.^{2} of surface area for each 1000 Btu of boiler input. The free open area of the adjoining room vented to the outdoors must be sized at 1 in.^{2} for each 5000 Btu of appliance input.

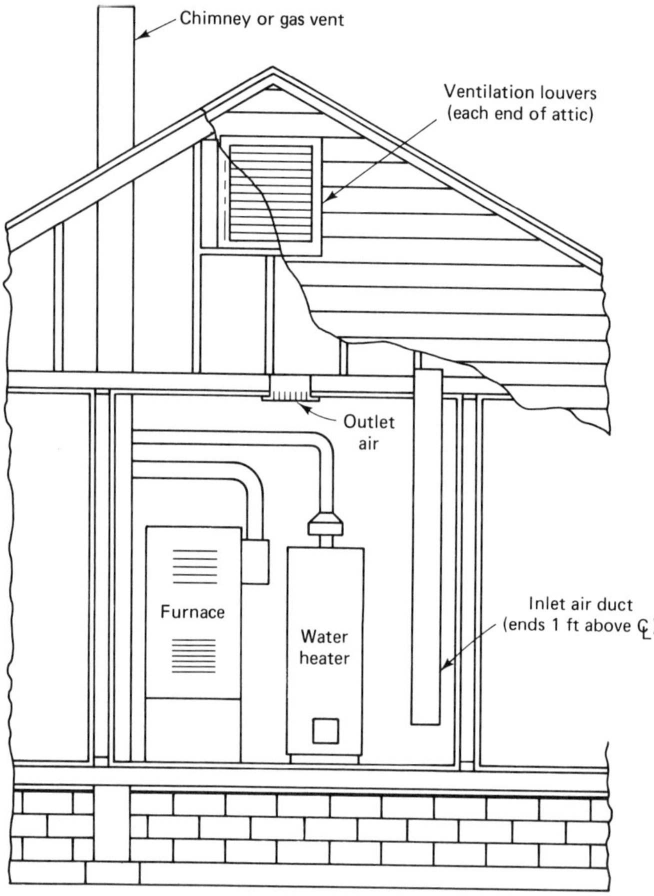

Figure 5.27 Boiler installed in confined spaces, all air drawn in from outdoors through ventilated attic. Note vertical air inlet duct. (Copyright 1984 by the American Gas Association and the National Fire Protection Association. Used by permission of the copyright holder.)

Some burners, such as pulsed combustion units for example, require special combustion air supply provisions. The unit illustrated in Figure 5.29 is factory equipped with materials for directly venting the appliancc to the outdoors. When the appliance is installed following the guidelines above, efficient combustion of the fuel gas can take place while maintaining proper ventilation of the unit and air circulation within the home.

Gas Delivery Systems

Prior to the installation of any gas-fired boiler, a piping layout should be made, indicating the length of pipe that will be required for the entire installation. Once

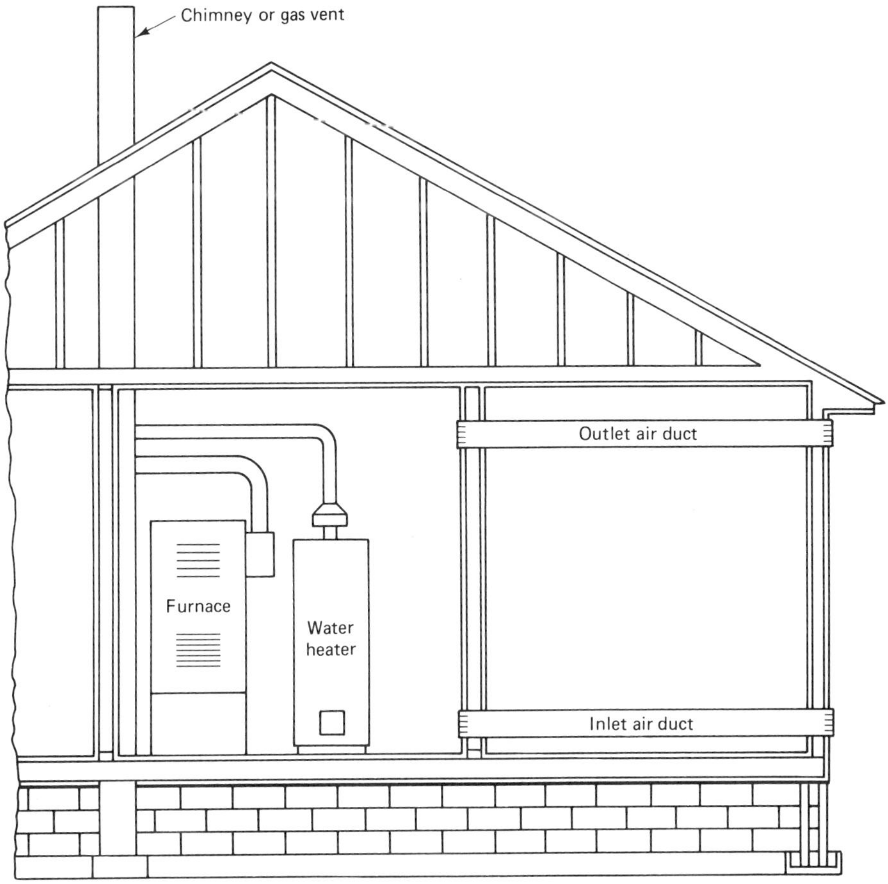

Figure 5.28 Boiler located in confined space. Note horizontal air duct to bring in outside combustion air. (Copyright 1984 by the American Gas Association and the National Fire Protection Association. Used by permission of the copyright holder.)

the layout has been finished and the Btu input of all appliances within the residence has been determined, the size of the gas delivery pipe for the system can be determined. For example, let us assume that we are going to install a 125,000-Btu natural-gas-fired boiler in a new home. In addition to the boiler, a 40,000-Btu hot water heater will also be installed. The layout of the installation is illustrated in Figure 5.30.

The first step in designing the gas delivery system is to determine the amount of gas, in cubic feet per hour (ft^3/hr) required by each appliance. This is done by using the formula

$$\frac{\text{consumption rating of appliance}}{\text{Btu/ft}^3 \text{ of gas}} = \text{maximum applicable gas demand}$$

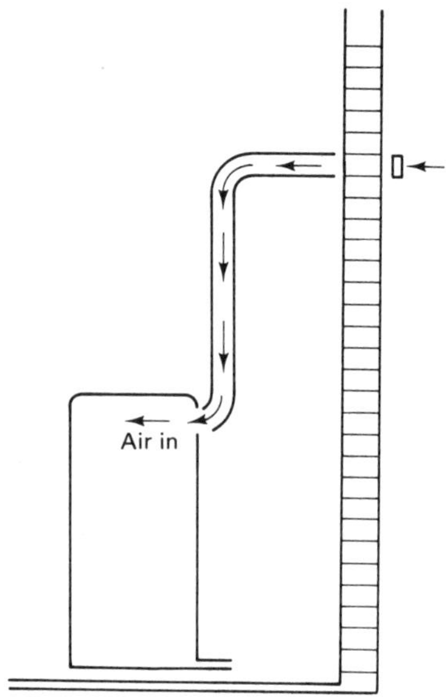

Figure 5.29 Combustion air provision for pulsed combustion boilers. (Courtesy of Hydrotherm, Inc.)

Therefore, for the heating boiler, the equation reads

$$\frac{125{,}000 \text{ Btu/hr boiler input}}{1000 \text{ Btu/ft}^3} = 125 \text{ ft}^3\text{/hr of gas required for the boiler}$$

For the domestic hot water heater, the equation reads

$$\frac{40{,}000 \text{ Btu/hr heater input}}{1000 \text{ Btu/ft}^3} = 40 \text{ ft}^3\text{/hr of gas required for the water heater}$$

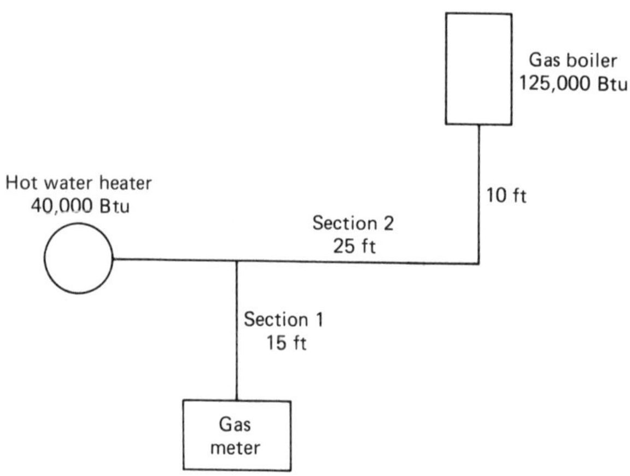

Figure 5.30 Gas piping schematic (typical).

To determine the pipe sizes required for this installation, assuming the use of iron pipe as the piping material, we use the longest total length of pipe run in our layout, 50 ft, the distance from the gas meter to our gas-fired boiler, in conjunction with the figures derived from Table 5.2. Using the column marked 50 ft, the boiler outlet that requires 125 ft^3 of gas per hour also requires a pipe size of $\frac{3}{4}$ in.

The water heater gas outlet, which requires 40 ft^3 of gas per hour, requires a $\frac{3}{8}$-in. pipe that T's off the main feed line at the junction of sections 1 and 2 illustrated in Figure 5.30. Section 1, which supplies gas to both the boiler and the hot water

TABLE 5.2 MAXIMUM CAPACITIES OF PIPE (FT^3 GAS/HR), SPECIFIC GRAVITY 0.60

	Nominal iron pipe size (ins.)			
Length (ft)	$\frac{1}{2}$	$\frac{3}{4}$	1	$1\frac{1}{4}$
	Based on a pressure drop of 0.3 in. water			
10	132	278	520	1050
20	92	190	350	730
30	73	152	285	590
40	63	130	245	500
50	56	115	215	440
60	50	105	195	400
70	46	96	180	370
80	43	90	170	350
90	40	84	160	320
100	38	79	150	305
	Based on a pressure drop of 0.5 in. water			
10	175	360	680	1400
20	120	250	465	950
30	97	200	375	770
40	82	170	320	660
50	73	151	285	580
60	66	138	260	530
70	61	125	240	490
80	57	118	220	460
90	53	110	205	430
100	50	103	195	400
125	44	93	175	360
150	40	84	160	325
175	37	77	145	300

Source: Burnham Corp.

Note: For all practical purposes, each 90° elbow can be considered as the following equivalent in length of straight pipe: $\frac{1}{2}$ in., 1.6 ft; $\frac{3}{4}$ in., 2.1 ft; 1 in., 2.6 ft; $1\frac{1}{4}$ in., 3.5 ft.

heater, totaling 165,000 Btu, should be sized as a 1-in. feed line (see Table 5.2). Note that the figures given in Table 5.2 are based on natural gas with a specific gravity of 0.60. If the specific gravity of the gas, a figure normally supplied by your local utility company, is other than 0.60, the multipliers given in Table 5.3 should be used to determine the correct pipe sizes for individual appliances.

Table 5.4 lists the capacity, in thousands of Btu per hour, for LP gas installations within the limits specified. To determine pipe sizes for piping materials and pressures other than those listed in Tables 5.2, 5.3, and 5.4, consult Appendixes A and B, the National Fuel Gas Code, or specific manufacturers' installation guides.

Once the pipe sizes required for the installation have been determined, the final configuration of the system can be done and preparations made to connect all appliances. In many instances where a new boiler of the same Btu capacity is replacing an older, outdated, or obsolete unit, the installer need only be concerned with a connection from the existing gas feed line to the new appliance.

Gas Piping Procedures

Only licensed technicians and those people familiar with gas piping procedures should become involved with gas-fired boiler piping. In some localities, only licensed technicians can perform this work. In others, the work must conform to existing codes and be inspected after completion, regardless of who performs the work.

The most commonly used gas piping materials are steel, wrought iron, and ductile iron pipe that conforms to applicable ANSI standards, as well as those set by the American Society for Testing and Materials (ASTM). In addition to the foregoing materials, seamless copper (type K or type L), aluminum alloy, or steel tube may be employed for gas piping as long as the gases are not corrosive to such materials [less than 0.3 grams (g) of hydrogen sulfide (H_2S) per 100 ft^3 of gas]. The specific standards for the corrosive effects of gas on metallic tubing may be found in ANSI/ASTM Standard D2385 or ANSI/ASTM Standard D2420. The use of cast iron pipe for gas-line connections is prohibited.

When threading pipe for specific installations, Table 5.5 should be followed to determine applicable thread lengths based on the specific size of the pipe and the number of threads per inch.

TABLE 5.3 MULTIPLIERS FOR NATURAL GAS, SPECIFIC GRAVITY OTHER THAN 0.60

Specific gravity	Correction factors
0.50	1.10
0.55	1.04
0.60	1.00
0.65	0.96
0.70	0.93

TABLE 5.4 PIPE CAPACITY (MBTU/HR) FOR LP GAS

Nominal iron pipe size (in.)	Length of pipe (ft)											
	10	20	30	40	50	60	70	80	90	100	125	150
$\frac{1}{2}$	275	189	152	129	114	103	96	89	83	78	69	63
$\frac{3}{4}$	567	393	315	267	237	217	196	185	173	162	146	132
1	1071	732	590	504	448	409	378	346	322	307	275	252
$1\frac{1}{4}$	2205	1496	1212	1039	913	834	771	724	677	630	567	511
$1\frac{1}{2}$	3307	2299	1858	1559	1417	1275	1181	1086	1023	976	866	787
2	6221	4331	3465	2992	2646	2394	2205	2047	1921	1811	1606	1496

Source: National Fuel Gas Code, ANSI Standard Z223.1. Copyright 1984 by the American Gas Association and the National Fire Protection Association. Used by permission of copyright holder.

Note: At 11 in. water column inlet pressure; based on a pressure drop of 0.5 in. water column.

TABLE 5.5 SPECIFICATIONS FOR THREADING METALLIC PIPE

Iron pipe size (in.)	Approximate length of threaded portion (in.)	Approximate number of threads to be cut
$\frac{1}{2}$	$\frac{3}{4}$	10
$\frac{3}{4}$	$\frac{3}{4}$	10
1	$\frac{7}{8}$	10
$1\frac{1}{4}$	1	11
$1\frac{1}{2}$	1	11
2	1	11
$2\frac{1}{2}$	$1\frac{1}{2}$	12
3	$1\frac{1}{2}$	12
4	$1\frac{5}{8}$	13

Source: National Fuel Gas Code, ANSI Standard Z223.1. Copyright 1984 by the American Gas Association and the National Fire Protection Association. Used by permission of copyright holder.

Pipe dope should be applied to all threaded connections to help ensure airtight connections. Teflon tape and Teflon pipe dope should not be used on threaded gas pipe connections. Teflon will tend to break down under the influence of fuel gases, leading to restrictions and clogging of valves and appliance orifices.

When installing gas pipe, it should be supported at adequate intervals to prevent induced stress and/or vibration of the system. Table 5.6 suggests maximum spacing distances for gas piping.

Direct connection of the gas feed line to the appliance should appear as illustrated in Figure 5.31. Note that a manual shutoff valve is installed upstream of the break-apart union. Also, a sediment trap with a sealed drip leg must be provided at all appliance connections as well as anywhere in the supply line that accumulation of moisture and sediment may be suspected. If the boiler gas line is supplied from the main gas feed line through a branch line that T's off the main, ensure that the T fitting is taken from the top or sides of the main, not from the bottom.

After the piping has been completed, air should be purged from the line and

TABLE 5.6 SUGGESTED SPACING OF PIPE SUPPORTS

Steel pipe, nominal size of pipe (in.)	Spacing of supports (ft)	Nominal size of tubing (in. O.D.)	Spacing of supports (ft)
$\frac{1}{2}$	6	$\frac{1}{2}$	4
$\frac{3}{4}$ or 1	8	$\frac{5}{8}$ or $\frac{3}{4}$	6
$1\frac{1}{4}$ or larger (horizontal)	10	$\frac{7}{8}$ or 1	8
$1\frac{1}{4}$ or larger (vertical)	every floor level		

Source: National Fuel Gas Code, ANSI Standard Z223.1. Copyright 1984 by the American Gas Association and the National Fire Protection Association. Used by permission of copyright holder.

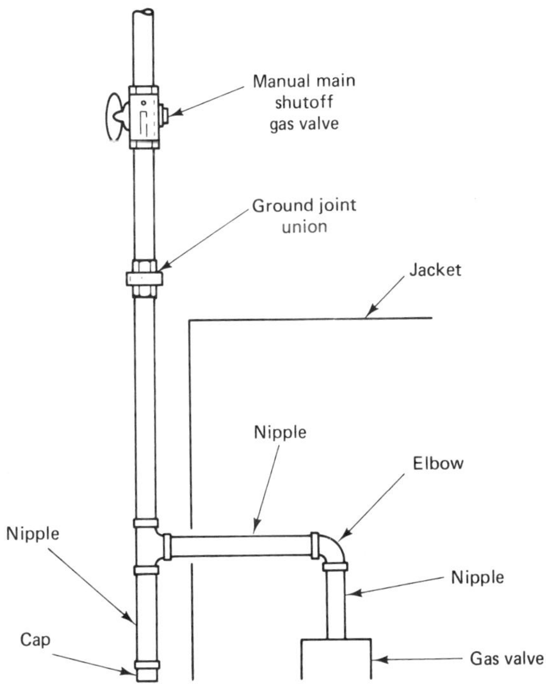

Figure 5.31 Gas supply piping to appliance (typical).

all connections should be checked for leaks. During this initial testing, the boiler *must* be disconnected from the gas line. A bubble solution of soap and water is generally used for this purpose, in addition to commercial gas leak detection fluids. If soap and water is used for lead detection, the pipe should be rinsed with water and thoroughly dried to prevent surface corrosion. After the gas has been turned on and the integrity of all joints established prior to ignition of the boiler, it is good practice to double check all joints.

Prior to energizing the boiler, all air must be purged from the gas lines. Purging should not be undertaken in any confined area where open flames or potential ignition sources exist. Adequate ventilation during purging should be provided at all times, and the rate of purging should be controlled to ensure that the procedure is safely accomplished. Purging is generally accomplished through the pilot light orifice by depressing the manual gas feed button to bypass the thermocouple gas shutdown mechanism. Under no circumstances should excess gases be voided into the combustion chamber.

Measuring Gas Input

Regulator adjustment. To adjust gas pressure at the appliance regulator, a manometer is connected into the gas manifold or suitable tapping on the appliance regulator. Pressure is then adjusted following the regulator adjustment procedure to obtain output pressure specified by the appliance manufacturer.

Gas meter calculations. In natural gas installations, it is important to ensure that the gas input into the boiler is delivered at a rate that is accurate enough to yield the nameplate Btu rating for which the appliance was designed. This ensures that the appliance will operate within the safety parameters set by the appliance manufacturer.

The Btu rating of the boiler is a function of three variables:

1. Specific gravity of the fuel gas
2. Heating value per cubic foot of the fuel gas
3. Volume of gas supplied to the appliance per unit of time

To calculate appliance gas consumption, a variety of methods can be used. Two methods that are fast and reliable are presented here. One procedure involves the use of a gas meter rate chart; the other procedure relies on manually timing the meter and interpolating the results mathematically.

To perform these calculations, both the specific gravity of the gas and its heating value per cubic foot should be obtained from the local utility company. When performing these calculations, all gas appliances other than the boiler being tested should be turned off to ensure the accuracy of the results.

Gas rate chart procedure. The use of a gas rate chart simplifies the determination of appliance gas consumption. A chart of this type is shown in Table 5.7. To use the chart, the size of the gas meter test dial, either $\frac{1}{2}$, 1, 2, or 5 ft^3, is timed for one revolution. The number of cubic feet of gas consumed by the appliance per hour is then read off the chart, based on the time necessary for one revolution of the test dial.

Manual calculations for consumption determination. With the gas boiler operating, observe the gas meter with a stopwatch and determine the amount of time, in seconds, required to consume 1 ft^3 of gas. Divide this figure into 3600 to determine cubic foot consumption per hour. After determining the number of cubic feet per hour consumed by the boiler, multiply this figure by 1000, the heating value of the gas per cubic foot, to determine the Btu input of the boiler. This result should correspond to the nameplate rating on the unit. If the input is not the same, the appliance regulator should be adjusted until the figures match. Note that in the calculations above, 1000 Btu/ft^3 is used as the standard for determining the input rating on the boiler. If the figure as supplied by the utility varies from 1000 Btu/ft^3, the utility figure should be substituted.

TABLE 5.7 GAS FLOW RATE CHART (FT^3/HR)

Seconds for one revolution	Size of test dial		Seconds for one revolution	Size of test dial		Seconds for one revolution	Size of test dial	
	$\frac{1}{2}$ ft³	1 ft³		2 ft³	5 ft³		2 ft³	5 ft³
10	180	360	10	720	1800	62	116	290
11	164	327	11	655	1636	64	112	281
12	150	300	12	600	1500	66	109	273
13	138	277	13	555	1385	68	106	265
14	129	257	14	514	1286	70	103	257
15	120	240	15	480	1200	72	100	250
16	113	225	16	450	1125	74	97	243
17	106	212	17	424	1059	76	95	237
18	100	200	18	400	1000	78	92	231
19	95	189	19	379	947	80	90	225
20	90	180	20	360	900	82	88	220
21	86	171	21	343	857	84	86	214
22	82	164	22	327	818	86	84	209
23	78	157	23	313	783	88	82	205
24	75	150	24	300	750	90	80	200
25	72	144	25	288	720	92	78	196
26	69	138	26	277	692	94		192
27	67	133	27	267	667	96	75	188
28	64	129	28	257	643	98		184
29	62	124	29	248	621	100	72	180
30	60	120	30	240	600	102		176
31		116	31	232	581	104	69	173
32	56	113	32	225	563	106		170
33		109	33	218	545	108	67	167
34	53	106	34	212	529	110		164
35		103	35	206	514	112	64	161
36	50	100	36	200	500	116	62	155

(continued)

TABLE 5.7 *(continued)*

Seconds for one revolution	Size of test dial		Seconds for one revolution	Size of test dial		Seconds for one revolution	Size of test dial	
	$\frac{1}{2}$ ft^3	1 ft^3		2 ft^3	5 ft^3		2 ft^3	5 ft^3
37		97	37	195	486	120	60	150
38	47	95	38	189	474	125		144
39		92	39	185	462	130		138
40	45	90	40	180	450	135		132
41			41	176	439	140		129
42	43	86	42	172	429	145		124
43			43	167	419	150		120
44	41	82	44	164	409	155		116
45	40	80	45	160	400	160		113
46		78	46	157	391	165		109
47	38		47	153	383	170		106
48		75	48	150	375	175		103
49			49	147	367	180		100
50	36	72	50	144	360			
51			51	141	353			
52		69	52	138	346			
53	34		53	136	340			
54		67	54	133	333			
55			55	131	327			
56	32	64	56	129	321			
57			57	126	316			
58	31	62	58	124	310			
59			59	122	305			
60	30	60	60	120	300			

Source: Burnham Corp.

VENTING SYSTEMS

The purpose of all venting systems on gas appliances is to remove safely the byproducts of combustion from the boiler combustion chamber to the outside of the residence to prevent contamination of interior living spaces. There are several factors to be considered in designing and installing an efficient and safe venting system. In gas boilers utilizing atmospheric combustion burners, natural draft created by the lighter-than-air flue gases is used to funnel the gases through an essentially vertical path for removal. In boilers using induced-draft technology, the venting system provides the same function as its noninduced counterpart, although the mechanics of the system are modified for the operation of the induced-draft system. Each of the common types of vent systems used in atmospheric combustion, induced-draft condensing and noncondensing units and pulsed combustion units, is examined below in greater detail.

Atmospheric Combustion Appliances

The vent system configuration for atmospheric combustion boilers is illustrated in Figure 5.32. The basic components of the system include the (1) draft hood assembly, (2) the vertical riser assembly, (3) the horizontal connectors, and (4) the chimney.

The draft hood is used on all atmospheric combustion burners. Two common types of draft hoods are illustrated in Figure 5.33. Note from the construction of the draft hood that dilution air is drawn in from around the appliance. This process both dilutes the flue gases with normal room air and lowers the flue gas temperature. Also, under high-draft conditions, the hood will act to stabilize the chimney draft as well as to divert momentary downdrafts away from the combustion chamber of the boiler (Figure 5.34).

Where lateral pipe runs are required for chimney connections, they should be kept as short as possible. All horizontal runs should be sloped upward with a minimum rise of 1 in. for every 4 ft of run. When the vent connector terminates in a masonry chimney, a masonry thimble or slip joint should be used with the vent pipe connector. Any space between the thimble and vent pipe connector should be sealed with high-temperature furnace cement. The thimble should be inserted into the chimney so that it is flush with the inside flue liner and located above the bottom of the chimney to prevent blockage from accumulated sediment and effluent byproducts. This installation arrangement is illustrated in Figure 5.35.

Several requirements and precautionary measures need to be observed when venting appliances. Most building codes usually prohibit venting the gas boiler into any chimney that also serves a solid-fuel appliance. Research local applicable codes before proceeding.

Although single-wall vent may be used for venting gas appliances, type B double-wall gas vent pipe is recommended. Permissible clearances for common types of vent pipes are given in Table 5.8.

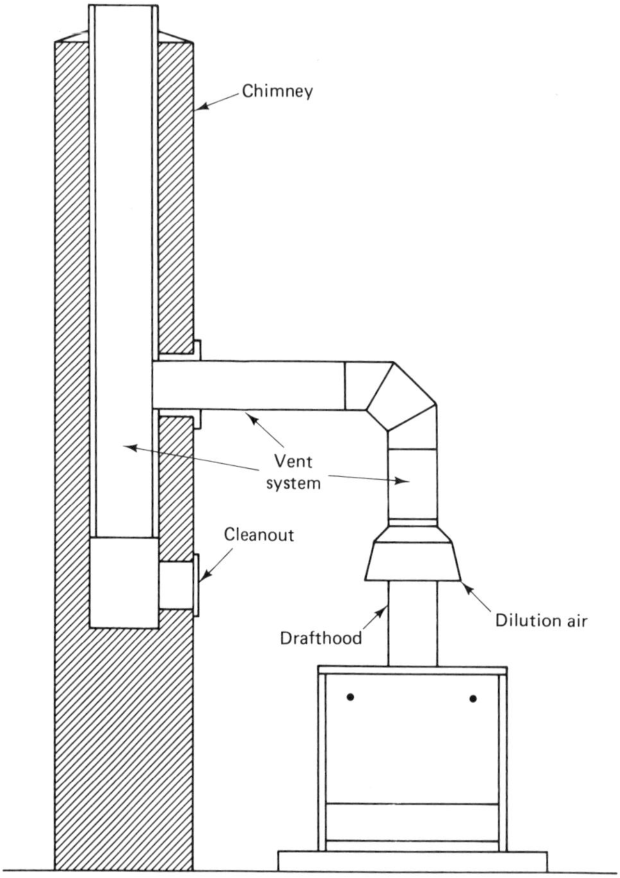

Figure 5.32 Vent system used in atmospheric combustion boilers.

The instructional manuals that are supplied with most domestically manufactured gas-fired boilers usually outline safe venting procedures to be followed for the specific appliance that comply with all applicable codes.

Vent Dampers

Electrically operated vent dampers that close off the majority of the cross-sectional area of the vent pipe during the off-cycle periods of the boiler have been coming into widespread use. A majority of these devices are installed when older, less efficient boilers are replaced with high-efficiency units. During operation of the heating system, the vent damper remains open to allow unrestricted removal of flue gases. During off-cycle periods, however, heat losses that would occur due to the natural aspiration of the chimney are all but eliminated when the automatic damper is closed. A typical electrically operated vent damper is illustrated in Figure 5.36.

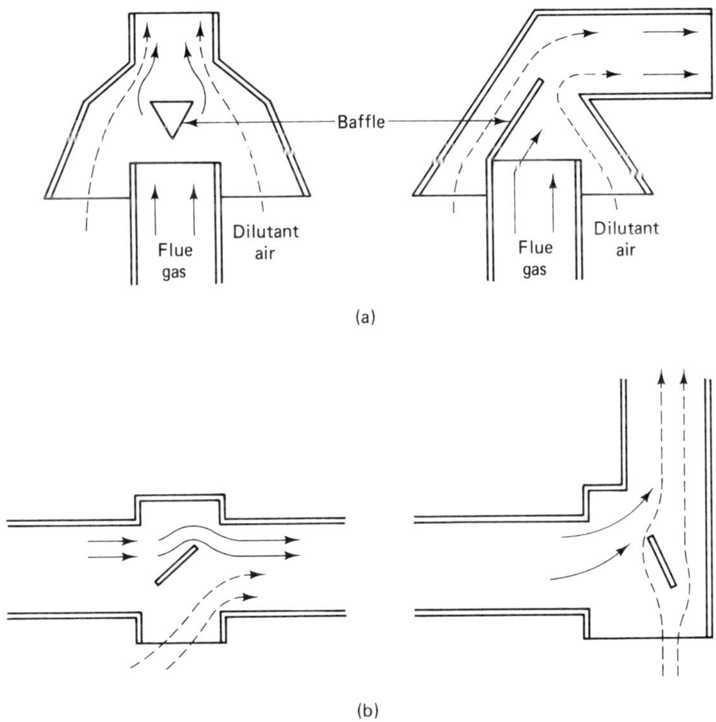

Figure 5.33 Design features of draft hoods: (a) vertical draft hood (vertical to horizontal); (b) horizontal draft hood (horizontal to vertical). (Courtesy of Robertshaw Controls Company.)

Vent dampers can be used on both standing pilot and intermittent spark ignition systems. Since the damper does not completely seal off the vent pipe, there is sufficient area available for venting the accumulated pilot light gases on standing pilot ignition systems. Electric vent dampers are installed above the appliance draft hood and connected into the control module as illustrated in Figure 5.37.

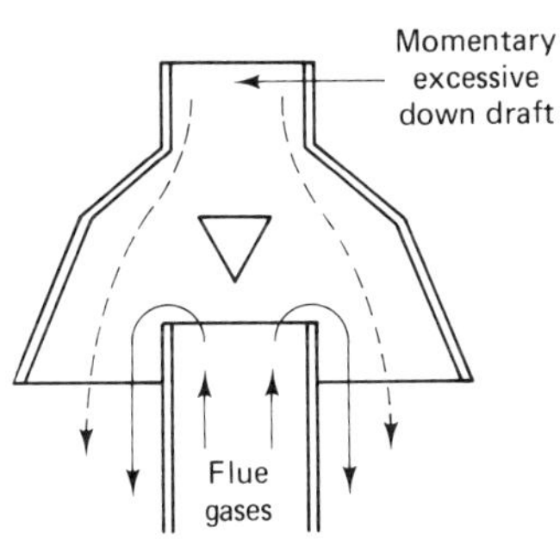

Figure 5.34 Stabilizing effect of draft hood during monentary downdrafts. (Courtesy of Robertshaw Controls Company.)

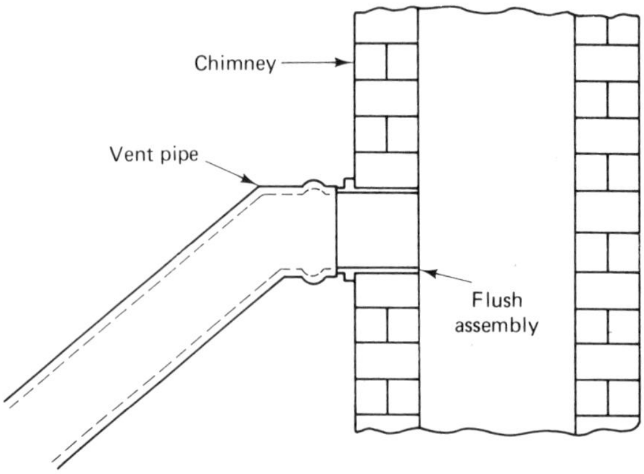

Figure 5.35 Chimney breeching connection.

Sequence of operation of vent dampers. During boiler cycling, the electric vent damper operates as a normally open electrical switch that is in the open position when no power is being applied to it. In this position, the vent damper itself is open. During the boiler off-cycle periods, power is applied to the damper control mechanism, keeping the vent closed. The main gas valve control wiring runs through a

TABLE 5.8 VENT PIPE CLEARANCES TO COMBUSTIBLE SURFACES

	Minimum distance from combustible material (in.)			
Equipment	Listed type B gas vent material	Listed type L vent material	Single-wall metal pipe	Factory-built chimney sections
Listed equipment with draft hoods and equipment listed for use with type B gas vents	As listed	As listed	6	As listed
Residential boilers and furnaces with listed gas conversion burner and with draft hood	6	6	9	As listed
Residential appliances listed for use with type L vents	Not permitted	As listed	9	As listed
Residential incinerators	Not permitted	9	18	As listed
Listed gas-fired toilets	Not permitted	As listed	As listed	As listed
Unlisted residential appliances with draft hood	Not permitted	6	9	As listed
Residential and low-heat equipment other than those above	Not permitted	9	18	As listed
Medium-heat equipment	Not permitted	Not permitted	36	As listed

Source: National Fuel Gas Code, ANSI Standard Z223.1. Copyright 1984 by the American Gas Association and the National Fire Protection Association. Used by permission of copyright holder.

Note: These clearances shall apply unless the listing of an appliance specifies different clearances, in which case the listed clearances shall apply.

Figure 5.36 Electric vent damper. (Courtesy of Slant/Fin Corp.)

switch in the vent damper (see Figure 5.37). Upon a call for heat from a room thermostat, the vent damper is deenergized, allowing the damper to open up the cross-sectional area of the vent pipe. As soon as the damper is fully open, an end switch in the damper housing closes. When this switch closes, it directs power to energize the main gas valve which supplies gas to the boiler. Most vent dampers are equipped with high-temperature sensing switches, so that in the event of excessively high flue gas temperatures, the sensing switch will keep the damper open even during the off-cycle periods of the boiler, until the stack temperature falls to a preset safe level.

Induced-Draft Condensing Boilers

Because induced-draft condensing boilers develop flue gas temperatures below the dew point, producing liquid effluents in the process that are corrosive to conventional venting systems and chimney materials, as well as venting under pressure developed by the draft-inducing fan, the vent systems on these boilers differ from their noncondensing counterparts in most respects.

Most condensing boilers vent in one of two ways: direct exhaust through the building wall or through an existing chimney by using a special appliance vent pipe that runs the length of the chimney, from the bottom to the chimney top or cap.

Direct-venting procedures. When directly venting a condensing boiler through existing building walls, stainless steel flue pipe must be used throughout the system. This type of venting configuration is illustrated in Figure 5.38. When installing this type of system, certain prerequisites must be met. These are:

1. Contrary to common venting practices, the crimped ends of the pipe should face toward the boiler to facilitate condensate drain-back.

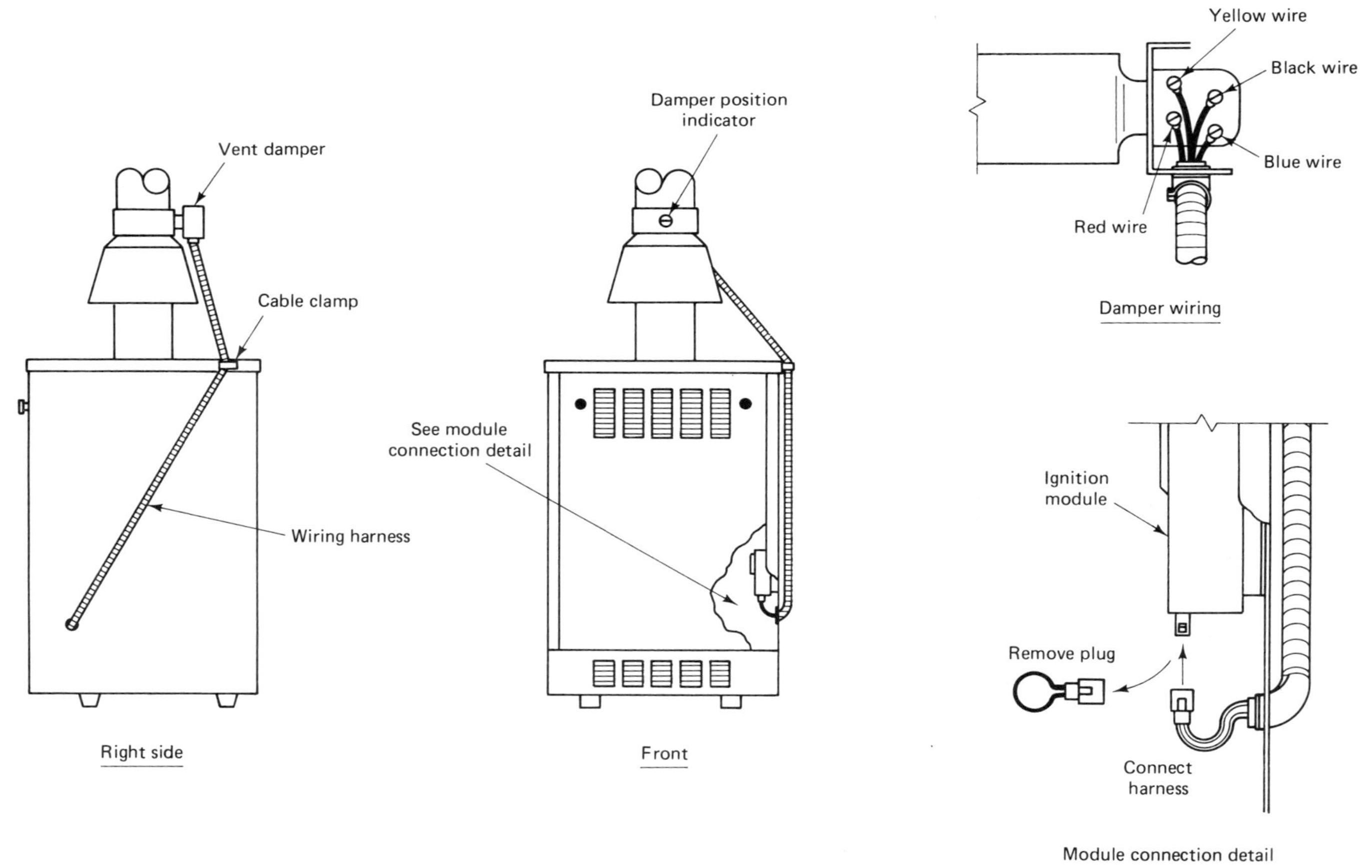

Figure 5.37 Electrical connections between vent damper and boiler. (Courtesy of Burnham Corp.)

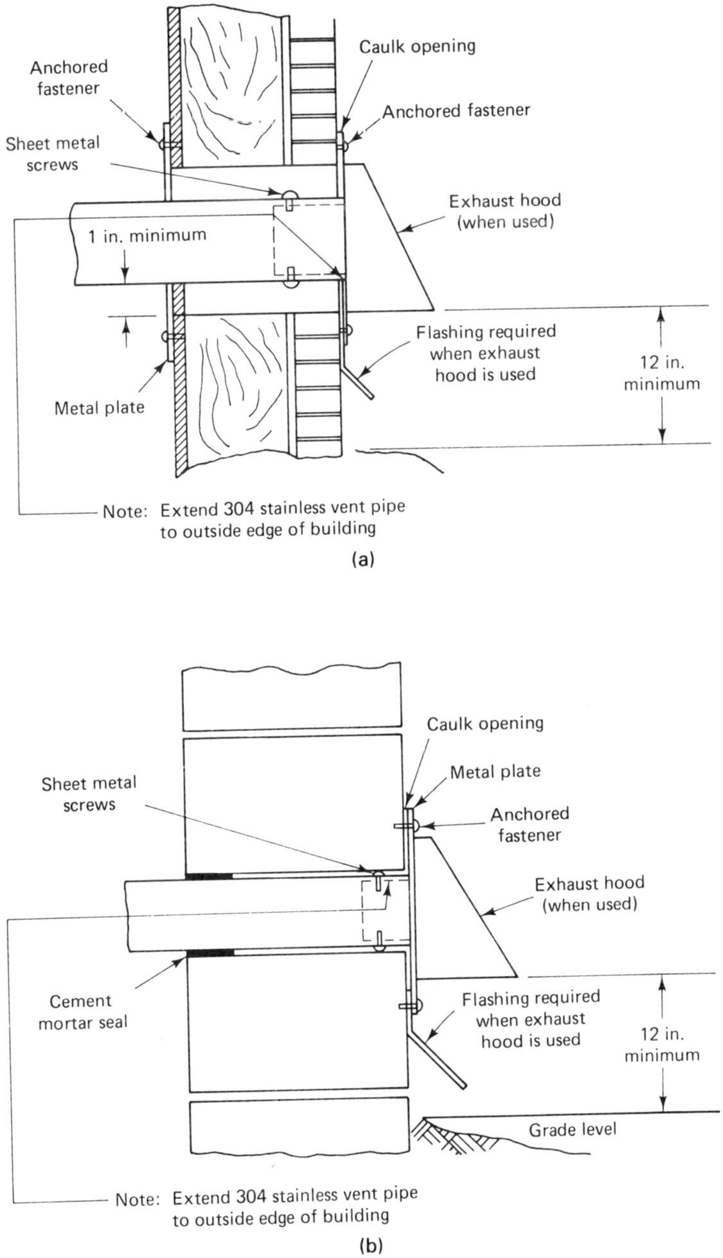

Figure 5.38 Venting a condensing boiler through (a) combustible and (b) noncombustible walls. (Courtesy of Weil-McLain, a Marley Co.)

2. All pipe seams should be placed in either the top or side position of the pipe. Do not install pipe seams on the bottom of horizontal pipe runs.
3. All pipe joints should be sealed with silicone sealant.

Vent termination shields, which protect the walls from the heat of the vent pipe, should conform to the specifications illustrated in Figure 5.39. When designing the venting terminations, the following factors should be kept in mind:*

1. Exhaust gases from the boiler will form a white plume during the winter months. Since this plume can cause condensation on windows in the immediate vicinity, locate vent termination hoods to minimize this possibility. Also, a buildup of ice on the exterior of the building and adjacent shrubbery may occur during severely cold weather. Examine the building and shrubbery within the immediate vicinity of the venting system for possible relocation of shrubs.
2. The vent should terminate at least 4 ft below or horizontally from, or 1 ft above any door, window, or air outlet into the residence. When the vent is located in the vicinity of public walkways, it should be placed a minimum of 7 ft above ground level.
3. The vent should be so located and guarded as to prevent accidental contact by people and household pets.

Induced-Draft Noncondensing Boilers

The venting procedures for induced-draft noncondensing boilers are similar in most respects to those recommended for standard atmospheric combustion units. The stack temperatures in these appliances are somewhat higher than their condensing counterparts. Standard masonry chimneys or type B gas vents are recommended for venting induced-draft noncondensing boilers. A typical venting configuration for a

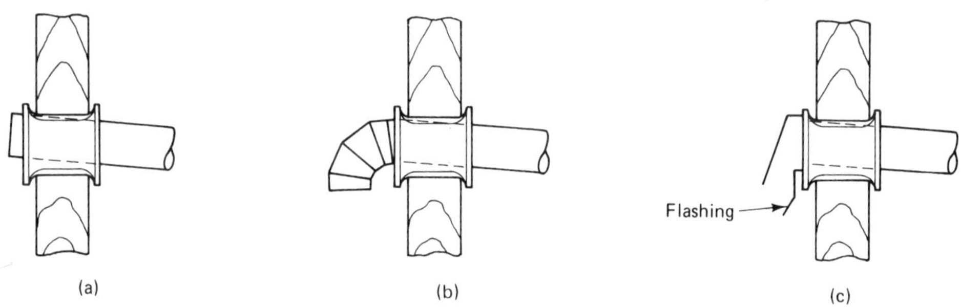

Figure 5.39 Methods of terminating vents in condensing boilers. (Courtesy of Weil-McLain, a Marley Co.)

*Adapted from *VHE Manual,* Weil-McLain, Michigan City, Ind., 1983.

noncondensing application is illustrated in Figure 5.40. Note that a draft hood is not used in this system.

Pulsed Combustion Boilers

Pulsed combustion boilers operating in the condensing mode generally produce exhaust gases in the range of 100 to 110°F. As such, these units use chlorinated polyvinyl chloride (CPVC) pipe.

Several variations of vent pipe configurations are common with these units. In one installation procedure, an existing chimney, if previously unused and of the proper length, can house both the air intake and exhaust piping (Figure 5.41). Note that to prevent contamination of intake combustion air, the exhaust piping should extend a minimum of 12 in. beyond the air intake pipe.

In applications where both the intake and exhaust pipe terminate in through-the-wall configurations, the guidelines in Figure 5.42 should be followed. Note that

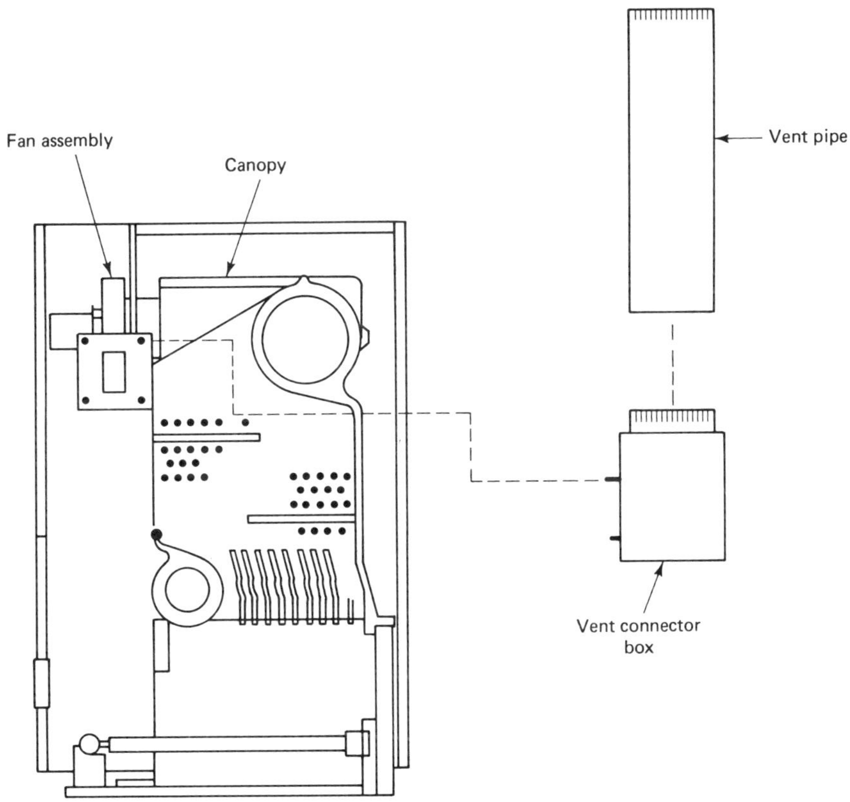

Figure 5.40 Venting configuration used in induced-draft noncondensing boiler. (Courtesy of Burnham Corp.)

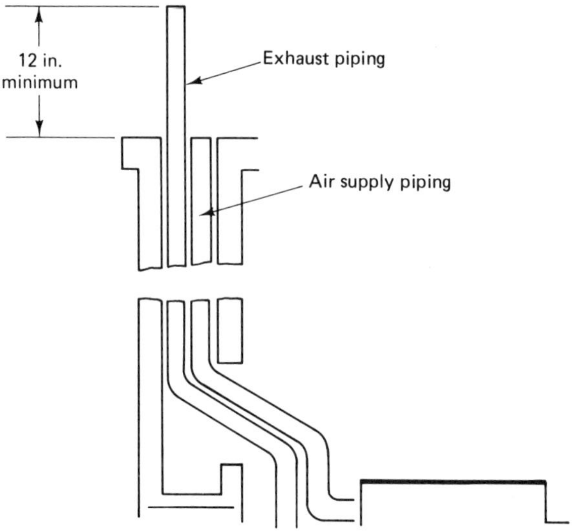

Figure 5.41 Intake and exhaust piping for induced-draft condensing boiler housed in masonry chimney. (Courtesy of Weil-McLain, a Marley Co.)

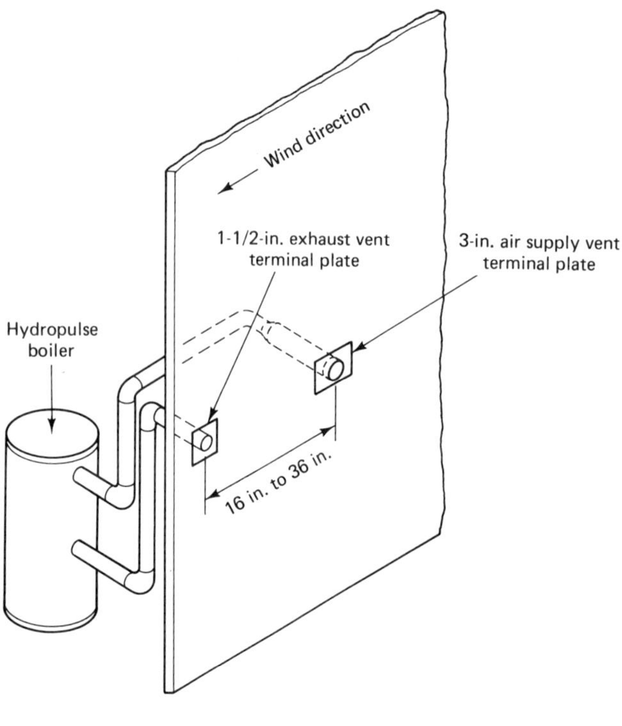

Figure 5.42 Through-the-wall intake and exhaust piping on pulsed combustion boiler. (Courtesy of Hydrotherm, Inc.)

Figure 5.42 emphasizes a minimum separation of 18 in. between intake and exhaust terminations. Also, the exhaust pipe should be positioned *downwind* from the air intake to prevent recirculation of exhaust gases through the air intake system. Downwind direction would be established by selecting the prevailing wind direction in a specific locality during the winter months.

In some instances, such as in multiple-family dwellings, mufflers may be required on pulsed combustion boilers if the venting system cannot be located so as to minimize objectionable noise. A typical muffler installation is illustrated in Figure 5.43.

Thermal and Pressure Safety Switches

In all gas-fired appliances, it is essential that the chimney remain free of obstructions to ensure efficient removal of all combustion by-products. Of particular concern is the removal of carbon monoxide, which is lethal in excessive amounts. If, for any reason, stack gases cannot exit through the venting system and chimney, lethal amounts of carbon monoxide will build up within the home.

Two methods to prevent this occurrence are currently used in modern gas-fired appliances: pressure switches and heat-sensitive thermal switches. On induced-draft boilers, a pressure-sensitive switch is mounted within the unit that detects excessive amounts of back pressure within the draft system. Activation of this switch automatically shuts off the gas feed system until the blockage is removed.

Field thermal switches, sometimes referred to as *spill switches,* detect blocked chimneys on atmospheric combustion units and are connected to the draft hood of the appliance (Figure 5.44). Blockages or downdrafts in the chimney that create a positive pressure in the smoke pipe or breeching adjacent to the draft hood will cause hot flue gases to spill out of the pipe and envelop the switch. Electrical contacts within the switch are normally closed but will open, causing the main gas valve to shut off the flow of all gas to the unit if the switch has been subjected to high-temperature flue gases for 30 to 60 seconds. The standard high-temperature settings on these switches is 140°F. Other temperature settings, as well as manual and automatic recycling of the switch contacts, are available. Figure 5.45 illustrates the wiring of a typical spill switch in series with the gas burner control for positive system shutdown in the event of chimney blockage.

Summary

When installing new heating systems, or when replacing older boilers with high-efficiency units, it is critical for safe operation of the heating system that all applicable code requirements are adhered to.

On replacement units, new vent pipe should always be used to replace older pipe regardless of the apparent condition of the original vent pipe. For maximum safety, only AGA-approved pipe should be used for vent connections.

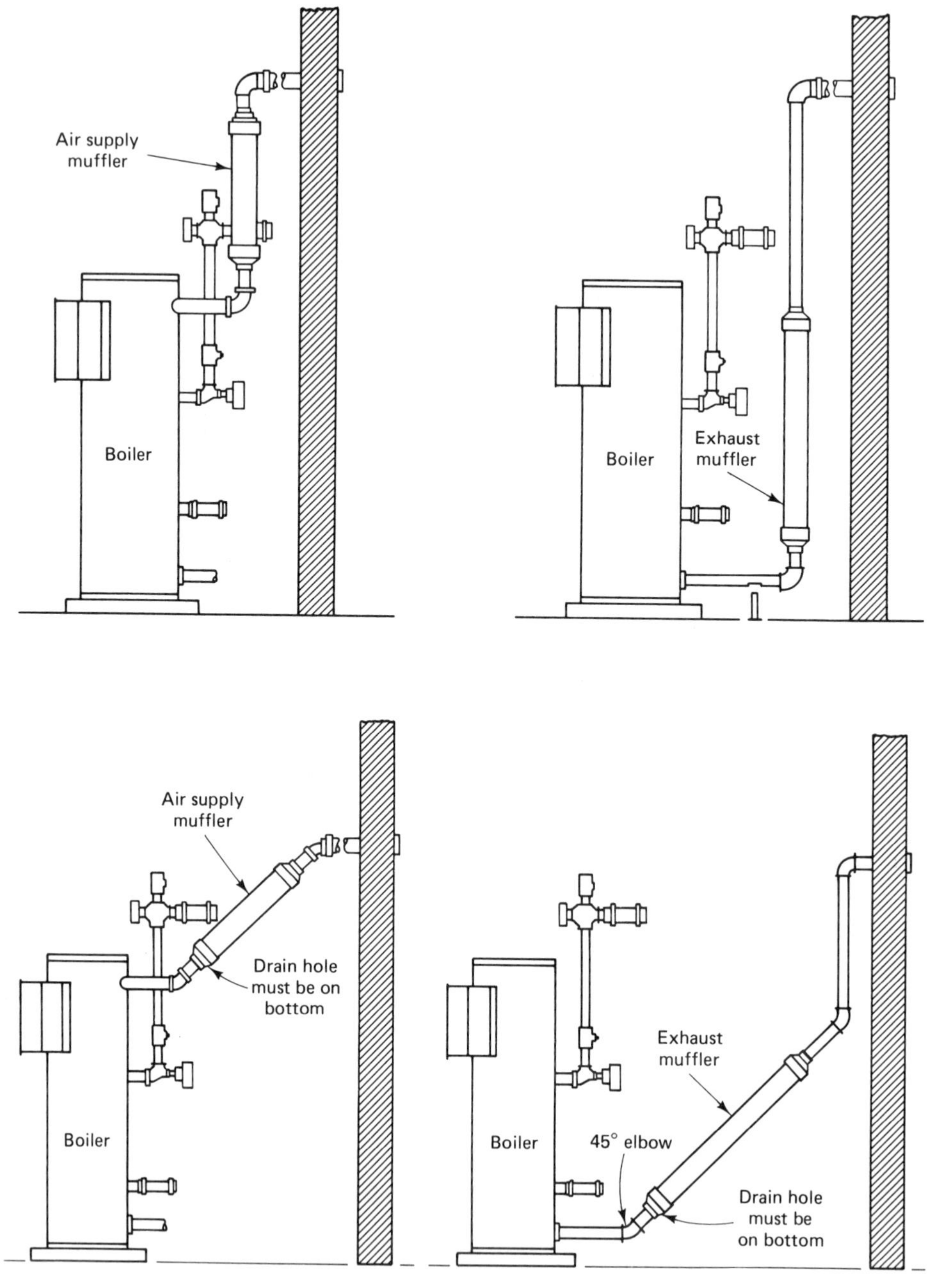

Figure 5.43 Muffler installation on pulsed combustion boiler. (Courtesy of Hydrotherm, Inc.)

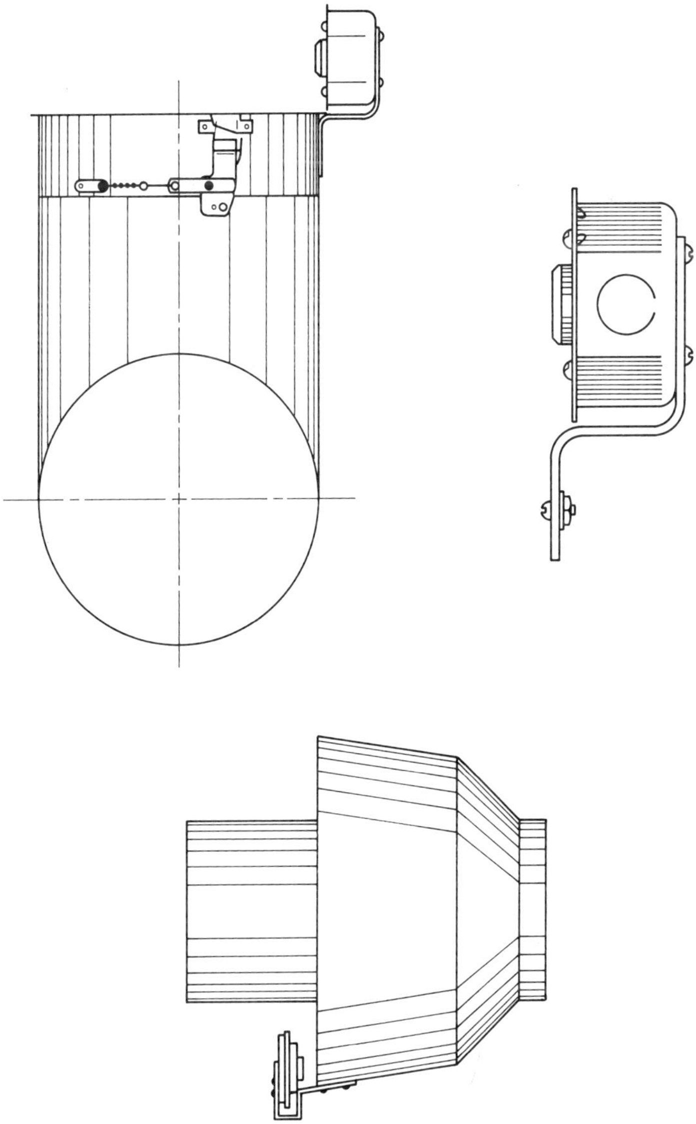

Figure 5.44 Spill switch mounting configuration on horizontal and vertical draft hoods. (Courtesy of the Field Controls Company.)

On all peripheral components, such as draft hoods, electric vent dampers, and factory-built chimneys, the installer should be certain that the components of each system are AGA approved and sized properly for the particular application. Following these guidelines, long-term trouble-free service and reliability should result from all components within the venting system.

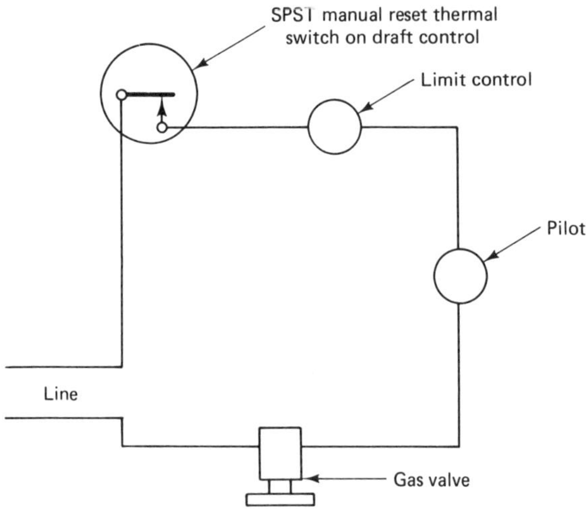

Figure 5.45 Wiring configuration of spill switch and pressure switches on gas-fired boilers. (Courtesy of the Field Controls Company.)

GAS-FIRED IGNITION AND CONTROL SYSTEMS

Most of the operating controls in residential gas-fired boilers are similar in function to oil and solid-fuel devices. A block sequence control diagram for gas-fired boilers is illustrated in Figure 5.46. In this operating sequence, upon a call for heat from a room thermostat, the gas valve is opened, allowing gas to flow through the valve and into the main burners. The pilot light functions to ignite the main burner assembly. The circulator on the boiler will operate on either an intermittent or a continual basis, depending on the configuration of the main relay control. In the intermittent mode, the circulator comes on only when the boiler has reached a preset low-limit water temperature. In the continuous mode, the circulator runs constantly throughout the heating season as long as there is power to the boiler control. The combination relay contains a high-limit safety cutoff and a low-voltage transformer that supplies electricity to all the relays in the system. If the boiler is located above the level of the baseboard radiation, a low-water cutoff should be installed in the system, usually located in the supply pipe directly above the boiler, which operates on the float principle. Should the level of the water in the system drop below the level of the float, it will automatically disconnect all electricity to the system, shutting down the boiler firing mechanism until the level of the water has been restored. The wiring configuration for this control is illustrated in Figure 5.47.

Most gas-fired boilers can operate in one of two pilot ignition modes: either as a standing pilot which is on continually, or as an intermittent ignition system which uses an electronic circuit to produce a spark to light a pilot that ignites the main burner assembly. Each of these ignition systems is examined below in greater detail.

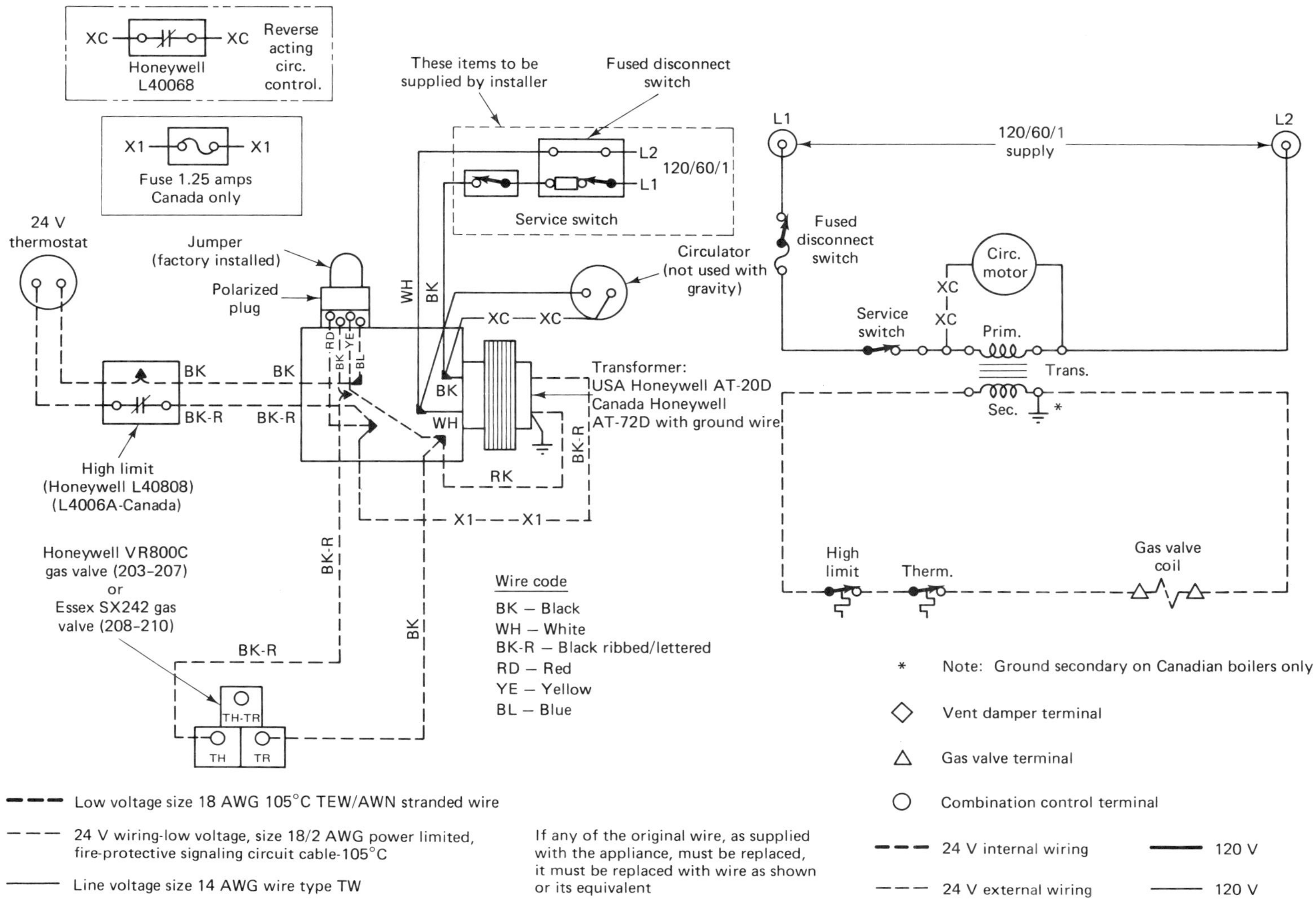

Figure 5.46 Control circuitry for typical gas-fired boiler, 24-V continuous or gravity circulation. (Courtesy of Burnham Corp.)

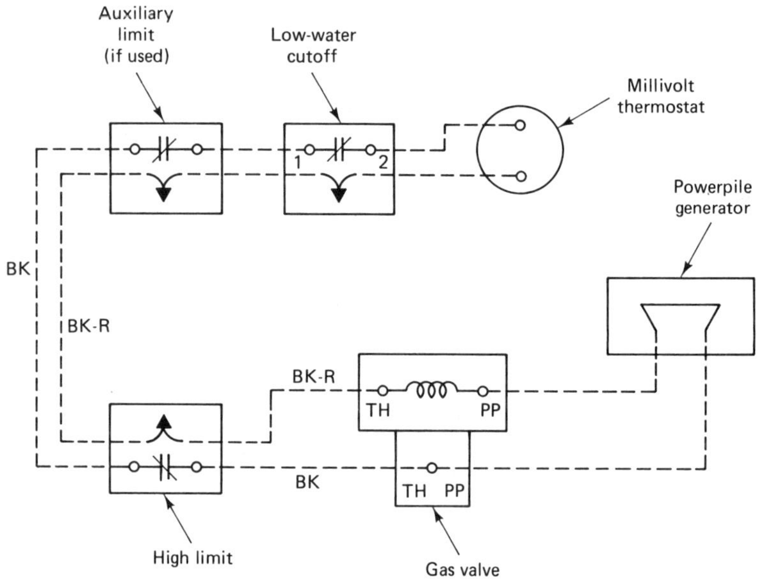

Figure 5.47 Wiring of low-water cutoff in control configuration of typical residential gas-fired boiler. (Courtesy of Burnham Corp.)

Standing Pilot Systems

A typical standing pilot ignition system is illustrated in Figure 5.48. There are a wide variety of standing pilot burners used in gas appliances. The standing pilot light performs a dual function in the system: It serves to ignite the main gas burners when there is a call for heat, and the pilot flame burns against a bimetallic thermocouple which produces energy that keeps the main gas supply valve open as long as the pilot light continues to burn.

During normal operation, gas passes through the automatic pilot valve prior to being routed to the main burner assembly. Should the pilot light be extinguished for any reason, or should the stability of the pilot flame be affected by chimney downdrafts, for example, both the pilot gas and main gas feed lines will be shut down. Known as 100% safety shutdown, this is required on all LP gas appliances and is used in most natural gas applications. Non-100% shutdown is sometimes used on natural gas appliances and allows pilot gas only to flow in the event of a pilot light outage. Since LP gas, unlike natural gas, is heavier than air and would accumulate at the bottom of the appliance rather than dispersing in air, 100% shutdown is mandatory for all LP appliances.

Autopilots. Most standing pilot valves are of the autopilot type, illustrated in Figure 5.49. In this system, the pilot flame is directed against a thermocouple,

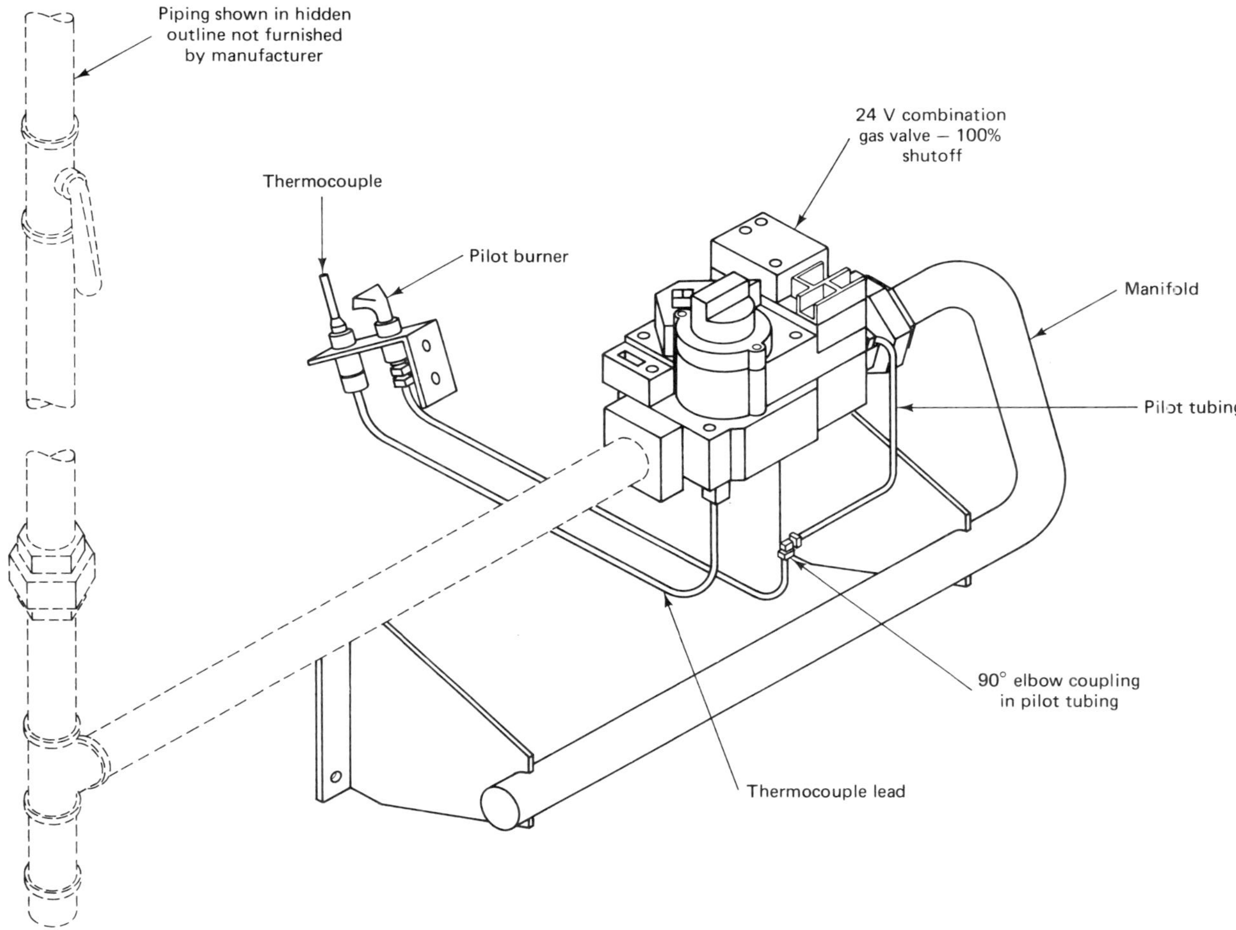

Figure 5.48 Standing pilot ignition system, 24-V combination gas valve, natural or LP gas. (Courtesy of Burnham Corp.)

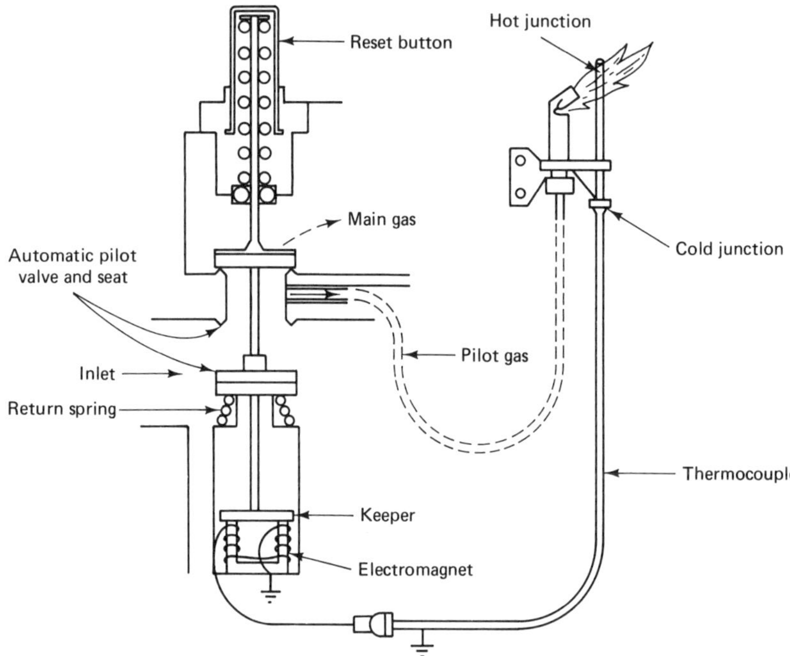

Figure 5.49 Autopilot ignition control valve and component arrangement. (Courtesy of Robertshaw Controls Company.)

which generates a small amount of electricity that is sufficient to energize the electromagnet in the pilot valve assembly.

During the pilot lighting procedure (Figure 5.50), the pilot button is depressed to hold the keeper against the electromagnet. Note that the interrupter valve is seated, which closes off the main gas port during the pilot lighting operation to ensure that no gas will flow to the main burners during this operation. The reset button must be held down sufficiently long (usually 30 seconds to 1 minute) to allow the pilot light to heat the thermocouple sufficiently for it to generate enough electrical power to energize the electromagnet that holds the keeper in place.

Under normal operating conditions, the pilot valve assembly functions as illustrated in Figure 5.51. After the pilot flame has been well established against the thermocouple, the reset button is released and main gas is free to flow to the burners when there is a call for heat.

Under conditions of loss of pilot flame against the thermocouple, generation of electricity from the thermocouple ceases, resulting in a loss of magnetism that releases the keeper. This closes the main gas and pilot valve assembly under the influence of the return spring (Figure 5.52).

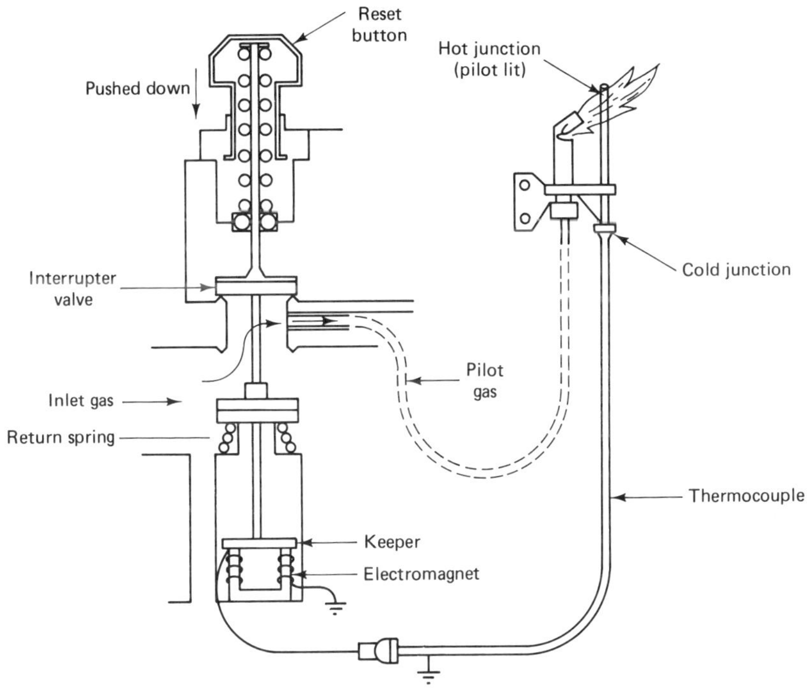

Figure 5.50 Autopilot configuration during pilot-lighting procedure. (Courtesy of Robertshaw Controls Company.)

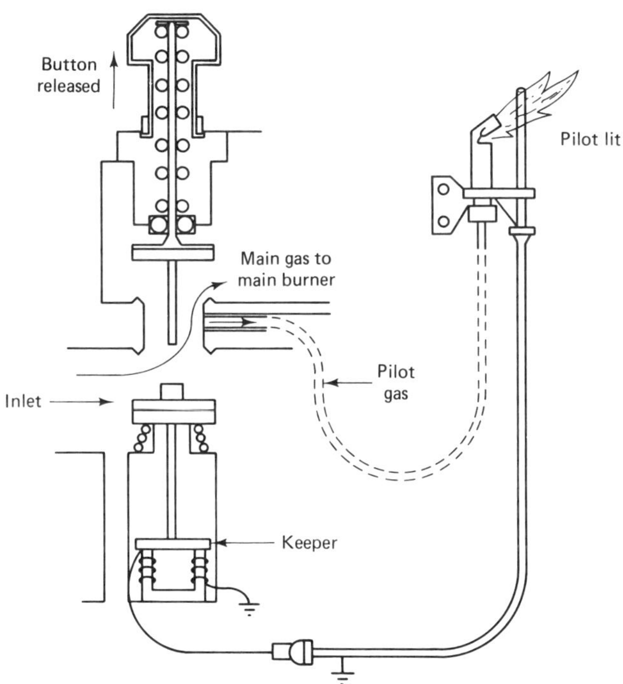

Figure 5.51 Normal operation of pilot valve assembly. (Courtesy of Robertshaw Controls Company.)

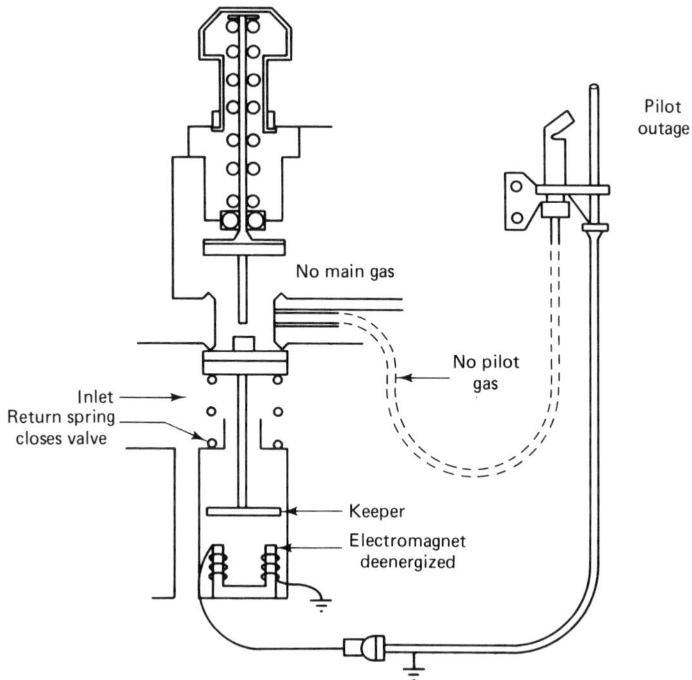

Figure 5.52 Loss of pilot flame in standing pilot ignition system. (Courtesy of Robertshaw Controls Company.)

Thermocouples. The action of the pilot valve is based on the operation of the thermocouple. A thermocouple is a device whose function is based on the principle that when two dissimilar metals are connected at each end and one of the ends in this junction is maintained at a higher temperature than the other, a flow of electrical current will take place. A simple iron/constantan thermocouple is illustrated in Figure 5.53. A conventional pilot light thermocouple of this type is illustrated in cross section in Figure 5.54.

Pilot burners may be either of the yellow or blue flame type. Blue flame, Bunsen-burner-type pilots premix combustion air and fuel gas prior to burning. Blue flame pilots are sometimes referred to as *aerated* pilots (Figure 5.55). A yellow flame pilot which relies on drawing secondary air that surrounds the flame for complete combustion is sometimes referred to as a *nonaerated* pilot.

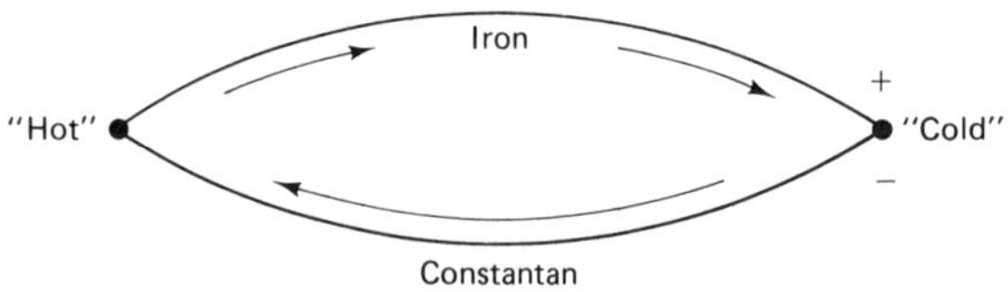

Figure 5.53 Simple thermocouple junction.

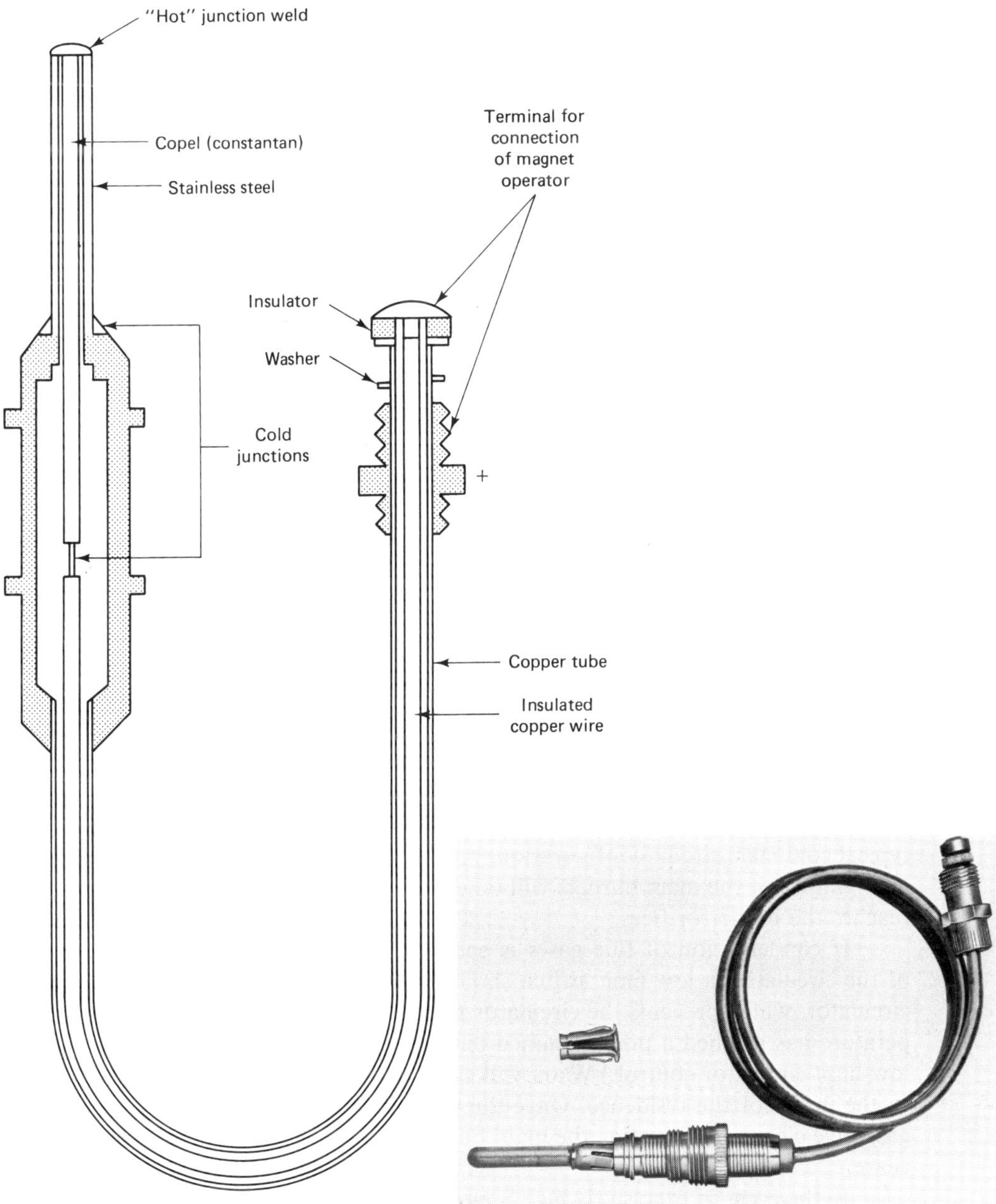

(a) (b)

Figure 5.54 Thermocouple. [(a) Courtesy of Robertshaw Controls Company; (b) courtesy of Honeywell, Inc.]

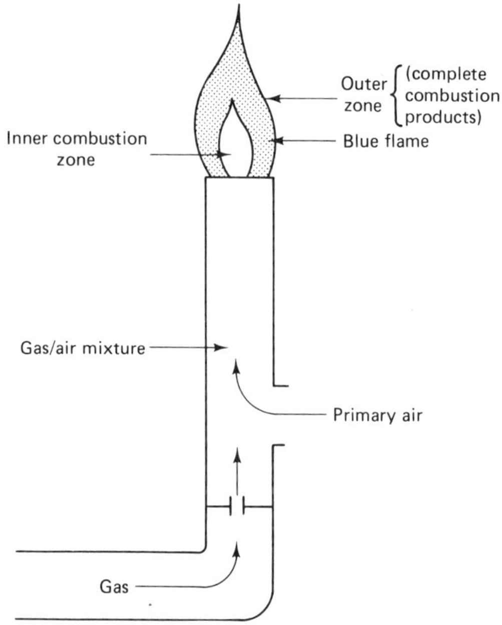

Figure 5.55 Blue flame pilot assembly. (Courtesy of Robertshaw Controls Company.)

BOILER OPERATING SEQUENCES

Standing Auto Pilot Ignition

The typical operating sequence for a low-voltage standing pilot ignition system gas-fired boiler is illustrated in Figure 5.56. A call for heat starts the circulator in a typical cold-start circuit configuration. At the same time, the gas valve is energized and ignition of the main burners will take place by the standing pilot once gas flow reaches the burner orifices.

If condensation of flue gases is encountered due to the cold-start operation of the circulator, a low-limit switch should be added to the wiring circuitry or the circulator, which prevents the circulator from operating until the boiler-water temperature has reached a predetermined temperature setting. (Figure 5.56 includes a low-limit circulator control.) Water will circulate through the system, radiating heat to the inside of the residence. Once the thermostat setting has been satisfied, the gas valve closes, deactivating the main burners, and operation of the circulator will cease.

In the event of excessively high boiler-water temperatures, the high-limit switch in the safety circuit of the boiler aquastat will open, closing the gas valve and shutting down the main burners. The circulator will continue to operate in this mode, as will the pilot light. Normal boiler operation will resume when the water temperature drops below the high-limit differential setting.

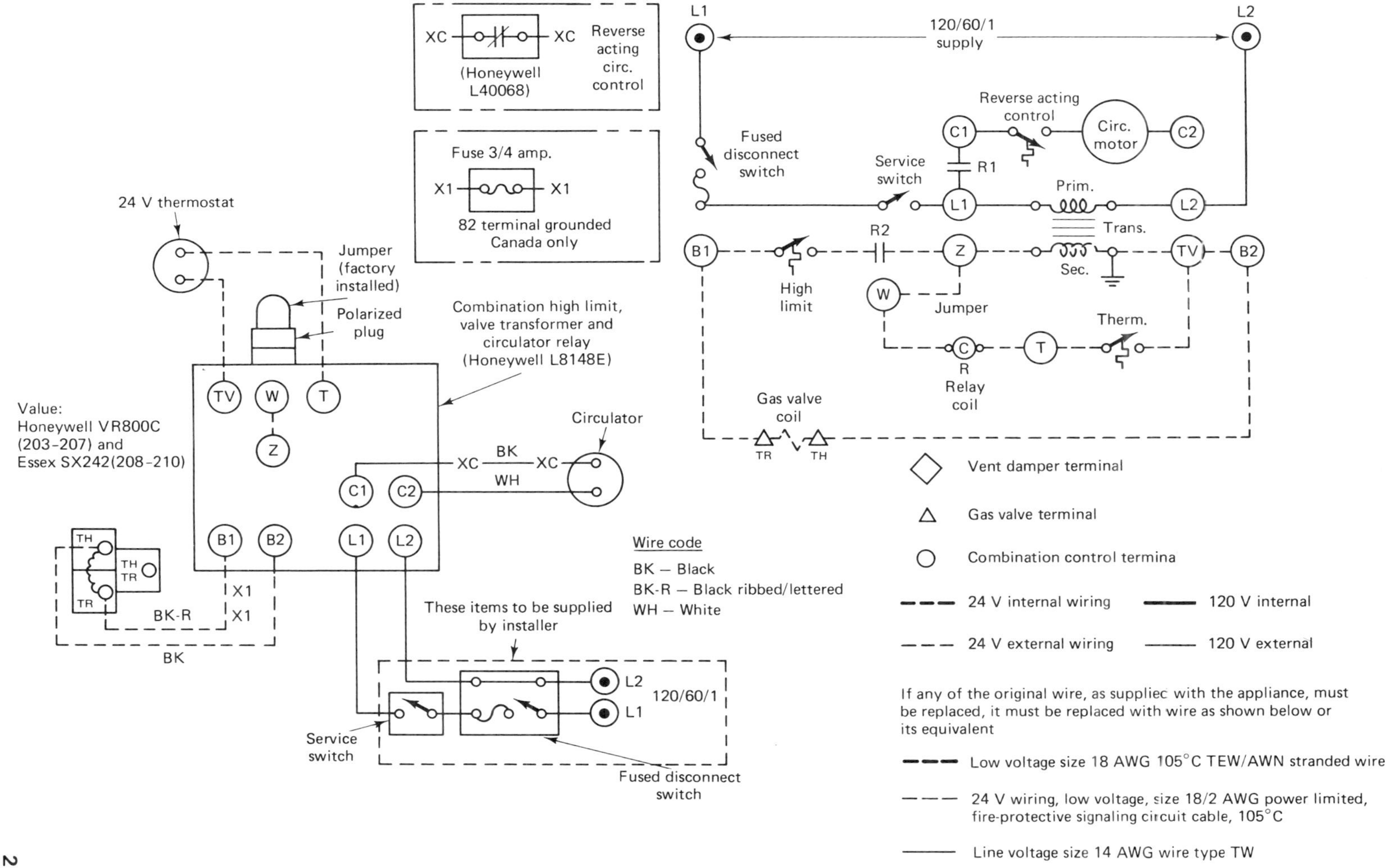

Figure 5.56 Boiler operating sequence with standing autopilot ignition and intermittent circulator operation. (Courtesy of Burnham Corp.)

Should a loss of pilot light occur at any time during this sequence, the gas valve will deenergize, resulting in 100% shutdown of the appliance.

Thermopile Pilot

A distinction is made between a thermocouple pilot system and a thermopile pilot system. A thermopile is an arrangement where several thermocouples are electrically connected in series to provide sufficient voltage to power all gas valve operation independently of external electrical power. A thermopile arrangement is illustrated in Figure 5.57. Note that the entire gas valve arrangement is wired in series with the thermopile generator. A manual auto switch is provided in this arrangement. This switch is normally set in the "auto" position. However, during loss of external electrical power, the switch is moved to the "manual" position to allow continued operation of the appliance.

An additional safety feature of most thermopile generators is the inclusion of an auxiliary limit control in the generator circuit. In the event that the boiler-water temperature exceeds a preset limit, the limit control will automatically interrupt the electrical current in the thermopile generator causing 100% gas shutdown in the

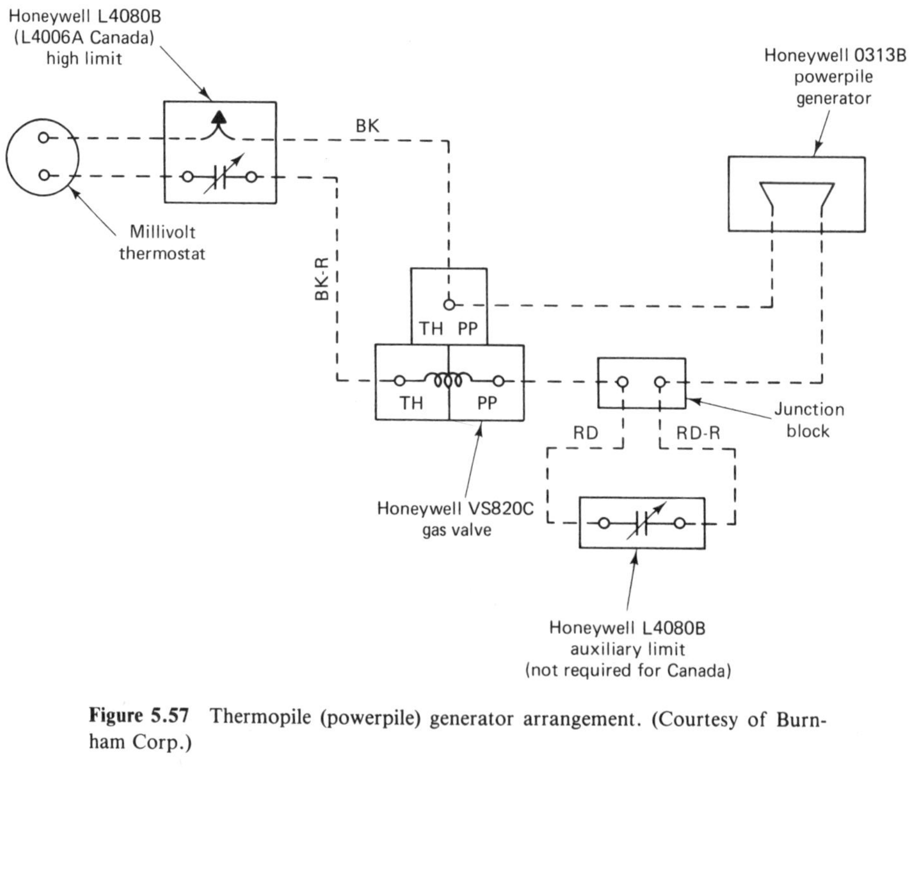

Figure 5.57 Thermopile (powerpile) generator arrangement. (Courtesy of Burnham Corp.)

system. When water temperature drops to a safe level, the auxiliary limit switch will again close. At this point, the pilot light must be relit manually to resume normal boiler operation.

Vent damper operating sequence. When electrically operated vent dampers are installed in boilers of either the thermocouple or thermopile ignition circuit, modifications in the circuitry are made to accommodate the damper motor and high-limit relay. Such a system schematic is illustrated in Figure 5.58. When the thermostat calls for heat, the damper motor is deenergized, allowing the damper to open. When the damper is fully opened, an end-switch relay in the damper mechanism closes the electrical circuit to the gas valve, which is wired in series through the damper motor switch relay. Closing the gas valve circuit allows ignition of the main burners to take place, and the boiler operates to provide space heating. Once the thermostat setting has been satisfied, the main gas valve will close, shutting down the burner assembly. Also, the damper motor will be reenergized, closing the damper to prevent off-cycle heat losses. The stack damper also contains a high-limit relay safety switch which will keep the damper open in the event of excessively high stack temperatures. The pilot ignition system functions with the damper as it does normally, serving to ignite the main burners on a call for heat from the thermostat as well as shut down the gas feed system should a loss of pilot occur.

Intermittent Ignition Systems

While standing pilot ignition systems have been the norm for gas-fired boilers for generations, new high-efficiency appliances are equipped with an intermittent type of ignition system. Intermittent ignition is used almost exclusively in induced-draft condensing and noncondensing boilers. In an intermittent system design, the pilot light burns only during the boiler on-sequence, shutting off when the boiler is in the off-cycle. This arrangement allows for an increase in efficiency since the boiler pilot light burns only when it is required, that is, when the gas system is firing. A typical intermittent pilot ignition system is illustrated in Figure 5.59. Pilot ignition is provided by a solid-state ignition electrode (Figure 5.60).

Upon a call for heat from the room thermostat, voltage is applied to the ignition electrode. A spark jumps a preset gap between the electrode and the grounded pilot burner. Simultaneously, the gas valve provides gas to the pilot burner, which results in a stable pilot flame that is directed against the electronic sensing probe. Once flame presence has been proven by the flame sensor, the ignition terminal is deenergized and the main gas valve is energized to allow ignition of the main gas burners.

Intermittent ignition pilot failure. Pilot failure in the intermittent configuration can occur during either initial startup or during the normal operating cycle of the boiler. Should a loss of pilot flame occur, the main gas valve is quickly deenergized, generally in less than 1 second. For a short period of time thereafter

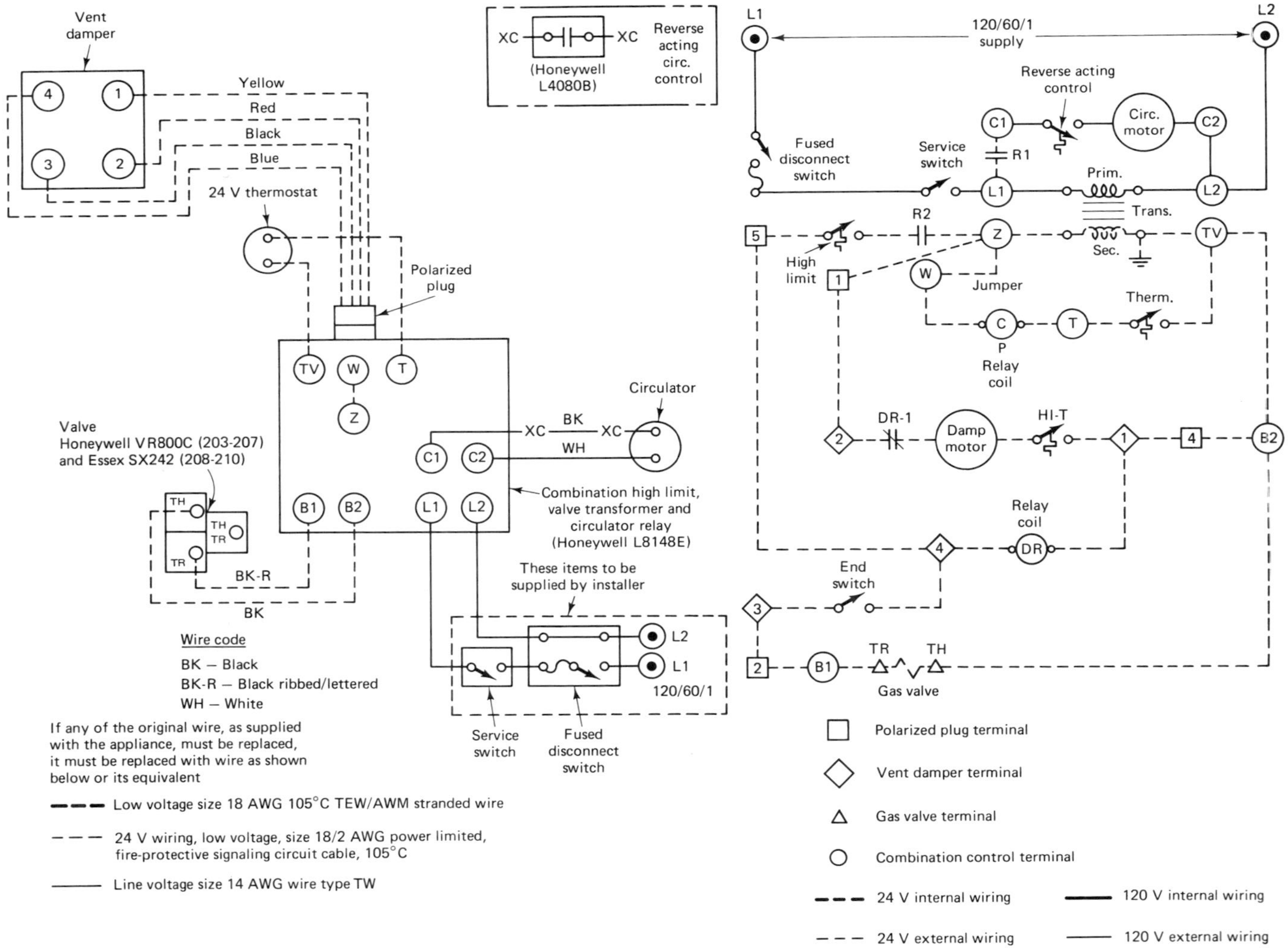

Figure 5.58 Ignition system sequence with electric damper relay. (Courtesy of Burnham Corp.)

(generally 30 seconds), the ignition terminal will attempt to relight the pilot flame. Should this attempt prove unsuccessful, the flame-sensing circuit will lock out the control module on safety shutdown. In this instance, the module must be manually reset before the ignition terminal can again be energized. Should the unit not function properly, most manufacturers supply troubleshooting guides which cover general service procedures that can be performed by the homeowner to help in pinpointing and correcting system malfunctions.

MAINTENANCE PROCEDURES

Although natural and LP gas boilers are cleaner burning than their oil and solid-fuel counterparts, annual cleaning of the boiler and periodic inspection of the system components is necessary for proper, safe, and efficient operation. The maintenance and cleaning procedures focus on the following areas: venting system, boiler flue passages, main burner and combustion chamber, pilot ignition system, general lubrication, gas feed line inspection, burner flame adjustment and efficiency testing, and proper boiler control sequence. Each of these areas is discussed below.

Venting System

The vent system of the boiler, including the chimney, should be inspected on an annual basis. Initial visual inspection of all appliance vent pipes should focus on the following:

1. Check piping for any deterioration and external corrosion. Any suspect pipe should be replaced with code-approved vent pipe.
2. Check all vent pipe connections and joints. A minimum of three screws should be used to connect each adjoining pipe section.
3. Remove pipe sections and inspect for any obstructions and accumulations of soot and sediment. A wire brush properly sized should be used to clean the pipe internally.
4. If an electric vent damper is installed in the system, check its operation to ensure proper rotation of the unit. Ensure that all electrical connections are tight and that no deterioration is present in the damper plate or motor housing.
5. Inspect the chimney for integrity of all masonry joints. If a factory-built chimney such as a type B gas vent chimney is used, inspect all pipe sections for corrosion in hidden snap-together joints.
6. Inspect all horizontal pipe sections for sags. Low spots, in particular, can act as condensate traps and should be given particular attention. Install insulated pipe supports where necessary to maintain proper position of pipe.

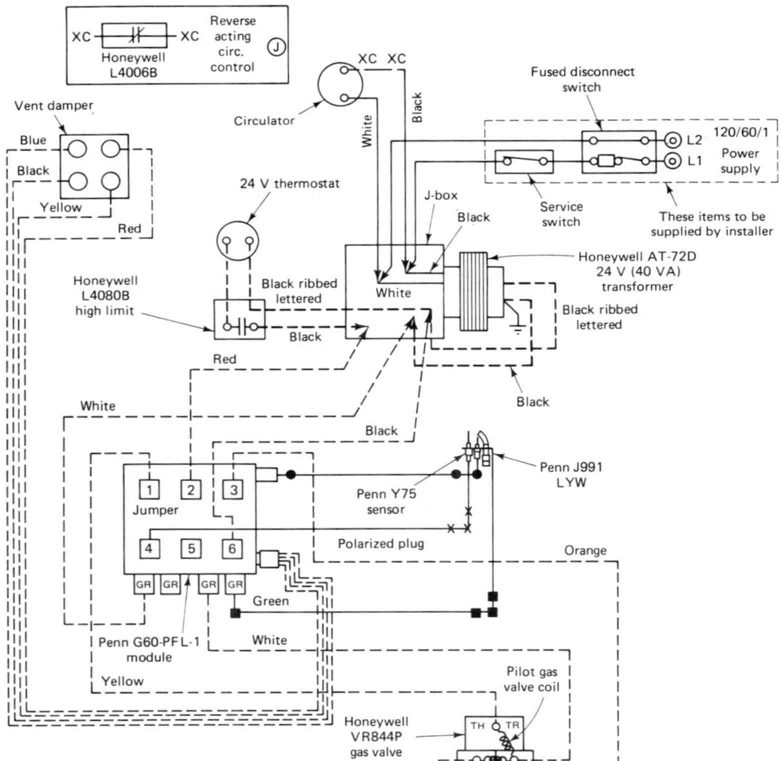

If any of the original wire, as supplied with the appliance, must be replaced, it must be replaced with wire as shown below or its equivalent

- – – – Low voltage size 18/2 AWG low-energy safety control wire
- –●–●– Ignition-Prestolite HAZ AWM 230°C
- –×–×– Sensing probe wire-Penn Y57AH
- –■–■– Ground-continental SRGX600 16 AWG 200°C wire or equal
- – – – – Low voltage size 18 AWG 105°C TEW/AWM stranded wire
- ——— Line voltage size 14 AWG wire type TW

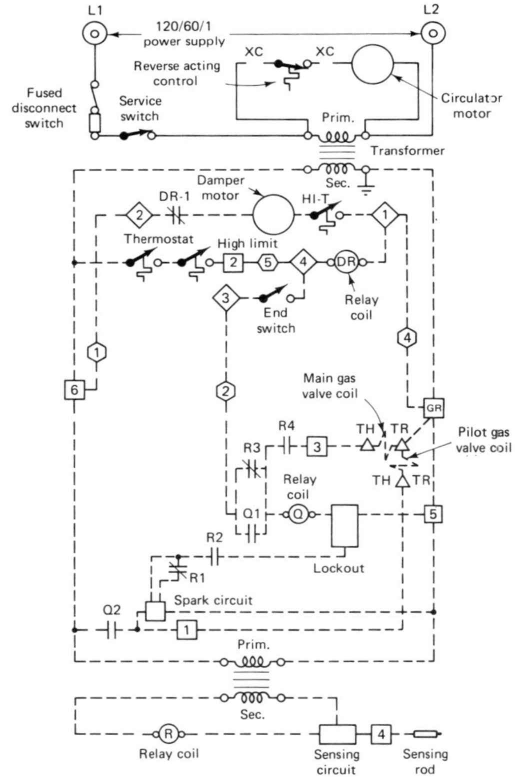

- △ Gas valve terminal
- □ Module terminal
- ◇ Stack damper terminal
- ○ Combination control terminal
- ⬡ Polarized plug terminal
- – – – – 24 V internal wiring
- – – – – 24 V external wiring
- ——— 120 V external wiring

(a)

The EI Control System utilizes a solid-state ignition control which lights the pilot burner by spark. Pilot gas is ignited and burns during each running cycle (intermittent pilot). Main burner and pilot gas are extinguished during the "off" cycle.

This system permits the main gas valves to open only when the pilot burner is proven to be lit.

Should a loss of flame occur, the main valve closes and the spark reoccurs with 0.8 second. The ignition control has an internal 100% lockout function to completely shut down the system should the pilot gas fail to ignite within approximately 30 seconds. To initiate a reignition trial when lockout occurs, the power must be interrupted for 30 seconds.

Sequence of Operation

Normal Operation

1. When THERMOSTAT calls for heat, the VENT DAMPER is deenergized, allowing the vent damper to open. When the vent damper reaches the fully open position, an end switch in the vent damper is closed which energizes the G60 module. On boilers with circulators that are wired to run continuously, the circulator operates as long as the switch providing power to the boiler is closed.
2. A component check circuit in the G60 is activated which checks the electronic network in the G60.
3. Terminals 1, GR, and the ignition terminal are energized. Terminals 1 and GR power the pilot valve in the VR844 DUAL COMBINATION GAS VALVE supplying gas to the pilot. The ignition terminal supplies voltage to the ignition electrode electronically and the ignition terminal is deenergized.
4. The sensing circuit between the J991 pilot burner and the G60 proves the presence of the pilot flame electronically and the ignition terminal is deenergized.
5. Terminals 3 and GR of the G60 are energized and supply power to the main gas valve in the VR844 DUAL COMBINATION GAS VALVE. The gas valve is energized allowing main gas flow and ignition of main burners.
6. On boilers equipped with circulators that are wired to run continuously and, where condensation of flue gas is encountered in boiler flues, a REVERSE ACTING CIRCULATOR CONTROL should be installed to stop the circulator before the boiler water temperature drops to that at which flue gas condensation may occur.
7. The burners will operate until the THERMOSTAT is satisfied.
8. After the THERMOSTAT is satisfied the GAS VALVE will be deenergized. The main burner and the pilot burner will be extinguished. If boiler is equipped with circulator, circulator will continue to run.

Safety Shutdown

1. Safety Switch Circuit
   In the event excessive boiler water temperature is developed, the high limit will open, interrupting power to the G60 IGNITION MODULE and the DUAL COMBINATION GAS VALVE. Main burners and pilot burner will be extinguished immediately. Normal operation will be resumed when the boiler water temperature drops to a point where the high limit switch closes.
   On boilers equipped with vent damper, if excessive stack temperatures are developed, a temperature sensing switch in the vent damper will keep the damper from closing until the stack temperature drops to a preset safe level.
2. Pilot Failure
   a. Pilot failure can occur during the start-up or the operating cycle of the boiler. Any pilot failure on the J991 electronic pilot after ignition of pilot flame, will close the main gas valve in 0.8 seconds.
   b. For approximately 30 seconds after failure of the J991 pilot, the module through the ignition terminal will try to reestablish pilot flame. If no pilot flame can be sensed by the flame rod circuit, terminals 3 and GR are deenergized and the module will lock out on safety. To reset G60, open the service switch for 30 seconds.

(b)

Figure 5.59 Intermittent (electronic) pilot ignition system with vent damper gravity flow or continuous circulation: (a) diagram; (b) description. (Courtesy of Burnham Corp.)

Boiler Flue Passages

All boiler flue passages should be inspected annually for soot accumulation. Excessive soot deposits can greatly reduce the efficiency of heat transfer within the boiler. Also, in warm, humid weather, condensate in conjunction with soot accumulations can cause corrosive formations within the boiler that can deteriorate metallic surfaces and internal water seals.

Most boilers come equipped with flue brushes that are sized for the particular

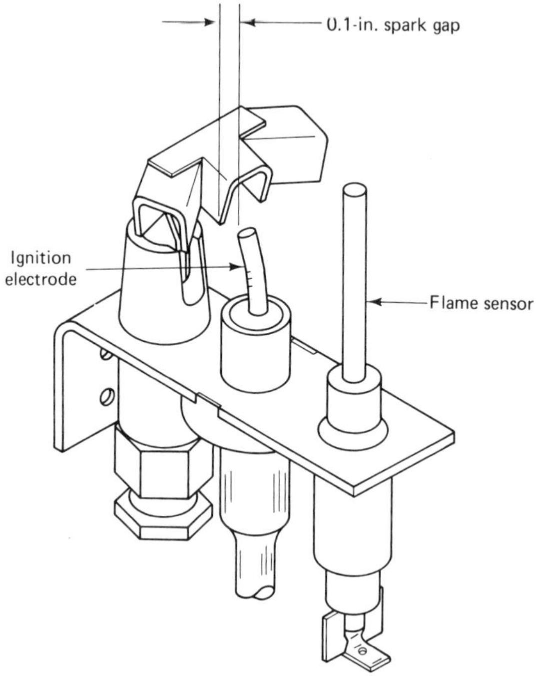

Figure 5.60 Solid-state ignition electrode.

unit. When cleaning flue passages, the main burners should be removed. The boiler is then brushed and the residue removed with a vacuum cleaner. For best results, annual inspection and cleaning should be done in the spring after the heating season. This ensures removal of soot accumulations prior to the onset of warm, humid weather, which can accelerate corrosion in a dirty, soot-laden boiler.

Main Burners and Combustion Chamber

Main gas burners should be removed annually, following the manufacturer's recommendations, and cleaned with a soft bristle brush and vacuumed (Figure 5.61). All orifices should be checked to make sure that they are free of dirt and lint.

Pilot Ignition Systems

The pilot ignition system requires a series of checks to ensure proper operation. The following guidelines are recommended:

1. With the boiler in operation, observe the position of the pilot flame against the thermocouple. Ensure that the tip of the flame is against the hot edge of the thermocouple junction. A typical flame setting is illustrated in Figure 5.62.

Figure 5.61 Main gas burners should be removed, inspected, and cleaned annually. (Courtesy of Slant/Fin Corp.)

2. If a thermopile generator is used, check the millivolt electrical output on the thermocouple. Compare the open-circuit-generated voltage with the manufacturer's specifications, and replace if necessary.
3. Check the pilot filter, located upstream of the pilot light adjustment, to ensure that it is clean.
4. Ensure that both the pilot light and thermocouple are firmly seated in their respective mounting brackets.
5. Check for damage or leaking gas connections at the pilot pressure regulator and other connections within this assembly. Remember to use a soapy water solution for this test and to clean all surfaces thoroughly after testing.
6. Ensure that draft conditions within the venting system do not adversely affect the pilot flame. If necessary, a draft protection hood on the pilot may be installed.

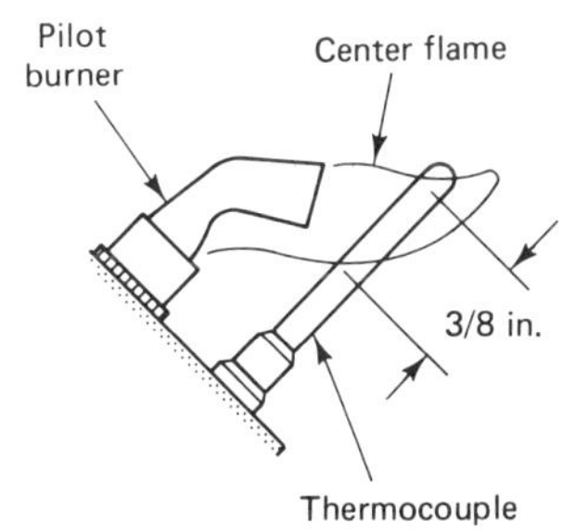

Figure 5.62 Proper flame setting for constant-burning pilot.

7. On aerated, blue-flame-type pilots, clean the air intake and orifice periodically if any dirt is present.

Main Burner Flames

Main burner flame adjustment should be done according to information supplied by the manufacturer of the boiler. The main burner flames on atmospheric burners will most certainly differ from those on pulsed combustion and induced-draft units. Most boilers feature a primary air shutter that is used to meter the amount of combustion air that is premixed with the incoming fuel. Some high-efficiency boilers cannot be field adjusted to change the main burner flame. Should any difficulties be encountered with these units, it is necessary to consult the factory for specific remedies.

Efficiency Testing

Boiler efficiency testing with gas-fired units is similar to the testing procedures used in oil-fired units. While different figures for CO_2 levels and stack temperatures and draft requirements will be encountered, the testing procedures are almost identical, using similar equipment in both systems. Manufacturers' specifications should be consulted to determine the ideal settings for each boiler.

General Lubrication

Lubrication of gas-fired boilers is usually limited to the circulator pumps used in the system. These pumps should be inspected and lubricated on an annual basis. Water-lubricated circulators do not require any lubrication—simply inspect them to make sure that no leaks are present in the vicinity of the circulator. Check the manufacturer's specifications for any additional service requirements.

Gas Feed Line

After all cleaning has been completed and all boiler components have been reassembled, the gas feed lines should be turned on and the system checked for leaks. Follow the procedures outlined earlier in this chapter on an annual basis to ensure that the integrity of the gas feed system remains intact. Replace any piping that is suspect. Also, an annual inspection of the gas feed system should include disassembly of sediment traps to drain any accumulated condensate.

Burner Flame Adjustment and Efficiency Testing

Once all components are back in place and the gas feed system has been inspected for integrity and purged, the burner flames should be checked and adjusted if necessary.

Boiler Control Sequence

After all service and maintenance procedures, including flame adjustments and efficiency testing, have been completed, the boiler should be run through all the operating sequences to ensure that all controls are functioning properly.

This sequence checking should involve an inspection of the ignition system for both standing pilot and intermittent systems. Of utmost importance is proper operation of the safety shutdown. This includes the high-limit shutdown as well as all gas shutdown provisions that occur on loss of the pilot light. If the boiler contains a low-water cutoff, it should be annually cleaned and inspected. If a thorough plan of general inspection and maintenance is followed, most gas-fired units can be expected to provide many years of trouble-free, reliable heating.

6

Steam Heating Systems

Steam heating was the most common type of central heating system used in both single- and multifamily residences throughout the early and middle twentieth century. Simple in design and matched with durable hardware, many of the systems installed more than 50 years ago are still in use today. Rising fuel prices in many parts of the United States, particularly in the northeast, where these systems predominate, has lead to a slowdown and in some areas a virtual halt in their use in new residential construction, in favor of less expensive, more efficient baseboard hydronic systems.

The principle of operation of a residential steam heating system is simple. Water is heated in a boiler that is located at the lowest part of the system. The water turns to low-pressure steam (under 15 psi), and the steam from the boiler rises and is channeled through a piping system to the room radiators; in the radiators, the steam condenses to water, releasing energy equivalent to the latent heat of vaporization of the water, which in turn heats the home. The condensate then returns via the piping circuit to the bottom of the boiler, where the process is repeated on a cyclical basis until the thermostat has been satisfied. During this heating cycle, returning condensate maintains the water at a fairly constant level to make up for the steam being distributed to the system radiation.

Modern conventionally fueled steam heating boilers are available in many different configurations and heating capacities. A high-efficiency steam heating oil-fired boiler is illustrated in Figure 6.1. Fired with high-efficiency flame retention burners, these boilers can yield combustion efficiencies in excess of 80%. Note also

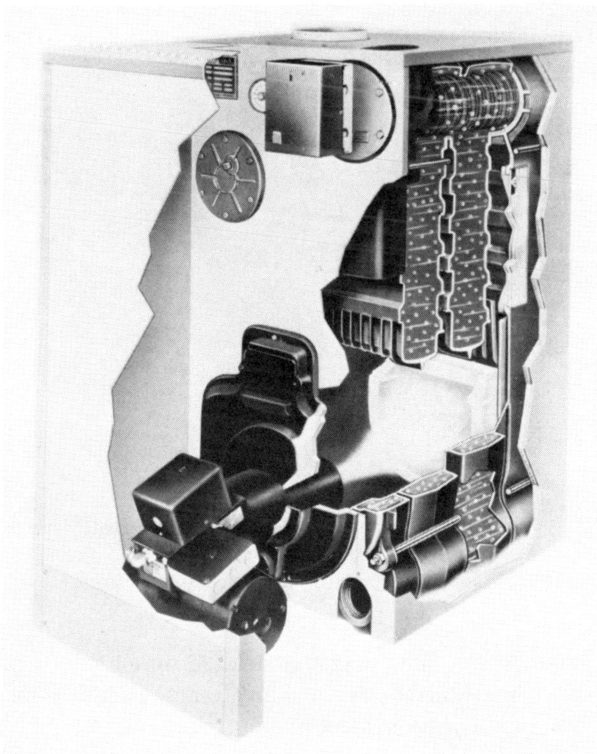

Figure 6.1 High-efficiency residential oil-fired steam boiler. (Courtesy of Burnham Corp.)

from Figure 6.1 that a tankless coil is incorporated into the boiler design for domestic hot water production.

Figure 6.2 illustrates a high-efficiency induced-draft gas-fired steam boiler suitable for many residential and small commercial applications. Using induced-draft combustion technology, these units can often offer combustion efficiencies in the neighborhood of 85%, a vast improvement over the inefficient steam boilers manufactured years ago that are still prevalent in many older heating systems that have not been updated.

HEAT DISTRIBUTION UNITS

In steam heating systems, two types of heat distribution units are commonly installed: radiators and convectors. The basic steam radiator is composed of a series of tubular sections or columns that are joined together by nipple connections at both the top and bottom of the units. During normal operation, room air circulates around the radiator, where it is heated by both radiation and convective currents. Most steam radiators are free-standing units and are usually located under windows or along exterior building walls for most efficient operation.

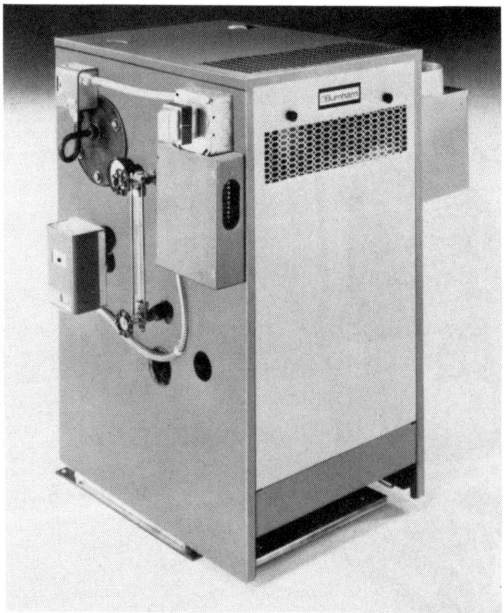

Figure 6.2 High-efficiency residential induced-draft gas-fired steam boiler. (Courtesy of Burnham Corp.)

Convectors, unlike radiators, are units that are enclosed on all sides with an air outlet located at the top or front of the housing and an air intake which is located at the bottom. Convectors are usually installed recessed into the wall, although they can protrude as well.

For the purposes of our discussion, we assume the use of radiators as the selected heat distribution unit. Within this framework we examine the basic types of residential steam heating systems and control configurations.

PIPING ARRANGEMENTS

Steam systems are classified according to the feed and return line piping configuration between the boiler and the radiators.

One-Pipe Steam Systems

In a one-pipe steam system, a single service main serves a dual purpose: It supplies the steam to the radiators as well as serving as a path for returning condensate to the boiler. A single-pipe steam system configuration is shown in Figure 6.3. Note that the flow of steam and condensate within the same pipe are in opposite directions. Steam is admitted to the radiator through a valve that can be opened or closed depending on heating preference. As steam enters the radiator, air from within is expelled through a radiator air vent located on the side opposite the steam intake. This arrangement is illustrated in Figure 6.4.

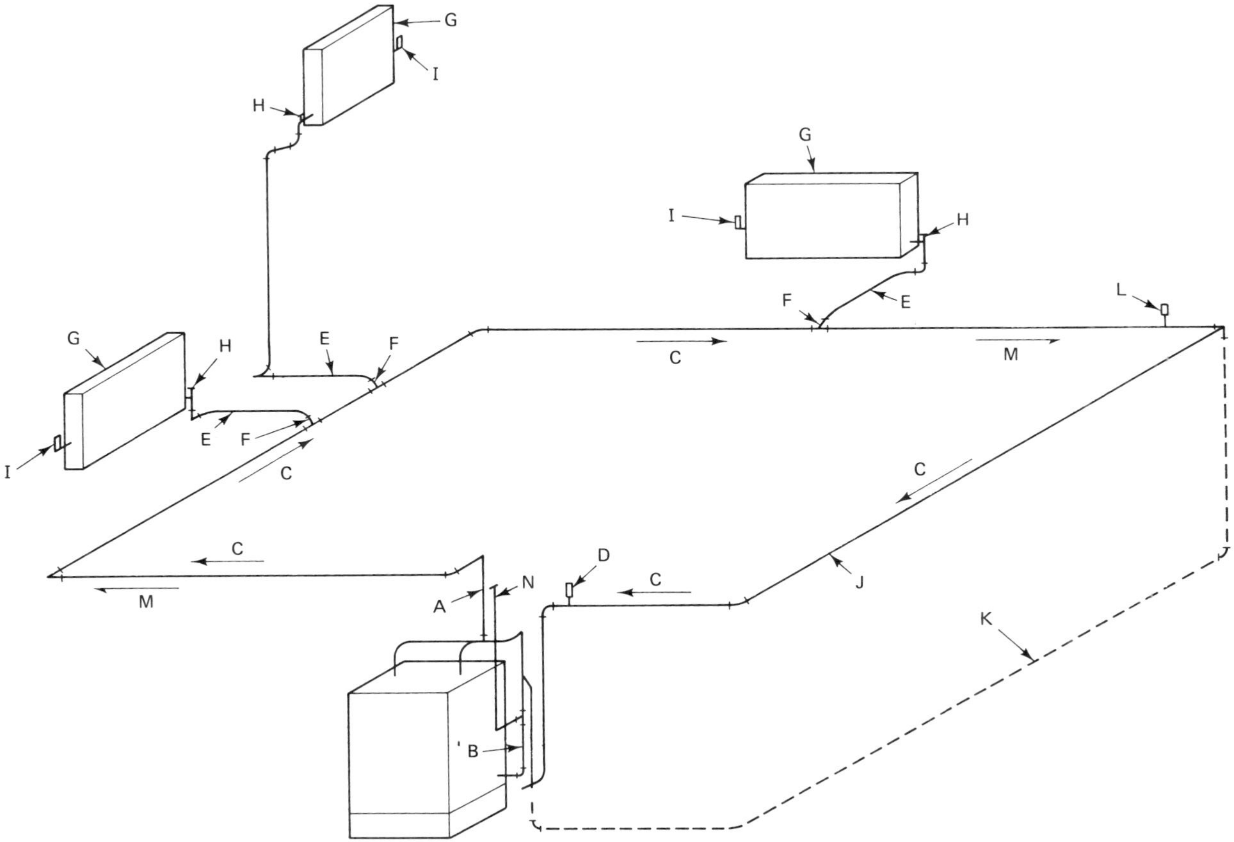

Figure 6.3 One-pipe steam heating system configuration. A, Supply pipe; B, return pipe; C, pipe pitches downward in direction of arrow not less than ¼ in. in 10 ft; D, main vent; E, branches; F, 45° ell; G, heat distributing units; H, radiator valve; I, air vent; J, dry return; K, wet return, if used; L, main vent located here when wet return is used; M, direction of flow of steam; N, makeup water line. (Courtesy of the Hydronics Institute.)

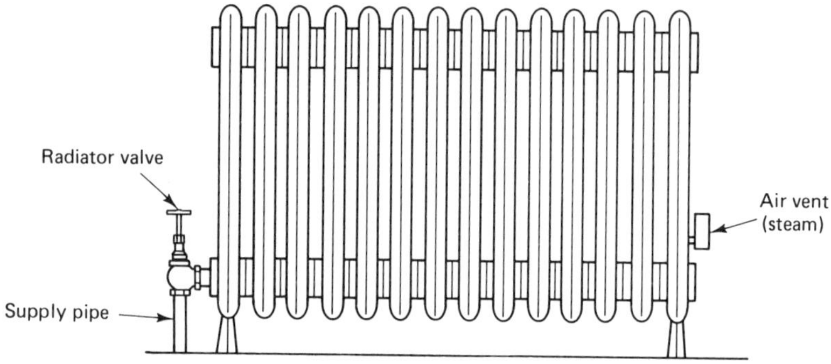

Figure 6.4 One-pipe steam radiator and air valve.

Two-Pipe Steam Systems

In a two-pipe steam system, each radiator is equipped with a separate feed line and condensate return line. A typical two-pipe circuit is illustrated in Figure 6.5. The operation of the two-pipe steam system is, apart from condensate return, identical to its single-pipe counterpart. Steam rises through the supply mains of the system into the radiator. As condensate is developed, it drains back down into the boiler through a separate return line that is attached to each radiator.

CONDENSATE RETURN CLASSIFICATION

In addition to classifying steam systems as either one- or two-pipe configurations, they are sometimes identified according to the type of return that is employed within the piping circuit, known as either a "wet" or a "dry" return.

Wet Return

A wet return is that part of the return main that is located below the water level of the boiler. A wet-return system is illustrated in Figure 6.6. When wet returns are employed, they should be manifolded together below the water level of the boiler, shown in detail in Figure 6.7. Note that main air vents are installed a minimum of 18 in. above the level of the water line. In a wet-return line, the pipe is completely filled with condensate.

Dry Return

Dry returns are those parts of the main return lines that are located above the water level of the boiler, illustrated in Figure 6.8. Dry return pipes contain a combination of steam, air, and condensate. Note that the air vent in a dry main is placed at the

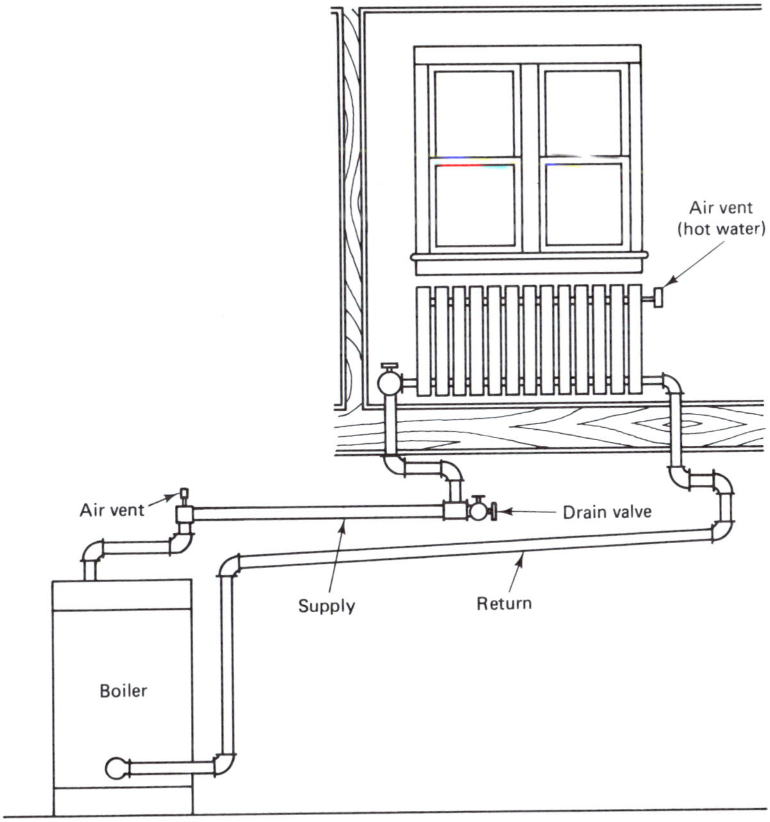

Figure 6.5 Two-pipe steam system. (Reprinted by permission of Small Homes Council, Building Research Council, University of Illinois, Technical Note 13.)

end of each return line run. Each separate return main requires its own separate air vent (see Figure 6.7).

SIZING AND INSTALLING THE SYSTEM

The sizing techniques used for determining steam boiler heating capacity in addition to the capacity of the steam radiator are similar in most respects to the techniques utilized in sizing hydronic systems. More detailed aspects of system sizing particular to steam systems are examined next.

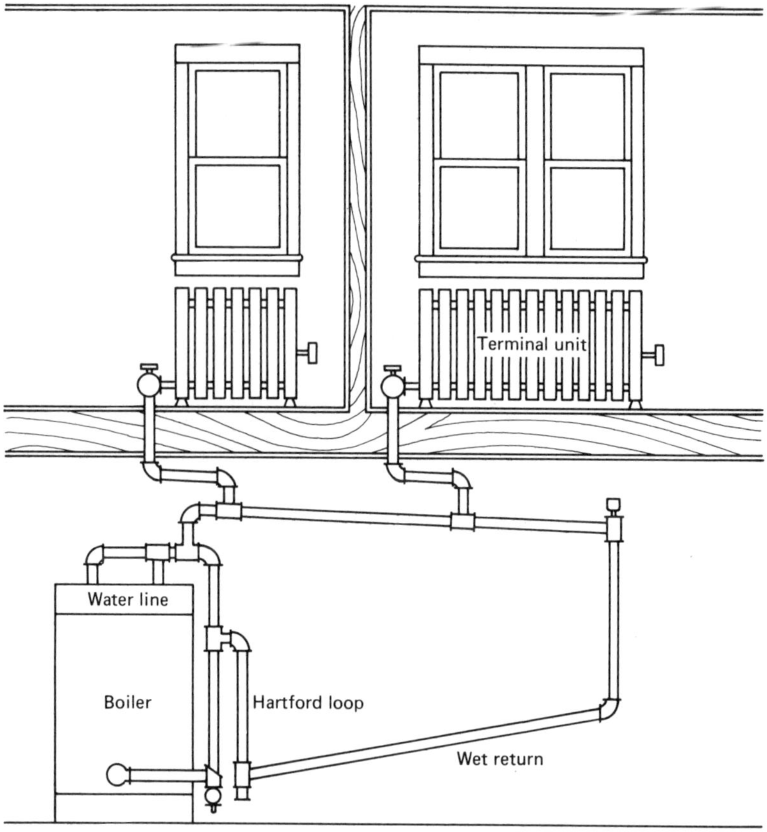

Figure 6.6 Wet returns on steam system. (Reprinted by permission of Small Homes Council, Building Research Council, University of Illinois, Technical Note 13.)

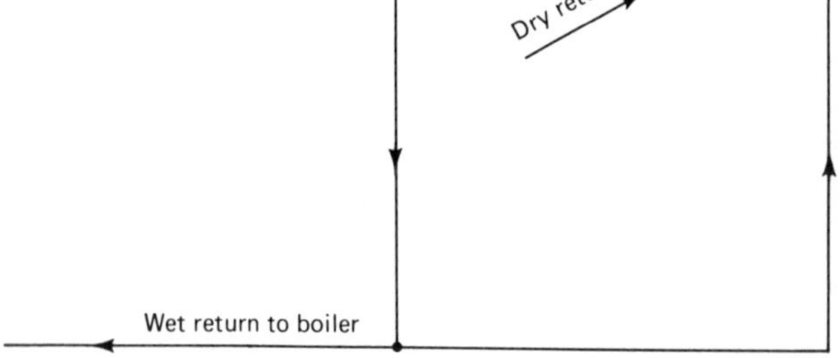

Figure 6.7 Manifolding of wet-return lines below water level of boiler.

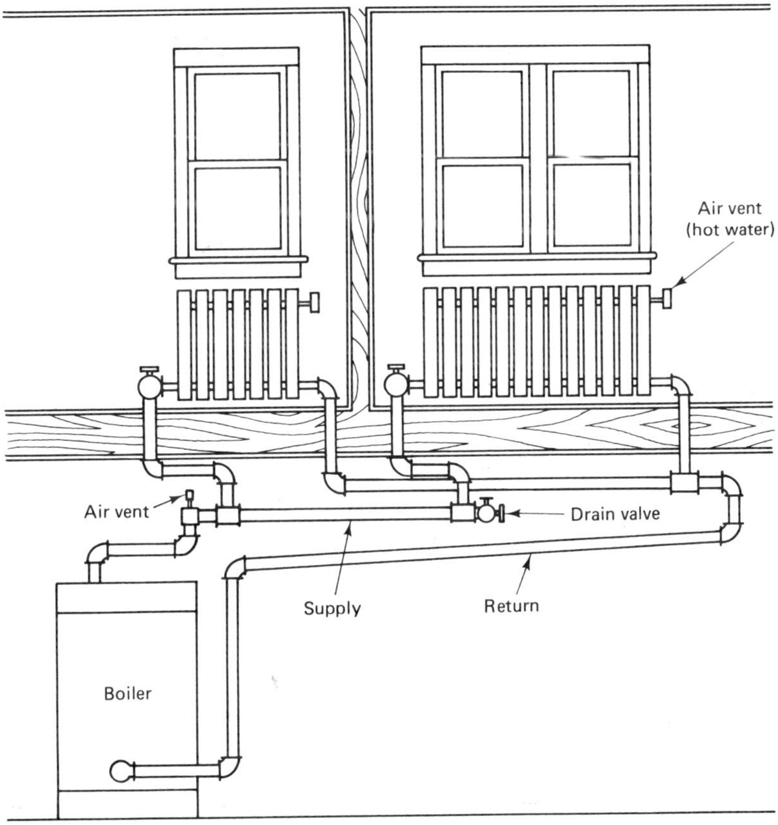

Figure 6.8 Dry-return configuration. (Reprinted by permission of Small Homes Council, Building Research Council, University of Illinois, Technical Note 13.)

Heat-Loss Analysis

A detailed heat-loss analysis must be undertaken for each room in the residence as a preliminary design step. After this has been completed, the size of the heating distribution units can be determined.

Sizing the Heat Distribution Units

When installing steam radiators, the Btu heat losses calculated from the residential heat-loss analysis must be coverted into a measurement known as *square feet of steam*. This term does not refer to the square-foot surface area of the radiation, but rather is an indication of the rating of the radiator. Square-feet steam figures are to be found in a conversion table shown in Table 6.1.

To illustrate how this table functions in sizing the radiators, let us assume that

TABLE 6.1 CONVERSION FIGURES FOR SQUARE FEET OF STEAM TO BTU

	Design water temperature (°F)									
Btu/hr	170	175	180	185	190	195	200	205	210	215
1,000	6.7	6.3	5.9	5.6	5.3	5.0	4.8	4.5	4.3	4.2
1,100	7.3	6.9	6.5	6.1	5.8	5.5	5.2	5.0	4.8	4.6
1,200	8.0	7.5	7.1	6.7	6.3	6.0	5.7	5.5	5.2	5.0
1,300	8.7	8.1	7.6	7.2	6.8	6.5	6.2	5.9	5.7	5.4
1,400	9.3	8.8	8.2	7.8	7.4	7.0	6.7	6.4	6.1	5.8
1,500	10.0	9.4	8.8	8.3	7.9	7.5	7.1	6.8	6.5	6.2
1,600	10.7	10.0	9.4	8.9	8.4	8.0	7.6	7.3	7.0	6.7
1,700	11.3	10.6	10.0	9.4	8.9	8.5	8.1	7.7	7.4	7.1
1,800	12.0	11.3	10.6	10.0	9.5	9.0	8.6	8.2	7.8	7.5
1,900	12.7	11.9	11.2	10.6	10.0	9.5	9.0	8.6	8.3	7.9
2,000	13.3	12.5	11.8	11.1	10.5	10.0	9.5	9.1	8.7	8.3
2,100	14.0	13.1	12.4	11.7	11.1	10.5	10.0	9.5	9.1	8.7
2,200	14.7	13.8	13.0	12.2	11.6	11.0	10.5	10.0	9.7	9.2
2,300	15.3	14.4	13.5	12.8	12.1	11.5	11.0	10.5	10.0	9.6
2,400	16.0	15.0	14.1	13.3	12.6	12.0	11.4	10.9	10.4	10.0
2,500	16.7	15.6	14.7	13.9	13.2	12.5	11.9	11.4	10.9	10.4
2,600	17.3	16.3	15.3	14.4	13.7	13.0	12.4	11.8	11.3	10.8
2,700	18.0	16.9	15.9	15.0	14.2	13.5	12.9	12.3	11.7	11.2
2,800	18.7	17.5	16.5	15.6	14.7	14.0	13.3	12.7	12.2	11.7
2,900	19.3	18.1	17.1	16.1	15.3	14.5	13.8	13.2	12.6	12.1
3,000	20.0	18.8	17.6	16.7	15.8	15.0	14.3	13.6	13.0	12.5
3,100	20.7	19.4	18.2	17.2	16.3	15.5	14.8	14.1	13.5	12.9
3,200	21.3	20.0	18.8	17.8	16.8	16.0	15.2	14.5	13.9	13.3
3,300	22.0	20.6	19.4	18.3	17.4	16.5	15.7	15.0	14.3	13.7
3,400	22.7	21.3	20.0	18.9	17.9	17.0	16.2	15.5	14.8	14.2
3,500	23.3	21.9	20.6	19.4	18.4	17.5	16.7	15.9	15.2	14.6
3,600	24.0	22.5	21.2	20.0	18.9	18.0	17.1	16.4	15.7	15.0
3,700	24.7	23.1	21.8	20.6	19.5	18.5	17.6	16.8	16.1	15.4
3,800	25.3	23.8	22.4	21.1	20.0	19.0	18.1	17.3	16.5	15.8
3,900	26.0	24.4	22.9	21.7	20.5	19.5	18.6	17.7	17.0	16.2
4,000	26.7	25.0	23.5	22.2	21.1	20.0	19.0	18.2	17.4	16.7
4,100	27.3	25.6	24.1	22.8	21.6	20.5	19.5	18.6	17.8	17.1
4,200	28.0	26.3	24.7	23.3	22.1	21.0	20.0	19.1	18.3	17.5
4,300	28.7	26.9	25.3	23.9	22.6	21.5	20.5	19.5	18.7	17.9
4,400	29.3	27.5	25.9	24.4	23.2	22.0	21.0	20.0	19.1	18.3
4,500	30.0	28.1	26.5	25.0	23.7	22.5	21.4	20.5	19.6	18.7
4,600	30.7	28.8	27.1	25.6	24.2	23.0	21.9	20.9	20.0	19.2
4,700	31.3	29.4	27.6	26.1	24.7	23.5	22.4	21.4	20.4	19.6
4,800	32.0	30.0	28.2	26.7	25.3	24.0	22.9	21.8	20.9	20.0
4,900	32.7	30.6	28.8	27.2	25.8	24.5	23.3	22.3	21.3	20.4
5,000	33.3	31.3	29.4	27.8	26.3	25.0	23.8	22.7	21.7	20.8
5,100	34.0	31.9	30.0	28.3	26.8	25.5	24.3	23.2	22.2	21.2
5,200	34.7	32.5	30.6	28.9	27.4	26.0	24.8	23.6	22.6	21.7

Source: The Hydronics Institute.

To use this table: Enter the Btu/hr column at the figure closest to the Btu/hr heat loss of each room. Read across to the column, which represents the design water temperature selected for the system, to find square feet of cast iron radiators.

TABLE 6.1 *(continued)*

Btu/hr	Design water temperature (°F)									
	170	175	180	185	190	195	200	205	210	215
5,300	35.3	33.1	31.2	29.4	27.9	26.5	25.2	24.1	23.0	22.1
5,400	36.0	33.8	31.8	30.0	28.4	27.0	25.7	24.5	23.5	22.5
5,500	36.7	34.4	32.4	30.6	28.9	27.5	26.2	25.0	23.9	22.9
5,600	37.3	35.0	32.9	31.1	29.5	28.0	26.7	25.5	24.3	23.3
5,700	38.0	35.6	33.5	31.7	30.0	28.5	27.1	25.9	24.8	23.7
5,800	38.7	36.3	34.1	32.2	30.5	29.0	27.6	26.4	25.2	24.2
5,900	39.3	36.9	34.7	32.8	31.1	29.5	28.1	26.8	25.7	24.6
6,000	40.0	37.5	35.3	33.3	31.6	30.0	28.6	27.3	26.1	25.0
6,100	40.7	38.1	35.9	33.9	32.1	30.5	29.0	27.7	26.5	25.4
6,200	41.3	38.8	36.5	34.4	32.6	31.0	29.5	28.2	27.0	25.8
6,300	42.0	39.4	37.1	35.0	33.2	31.5	30.0	28.6	27.4	26.2
6,400	42.7	40.0	37.6	35.6	33.7	32.0	30.5	29.1	27.8	26.7
6,500	43.3	40.6	38.2	36.1	34.2	32.5	31.0	29.5	28.3	27.1
6,600	44.0	41.3	38.8	36.7	34.7	33.0	31.4	30.0	28.7	27.5
6,700	44.7	41.9	39.4	37.2	35.3	33.5	31.9	30.5	29.1	27.9
6,800	45.3	42.5	40.0	37.8	35.8	34.0	32.4	30.9	29.6	28.3
6,900	46.0	43.1	40.6	38.3	36.3	34.5	32.9	31.4	30.0	28.7
7,000	46.7	43.8	41.2	38.9	36.8	35.0	33.3	31.8	30.4	29.2
7,100	47.3	44.4	41.8	39.4	37.4	35.5	33.8	32.3	30.9	29.6
7,200	48.0	45.0	42.4	40.0	37.9	36.0	34.3	32.7	31.3	30.0
7,300	48.7	45.6	42.9	40.6	38.4	36.5	34.8	33.2	31.7	30.4
7,400	49.3	46.3	43.5	41.1	38.9	37.0	35.2	33.6	32.2	30.8
7,500	50.0	46.9	44.1	41.7	39.5	37.5	35.7	34.1	32.6	31.2
7,600	50.7	47.5	44.7	42.2	40.0	38.0	36.2	34.5	33.0	31.7
7,700	51.3	48.1	45.3	42.8	40.5	38.5	36.7	35.0	33.5	32.1
7,800	52.0	48.8	45.9	43.3	41.1	39.0	37.1	35.5	33.9	32.5
7,900	52.7	49.4	46.5	43.9	41.6	39.5	37.6	35.9	34.3	32.9
8,000	53.3	50.0	47.1	44.4	42.1	40.0	38.1	36.4	34.8	33.3
8,100	54.0	50.6	47.6	45.0	42.6	40.5	38.6	36.8	35.2	33.7
8,200	54.7	51.3	48.2	45.6	43.2	41.0	39.0	37.3	35.7	34.2
8,300	55.3	51.9	48.8	46.1	43.7	41.5	39.5	37.7	36.1	34.6
8,400	56.0	52.5	49.4	46.7	44.2	42.0	40.0	38.2	36.5	35.0
8,500	56.7	53.2	50.0	47.2	44.7	42.5	40.5	38.6	37.0	35.4
8,600	57.3	53.8	50.6	47.8	45.3	43.0	41.0	39.1	37.4	35.8
8,700	58.0	54.4	51.2	48.3	45.8	43.5	41.4	39.5	37.8	36.2
8,800	58.7	55.0	51.8	48.9	46.3	44.0	41.9	40.0	38.3	36.7
8,900	59.3	55.6	52.4	49.4	46.8	44.5	42.4	40.5	38.7	37.1
9,000	60.0	56.3	52.9	50.0	47.4	45.0	42.9	40.9	39.1	37.5
9,100	60.7	56.9	53.5	50.6	47.9	45.5	43.3	41.4	39.6	37.9
9,200	61.3	57.5	54.1	51.1	48.4	46.0	43.8	41.8	40.0	38.3
9,300	62.0	58.1	54.7	51.7	48.9	46.5	44.3	42.3	40.4	38.7
9,400	62.7	58.8	55.3	52.2	49.5	47.0	44.8	42.7	40.9	39.2
9,500	63.3	59.4	55.9	52.8	50.0	47.5	45.2	43.2	41.3	39.6
9,600	64.0	60.0	56.5	53.3	50.5	48.0	45.7	43.6	41.7	40.0
9,700	64.7	60.6	57.1	53.9	51.1	48.5	46.2	44.1	42.2	40.4
9,800	65.3	61.3	57.6	54.4	51.6	49.0	46.7	44.5	42.6	40.8
9,900	66.0	61.9	58.2	55.0	52.1	49.5	47.1	45.0	43.0	41.2

(continued)

TABLE 6.1 *(continued)*

	Design water temperature (°F)									
Btu/hr	170	175	180	185	190	195	200	205	210	215
10,000	66.7	62.5	58.8	55.6	52.6	50.0	47.6	45.5	43.5	41.7
10,100	67.3	63.1	59.4	56.1	53.2	50.5	48.1	45.9	43.9	42.1
10,200	68.0	63.8	60.0	56.7	53.7	51.0	48.6	46.4	44.3	42.5
10,300	68.7	64.4	60.6	57.2	54.2	51.5	49.0	46.8	44.8	42.9
10,400	69.3	65.0	61.2	57.8	54.7	52.0	49.5	47.3	45.2	43.3
10,500	70.0	65.6	61.8	58.3	55.3	52.5	50.0	47.7	45.7	43.7
10,600	70.7	66.3	62.4	58.9	55.8	53.0	50.5	48.2	46.1	44.2
10,700	71.3	66.9	62.9	59.4	56.3	53.5	51.0	48.6	46.5	44.6
10,800	72.0	67.5	63.5	60.0	56.8	54.0	51.4	49.1	47.0	45.0
10,900	72.7	68.1	64.1	60.6	57.4	54.5	51.9	49.5	47.4	45.4
11,000	73.3	68.8	64.7	61.1	57.9	55.0	52.4	50.0	47.8	45.8
11,100	74.0	69.4	65.3	61.7	58.4	55.5	52.9	50.5	48.3	46.2
11,200	74.7	70.0	65.9	62.2	58.9	56.0	53.3	50.9	48.7	46.7
11,300	75.3	70.6	66.5	62.8	59.5	56.5	53.8	51.4	49.1	47.1
11,400	76.0	71.3	67.1	63.3	60.0	57.0	54.3	51.8	49.6	47.5
11,500	76.7	71.9	67.6	63.9	60.5	57.5	54.8	52.3	50.0	47.9
11,600	77.3	72.5	68.2	64.4	61.1	58.0	55.2	52.7	50.4	48.3
11,700	78.0	73.1	68.8	65.0	61.6	58.5	55.7	53.2	50.9	48.7
11,800	78.7	73.8	69.4	65.6	62.1	59.0	56.2	53.6	51.3	49.2
11,900	79.3	74.4	70.0	66.1	62.6	59.5	56.7	54.1	51.7	49.6
12,000	80.0	75.0	70.6	66.7	63.2	60.0	57.1	54.5	52.2	50.0
12,100	80.7	75.6	71.2	67.2	63.7	60.5	57.6	55.0	52.6	50.4
12,200	81.3	76.3	71.8	67.8	64.2	61.0	58.1	55.5	53.0	50.8
12,300	82.0	76.9	72.4	68.3	64.7	61.5	58.6	55.9	53.5	51.2
12,400	82.7	77.5	72.9	68.9	65.3	62.0	59.0	56.4	53.9	51.7
12,500	83.3	78.1	73.5	69.4	65.8	62.5	59.5	56.8	54.3	52.1
12,600	84.0	78.8	74.1	70.0	66.3	63.0	60.0	57.3	54.8	52.5
12,700	84.7	79.4	74.7	70.6	66.8	63.5	60.5	57.7	55.2	52.9
12,800	85.3	80.0	75.3	71.1	67.4	64.0	61.0	58.2	55.7	53.3
12,900	86.0	80.6	75.9	71.7	67.9	64.5	61.4	58.6	56.1	53.7

we are going to install steam radiators in a three-room house. Two of the rooms are calculated to have a heat loss of 10,000 Btu/hr, and the other has a heat loss of 8700 Btu/hr at the specified design temperature differential. The required square feet of steam radiation capability for each of the two large rooms with a 10,000-Btu/hr heat loss is 41.7; the room with the heat loss of 8700 Btu/hr requires 36.2 ft^2 of steam radiation capacity based on a boiler temperature of 215°F.

To convert the heat emission rates of cast iron radiators from square feet of steam to Btu, assuming a design water temperature of 215°F, the square-footage rating of the radiator is multiplied by 240. Thus, in this example, a radiator with a square-foot steam rating of 41.7 will produce 10,008 Btu/hr, the exact figure required to provide adequate space heating to the room.

Selecting the radiators. Once the required square feet of steam has been determined, a specification chart provided by the manufacturer of the radiator should be consulted to determine which units are appropriate for the installation and the size of the units that are required. A typical chart of this type is illustrated in Table 6.2.

It should be emphasized that only radiators specified by the manufacturer as being suitable for steam applications should be installed in the heating system. Trying to substitute units that are "felt" to be appropriate can be highly dangerous, in addition to being ineffective.

Sizing the Boiler

When selecting a steam boiler for a specific heating system, the boiler must have a net Btu rating equivalent to, or in excess of, the total square footage of radiation required in the residence. If it is anticipated that additions to the home will take place in the near future, the anticipated heat losses from these alterations should be estimated to ensure that the boiler has sufficient heating capacity to serve the enlarged system. In making this decision, a specification sheet provided by the manufacturer listing the Btu outputs of different boiler models must be consulted, as shown in Table 6.3.

For example, let us assume that we wish to install a steam system in a house with a calculated heat loss of 88,000 Btu/hr. Consulting the specification sheet, we find that boiler model V-33 with a 1.00-gal/hr burner capacity comes closest to our requirements. It should be observed that the net IBR rating for this unit, 90.8 MBtu/hr, is somewhat lower than the corresponding DOE heating capacity. Note, too, that steam ratings are somewhat lower in heating capacity than equivalent Btu ratings when the boiler is used for hot water systems. The net square-foot steam rating for the boiler should be sized properly to account for the total square footage of radiators to be served by the heating system.

Piping Considerations

Considerable thought must be given to piping steam systems. In most instances, the piping configurations in steam heating applications are much more critical, due to the nature of steam rather than water as the heating medium. Pipe that is cold one minute will, under the influence of steam that has expanded many times in volume, come up to a temperature of over 200°F the next. Some of the major considerations affecting steam piping practices are examined next.

System layout. Prior to undertaking any work on the heating system, a detailed scale drawing needs to be made of the entire system. The procedures recommended for this design phase are detailed in Chapter 2. A typical system layout is illustrated in Figure 6.9. This layout should include the square-foot rating for

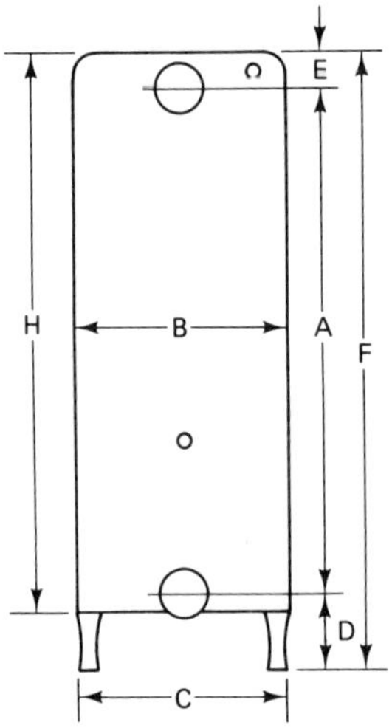

TABLE 6.2 MANUFACTURER'S SPECIFICATION SHEET FOR CAST IRON RADIATORS

a. Dimensions (in.)

Size	A	B and C	D	E	F	H[a]
25-6	$21\frac{7}{16}$	$6\frac{15}{16}$	$2\frac{1}{2}$	$1\frac{1}{16}$	25	$23\frac{13}{16}$
25-4	$21\frac{7}{16}$	$4\frac{7}{16}$	$2\frac{1}{2}$	$1\frac{1}{16}$	25	$23\frac{13}{16}$
19-4	$15\frac{7}{16}$	$4\frac{7}{16}$	$2\frac{1}{2}$	$1\frac{1}{16}$	19	$17\frac{13}{16}$

b. Ratings

Size (number of sections)	Length, $1\frac{3}{4}$ in. per section (in.)	Four-tube		Six-tube, 25 in. high
		19 in. high	25 in. high	
4	7	6.4	8	12
6	$10\frac{1}{2}$	9.6	12	18
8	14	12.8	16	24
10	$17\frac{1}{2}$	16.0	20	30
12	21	19.2	24	36
14	$24\frac{1}{2}$	22.4	28	42
16	28	25.6	32	48
18	$31\frac{1}{2}$	28.8	36	54
20	35	32.0	40	60
22	$38\frac{1}{2}$	35.2	44	66
24	42	38.4	48	72
26	$45\frac{1}{2}$	41.6	52	78
28	49	44.8	56	84
30	$52\frac{1}{2}$	48.0	60	90
32	56	51.2	64	96
34	$59\frac{1}{2}$	54.4	68	102
36	63	57.6	72	108
38	$66\frac{1}{2}$	60.8	76	114
40	70	64.0	80	120
42	$73\frac{1}{2}$	67.2	84	126
44	77	70.4	88	132
46	$80\frac{1}{2}$	73.6	92	138
48	84	76.8	96	144
50	$87\frac{1}{2}$	80.0	100	150
52	91	83.2	104	156
54	$94\frac{1}{2}$	86.4	108	162
56	98	89.6	112	168

Source: Burnham Corp.

Note: Center leg section furnished on 30 section and larger. Heating surface based on the standard heat emission of 240 Btu/ft^2/hr.

[a]Legless, overall height; tappings: top, 1 in., bottom, $1\frac{1}{4}$ in.

TABLE 6.3 SPECIFICATION SHEET FOR STEAM BOILERS

Boiler number	Burner capacity (gal/hr)	DOE heating capacity (MBtu/hr)	IBR net ratings: Water (MBtu/hr)	IBR net ratings: Steam (MBtu/hr)	Steam (ft²)	Net water rating (ft²)	DOE annual efficiency (%): Less vent damper	DOE annual efficiency (%): With vent damper
V-33	1.05	121	105.2	90.8	378	700	80.8	82.4
	1.15	133	115.7	99.8	416	770	80.8	82.4
V-34	1.55	178	154.8	133.5	556	1030	80.6	82.2
	1.70	194	168.7	145.5	606	1120	80.0	81.6
V-35	2.10	241	209.6	180.8	753	1395	80.3	81.9
	2.25	244[a]	212.2	183.0	763	1415	—	—
V-36	2.60	284[a]	247.0	213.1	888	1645	—	—
	2.80	302[a]	262.6	226.6	944	1750	—	—
V-37	3.15	342[a]	297.4	256.6	1069	1985	—	—
	3.40	370[a]	321.7	277.6	1157	2145	—	—
V-38	3.65	398[a]	346.1	298.6	1244	2310	—	—
	4.00	438[a]	380.9	328.6	1369	2540	—	—

Source: Burnham Corp.

[a]IBR gross output—not rated for DOE capacity or efficiency.

each radiator. This will aid in selecting circuit pipe sizes, an explanation of which follows.

Sizing the pipe:

Mains. The majority of residential steam mains are sized between 2 and 3 in. If it is suspected that larger-than-normal mains will be required due to the amount of square footage of radiation required for a specific installation, the boiler manufacturer should be consulted. If, for example, the pipe size to be used in the mains is 2 in. and the boiler tapping is 3 in., it is possible to bush down the tapping to 2 in.* The size of the steam mains should be listed at all applicable points in the piping layout drawings.

Branch Circuits. When sizing the branch circuits in a typical one-pipe steam system, Table 6.4 should be consulted. For example, Table 6.4 indicates that for a room that requires a total of 41.7 ft^2 of steam radiation, a feed branch circuit of $1\frac{1}{4}$ in. is necessary. Similarly, a room in which the required square footage of radiation is 36.2 should also be fed with a $1\frac{1}{4}$-in. branch circuit. It is good practice to

*Although reducing bushings may be used to reduce nipple sizes in the boiler, it is preferable to install at least 12 to 18 in. of piping sized to match the boiler tapping at the boiler takeoff, and then use reducing bushings or couplings. This procedure allows for maximum steam acceleration as it exits the boiler.

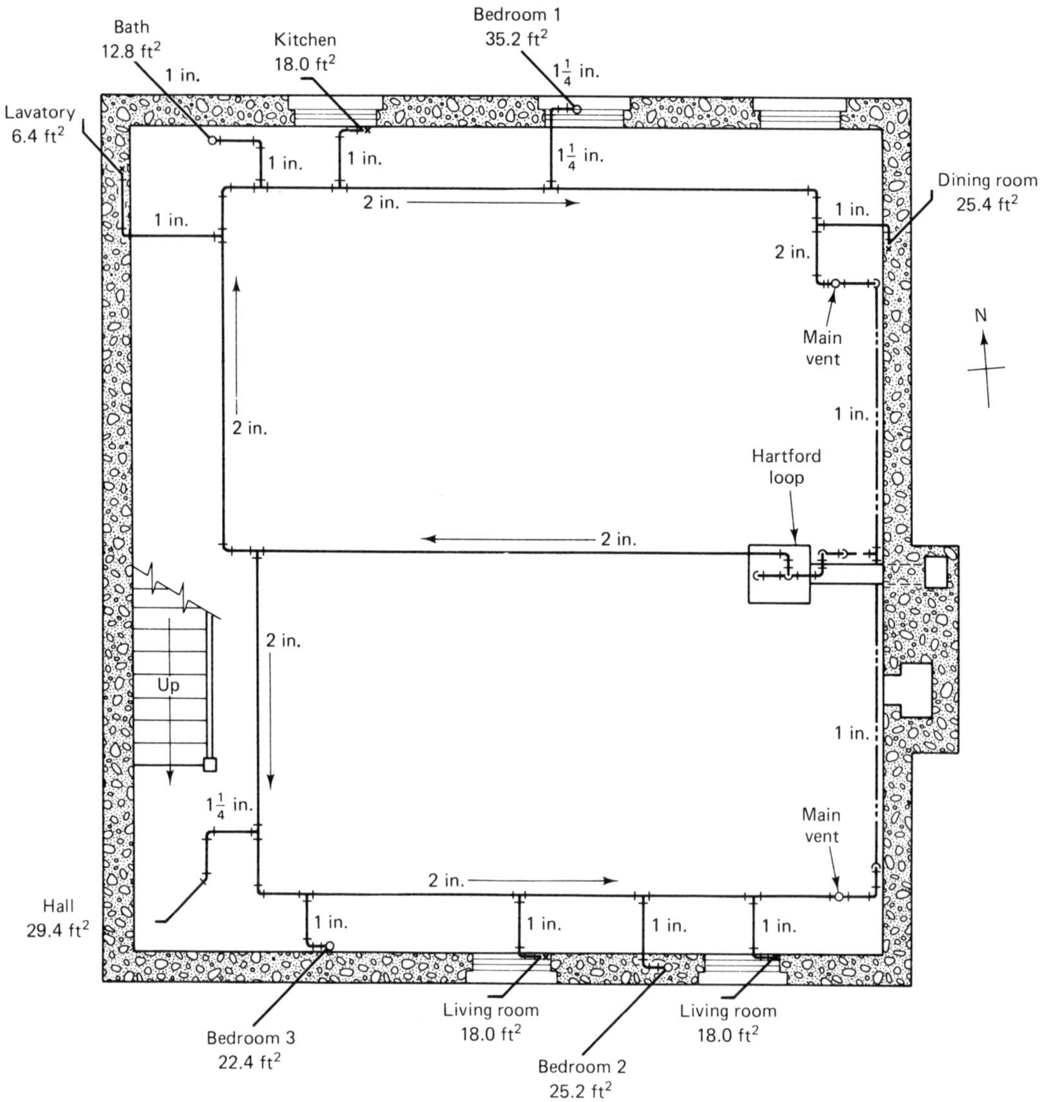

Figure 6.9 Detailed steam system pictorial. Pitch mains down in direction of arrows at least ¼ in. in 10 ft. Pitch branches up from main at least ½ in. in 10 ft. (Courtesy of the Hydronics Institute.)

list the branch circuit pipe sizes on the system layout as soon as they have been determined.

Steam supply mains. When steam supply mains are installed along with separate branch circuit piping, a minimum pitch of $\frac{1}{4}$ in. per 10 ft must be main-

TABLE 6.4 BRANCH CIRCUIT PIPE SIZING FOR ONE-PIPE STEAM HEATING SYSTEMS

Radiation (ft^2)	Branch size[a] (in.)	Main size[b] (in.)	Wet return size (in.)
1–28	1	—	—
29–64	$1\frac{1}{4}$	—	—
65–92	$1\frac{1}{2}$	—	—
95–168	2	—	—
432	—	2	1

Source: The Hydronics Institute.

Note: Based on pressure drop of 1 oz per 100 ft of equivalent length.

[a]If horizontal length of the branch exceeds 8 ft, increase the branch piping one size larger.

[b]Based on an equivalent length not exceeding 200 ft. To determine equivalent length, add to the actual length of main 4.3 ft for each elbow and 8 ft for each tee.

tained for proper condensate return. Adequate support bracing should be installed to prevent pipe stress and sag (Figure 6.10).

Branch circuits and swing joints. Due to the high operating temperatures encountered in steam systems, together with the rapid expansion and contraction of the piping during normal system cycling, provisions must be made to accommodate these dimensional changes. This is accomplished by the installation of swing joints at all strategic points. A typical configuration for a branch circuit–swing joint combination is illustrated in Figure 6.11.

The swing joint should be located at a point below the vertical riser to the radiator. Also, adequate space should be provided for all horizontal pipe runs and risers, especially where they pass through walls, floors, or framing members of the house. Failure to configure the piping with ample swing joints and clearances results in cracked and leaking pipe fittings, low spots in the mains, and eventual premature boiler replacement due to excessive amounts of makeup water being taken into the system to replace lost water.

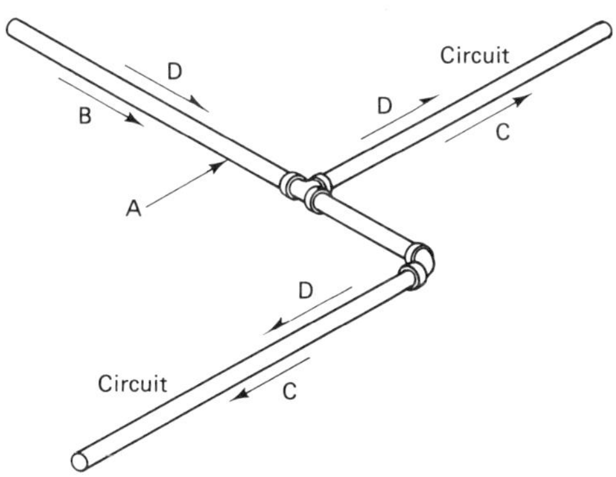

Figure 6.10 Steam supply mains and division into branch circuits. A, Trunk; B, pitch trunk downward in direction of arrow not less than ¼ in. in 10 ft; C, pitch circuits downward in direction of arrows not less than ½ in. in 10 ft; D, direction of flow of steam. (Courtesy of the Hydronics Institute.)

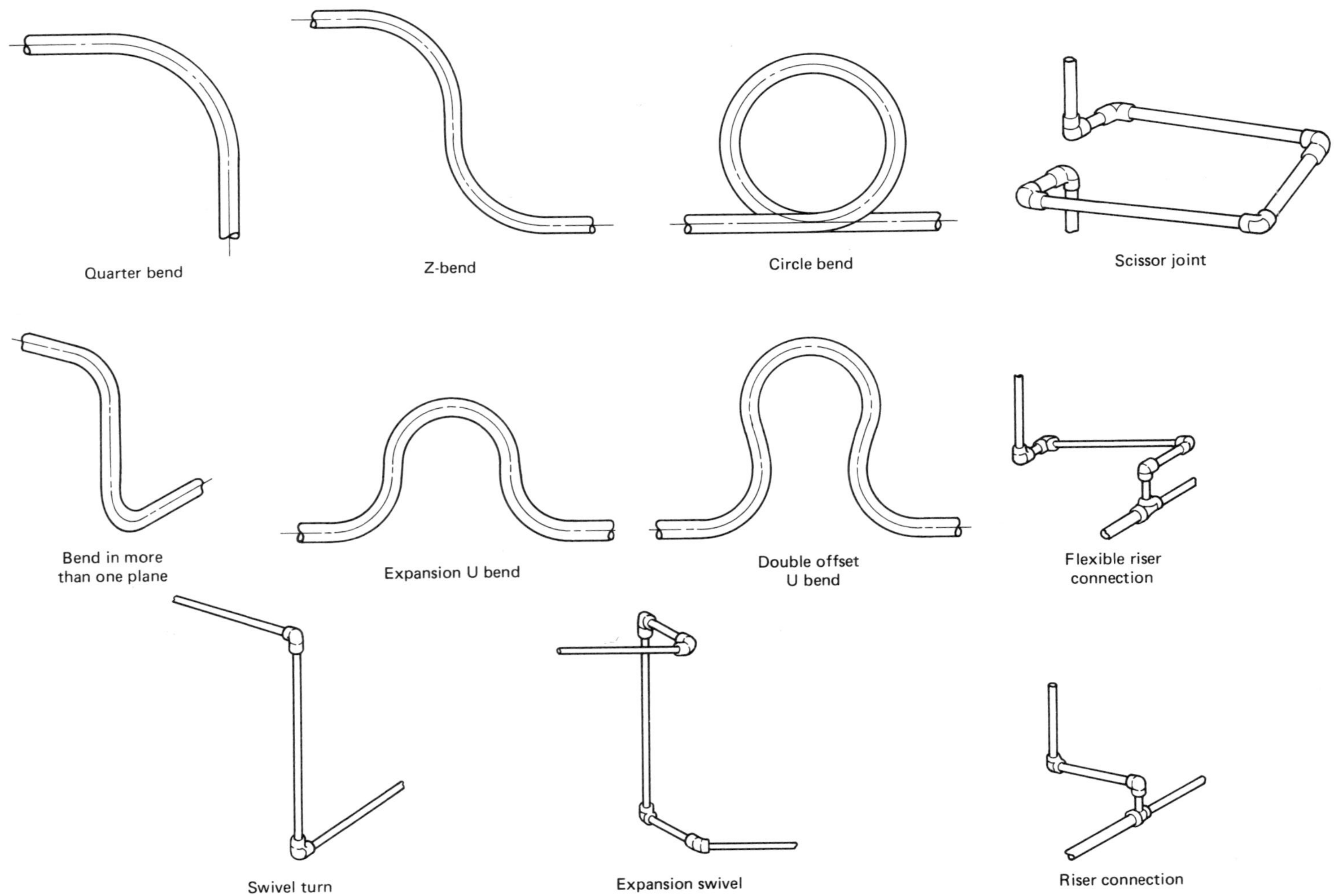

Figure 6.11 Branch circuit and swing joints. (Courtesy of Taco, Inc.)

Installing the Radiator

For best performance, radiators should be installed under windows. If this is not possible, placc radiators along exterior building walls where room design permits. In one- and two-pipe steam systems, the radiator supply pipe enters from a vertical takeoff riser in most branch circuit designs. Entrance of steam into the radiator is controlled by a valve that should be either fully opened or fully closed for proper operation. The radiator air vent is located on the bottom nipple assembly opposite the steam entrance pipe (Figure 6.12).

Installing the Boiler Piping and Controls

Steam boilers require specialized piping as well as controls that are specific to steam systems.

Hartford loop. A Hartford loop is a piping configuration used to connect the condensate return lines to the boiler. The purpose of this loop is to equalize the pressure between the water in the boiler and the return mains, preventing the boiler from pushing water into the mains. Figure 6.13 illustrates the Hartford loop piping arrangement. Note that the crossover equalizes the pressure between the boiler and the return line. Since water within the boiler is normally at a higher pressure than the condensate, the tendency without this equalization arrangement would be for the higher pressure in the boiler to force water back into the return main, as illustrated in Figure 6.14.

In connecting the Hartford loop, an inverted Y fitting should be used where possible; however, a T with a closed nipple can be substituted (Figure 6.15). The location of the Y or T fitting in the loop is important for proper operation of the equalization line. In steam systems using a gravity return for the condensate, the loop Y or T should be located 2 in. below the normal water level in the boiler. For systems employing a pump for condensate return, these fittings should be located 4 in. below the boiler-water level. Also, the equalizer line should be run with

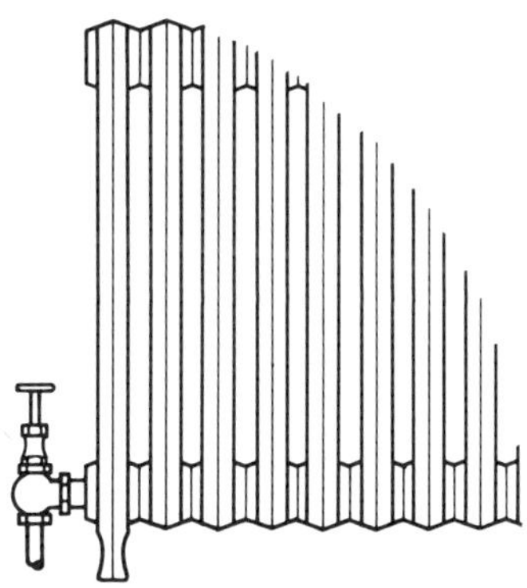

Figure 6.12 Radiator piping configuration.

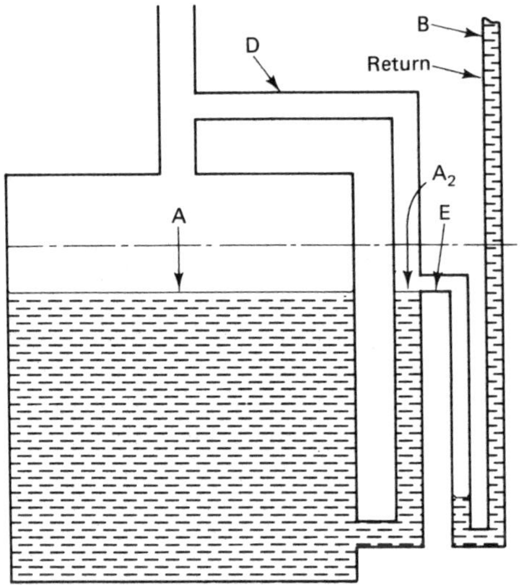

Figure 6.13 Hartford loop. The pressure at A is greater than that at B. Pressure at A and A_2 are equal due to equalizing line D, which should not be less than 1½ in. pipe size. If water line reaches level of close nipple E, no more water can be forced from boiler. Do not use longer than a close nipple at E and locate nipple 2 in. below normal water line. *Note:* An inverted Y may be used in place of close nipple at E. (Courtesy of the Hydronics Institute.)

a minimum of $1\frac{1}{2}$-in line. If two return pipings drain into the boiler, they should be manifolded as illustrated in Figure 6.16.

Supply headers. If two or more supply tappings are built into the boiler, and the use of both of them is recommended by the manufacturer, a header arrangement should be piped as shown in Figure 6.17.

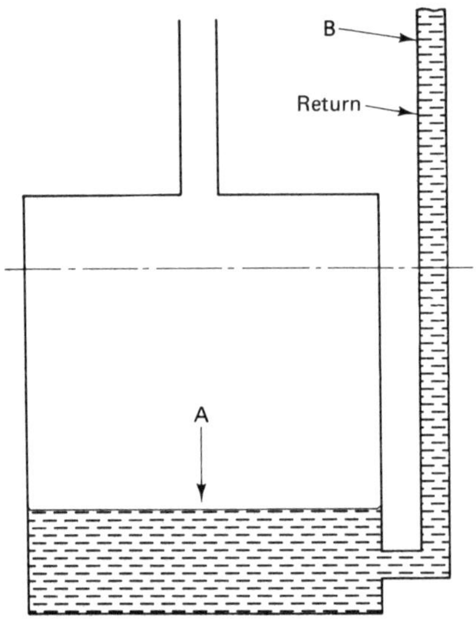

Figure 6.14 Steam boiler without Hartford loop. Pressure at A greater than that at B forces water from the boiler into the return. (Courtesy of the Hydronics Institute.)

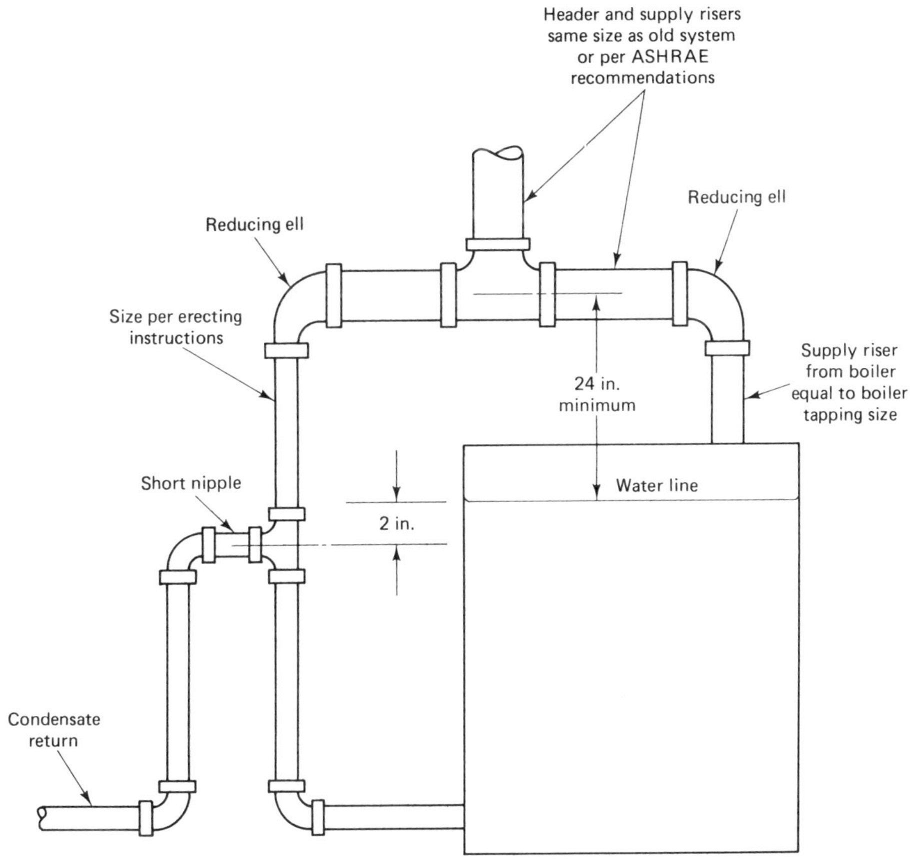

Figure 6.15 T fitting installed in Hartford loop (swing joints not shown). (Courtesy of Weil-McLain, a Marley Co.)

BOILER CONTROLS

After the piping has been installed, the boiler trim, consisting mainly of the operating controls, is added to the unit to complete the major aspects of system installation. These controls include the pressure operating control, steam pressure gauge, water gauge glass, automatic low-water cutoff, and feedwater makeup systems.

Pressure Control Switches

The operational cycling of the steam boiler is controlled by a specially designed pressure switch, shown installed on the boiler in Figure 6.18. Note that the switch is piped into the boiler using a special pigtail pipe fitting. This prevents the switch

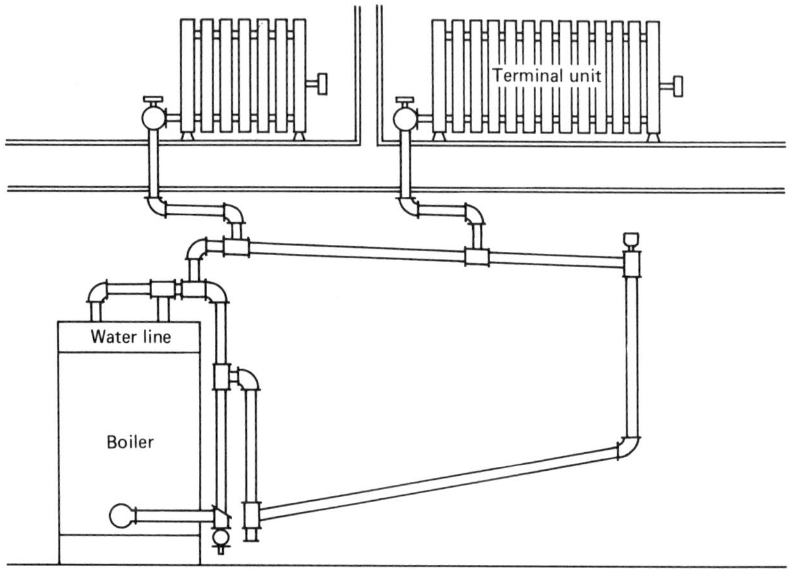

Figure 6.16 Manifolded return lines in steam boiler. (Reprinted by permission of Small Homes Council, Building Research Council, University of Illinois, Technical Note 13.)

from coming into direct contact with steam, which would cause premature failure of the switching mechanism. This control features a spring-activated pressure switch with an adjustable differential to turn the burner on and off depending on the pressure setting. To set the control for boiler operation, the cut-in pressure is selected and calibrated (Figure 6.19).

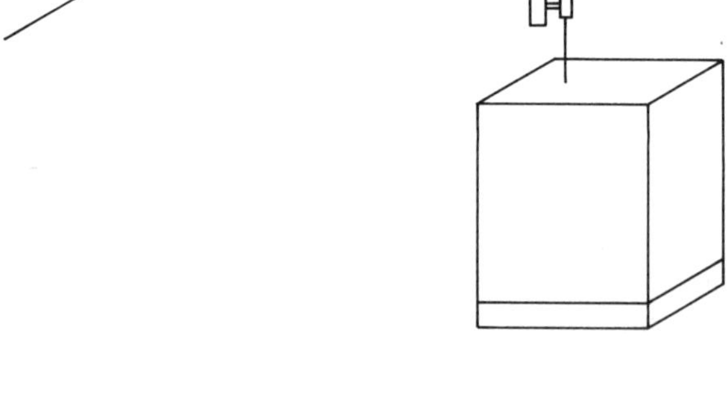

Figure 6.17 Manifolded supply mains in steam piping.

Figure 6.18 Pressure control switch for steam boiler. (Courtesy of Honeywell, Inc.)

The cutout is set by adjusting the differential wheel inside the control switch. Boiler cutout equals the cut-in pressure plus the differential pressure chosen. Normal settings are 3 lb cut-in with a differential of 2 lb (Figure 6.20). The pressure control switch is wired in series with the burner. A typical wiring arrangement is illustrated in Figure 6.21.

Assuming that the room thermostat is calling for heat, if pressure within the boiler is below the cut-in point, the switch connection to the burner will be closed, and ignition of the burner will begin. When the boiler reaches the cutout pressure, the switch opens the circuit to the burner. When the pressure drops to the cut-in point, the burner begins operating once more. This cycling will continue as long as the thermostat is calling for heat. The maximum working pressure for most residential steam systems is 15 psi.

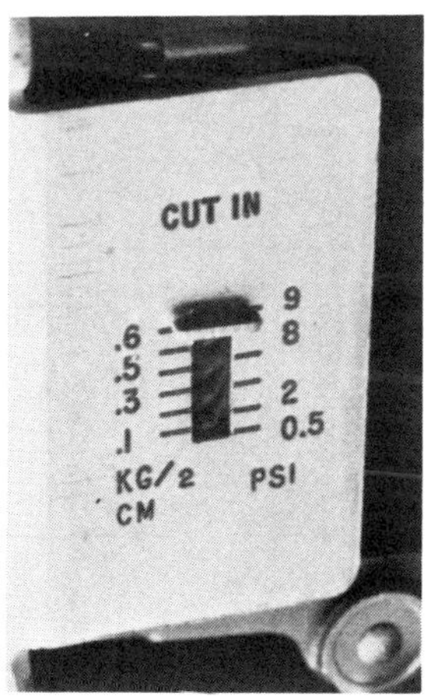

Figure 6.19 Setting pressure cut-in/cutout on steam regulator. (Courtesy of Honeywell, Inc.)

Figure 6.20 Pressure cut-in/cutout differential on steam pressure gauge. (Courtesy of Honeywell, Inc.)

Low-Limit Domestic Hot Water Aquastats

If the steam boiler contains a domestic hot water tankless coil, a low-limit aquastat is placed in the control circuitry of the boiler, the purpose of which is to maintain steam boiler water temperature at a level that is sufficient to provide for indirect water heating. A typical low-limit switch wiring arrangement is illustrated in Figure 6.22.

Steam Gauges and Relief Valves

Steam gauges and relief valves in steam systems function in a manner similar to that of their hydronic counterparts.

Steam pressure gauges in residential boilers indicate boiler pressure in psi. These gauges are inserted into threaded tappings provided in the casting of the boiler located above the normal water line. A steam pressure gauge of this type is illustrated in Figure 6.23.

The pressure relief valves used on steam boilers are similar in function to those relief valves used on hot water boilers; however, they are normally set to discharge

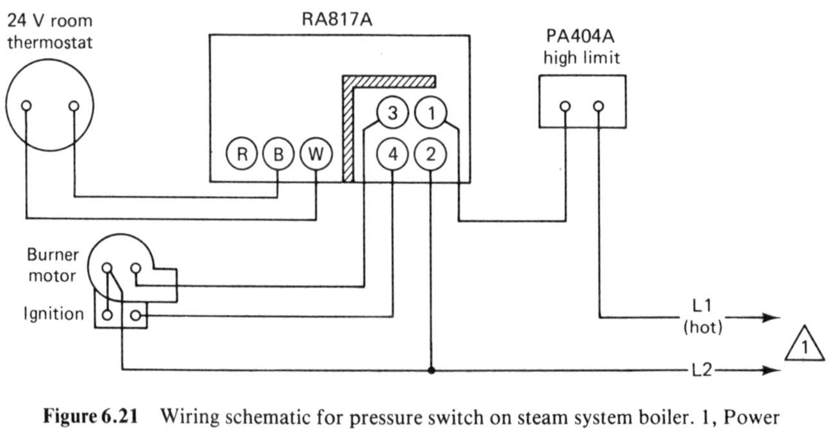

Figure 6.21 Wiring schematic for pressure switch on steam system boiler. 1, Power supply: provide disconnect means and overload protection as required. (Courtesy of Honeywell, Inc.)

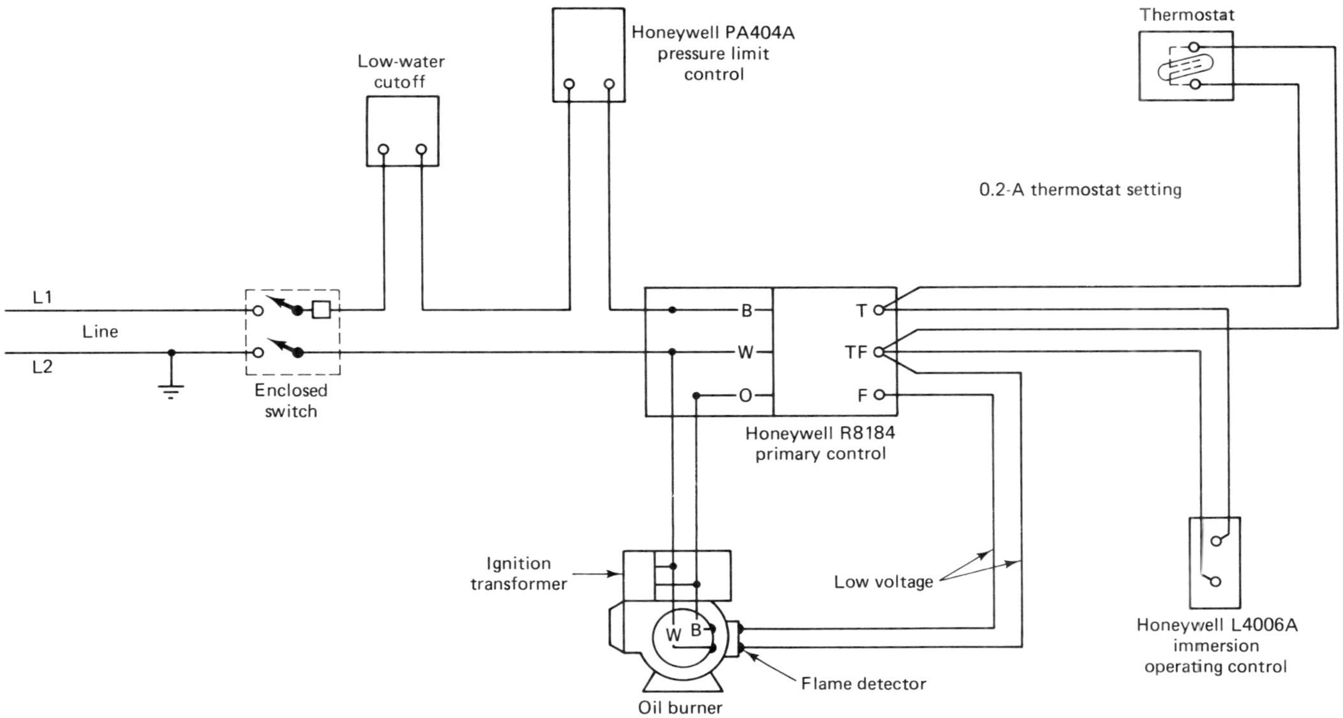

Figure 6.22 Low-limit switch wiring for domestic hot water coil in steam boiler. (Courtesy of Burnham Corp.)

Figure 6.23 Steam pressure gauge. (Courtesy of Dresser Industries—Instrument Div.)

at 15 psi, rather than the hydronic pressure setting of 30 psi. Piping should be provided from the discharge opening of the relief valve to either a floor drain or to within 6 in. of the surface of the floor. Local codes should be consulted for all applicable regulations concerning pressure relief valve discharge configurations.

Water Column Gauge Glass Assembly

Each steam boiler operates at a specified water level recommended by the manufacturer. If the water level in the unit is too low, insufficient water is available for conversion to steam, and in extreme cases can cause failure of the boiler due to overheating by the burner. If the water level in the boiler is too high, a condition known as *waterbound* or *waterlogging* can occur, causing excess water to circulate within the supply mains, creating water traps that will prevent steam from being delivered to the radiators.

For a visual inspection of the water level in the unit to be performed, all steam boilers are equipped with externally mounted water columns, consisting of a glass gauge assembly and try-cock valves. A typical water column assembly is shown in Figure 6.24. The valves serve to admit water into the glass gauge to display the water level of the boiler. They also serve to drain the assembly for servicing or glass replacement when necessary. This assembly is installed at the boiler-water-level line in tappings provided by the appliance manufacturer. Sometimes, the glass gauge is marked with two lines that indicate the high- and low-water-level operating limits. The lowest safe water operating level of the boiler must not be lower than the bottom of the glass gauge assembly.

Automatic Low-Water Cutoff

All steam boilers must be equipped with an automatic low-water cutoff mechanism. The device shuts off the burner should the water level fall to a predetermined low limit. Most of these shutoff devices are of the float type, as shown in Figure 6.25.

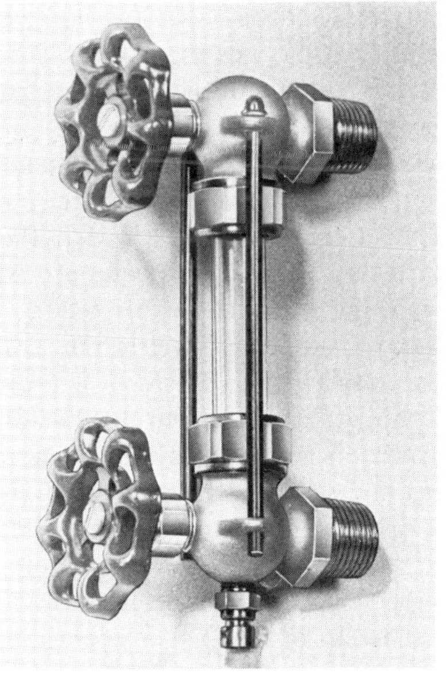

Figure 6.24 Water column and glass gauge assembly. (Courtesy of Conbraco Industries, Inc.)

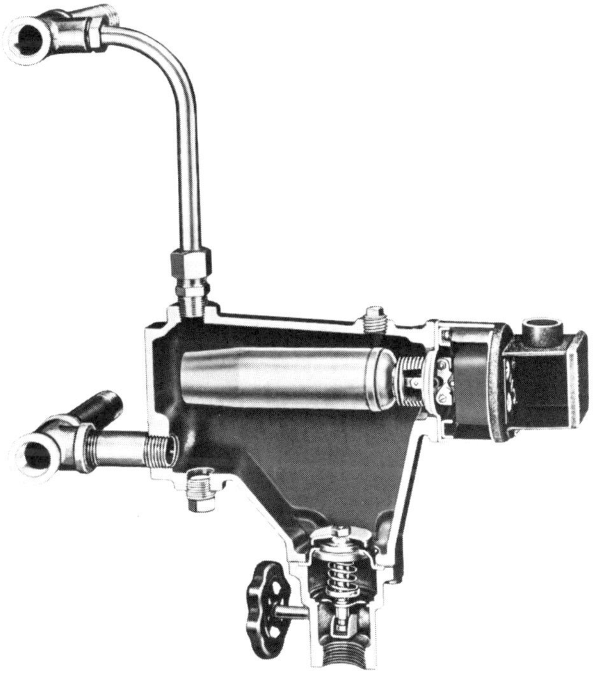

Figure 6.25 Low-level shutoff for steam boiler. When the water level drops, the float lowers, opening an electrical end switch, interrupting burner operation until the water level is restored. (Courtesy of ITT Fluid Handling Division.)

If the water level in the boiler drops to a low limit, a lowering of the float mechanism within the shutoff device opens the control circuit to the oil or gas ignition system, which is wired in series with the low-water cutoff.

The low-water cutoff requires periodic maintenance during the heating season. As boiler water circulates within the float chamber, sediment from the piping circuit can become deposited within. Although most low-water shutoff controls contain a large bowl for dirt and sediment accumulation, a weekly flushing of this control is recommended. To accomplish this, the cutoff is equipped with a blow-off valve at the bottom of the unit to void the sediment (Figure 6.26).

When flushing the control, the water level in the glass gauge should be observed and water added if necessary. Also, opening the blow-off valve to flush the unit checks the operation of the cutoff circuit. During blow-off, the float level should drop, shutting down the burner; hot water and steam will also come out of the blow-off valve, flushing away dirt and sediment from the unit.

When new boilers are installed in an existing steam heating system, it is common for scale and sediment to break loose from the piping and accumulate in the boiler and cutoff float chamber. In this instance, the cutoff should be purged several times a week during the first weeks of new boiler operation.

Feedwater Mechanisms

During the heating season, small amounts of water will be lost from the heating system due to the operation of main and radiator vents. These losses will result in a drop in the boiler-water level that must be made up for efficient and safe operation of the system. Maintaining the water level in the system can be accomplished by either a manual or an automatic control.

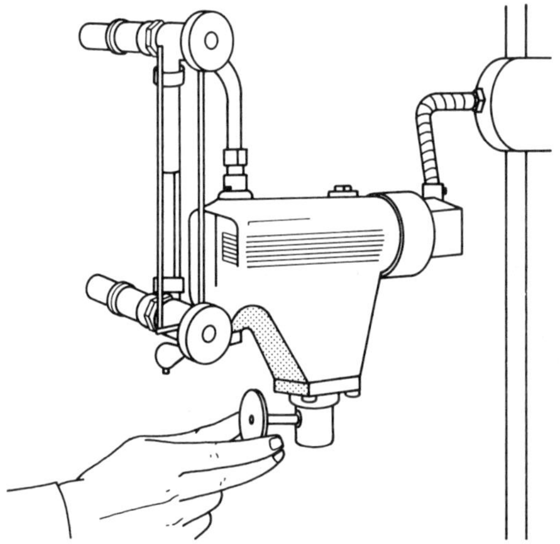

Figure 6.26 Flushing low-level cutoff. For best results, the cutoff should be flushed once a week during the heating season when the boiler is in continual operation. (Courtesy of ITT Fluid Handling Division.)

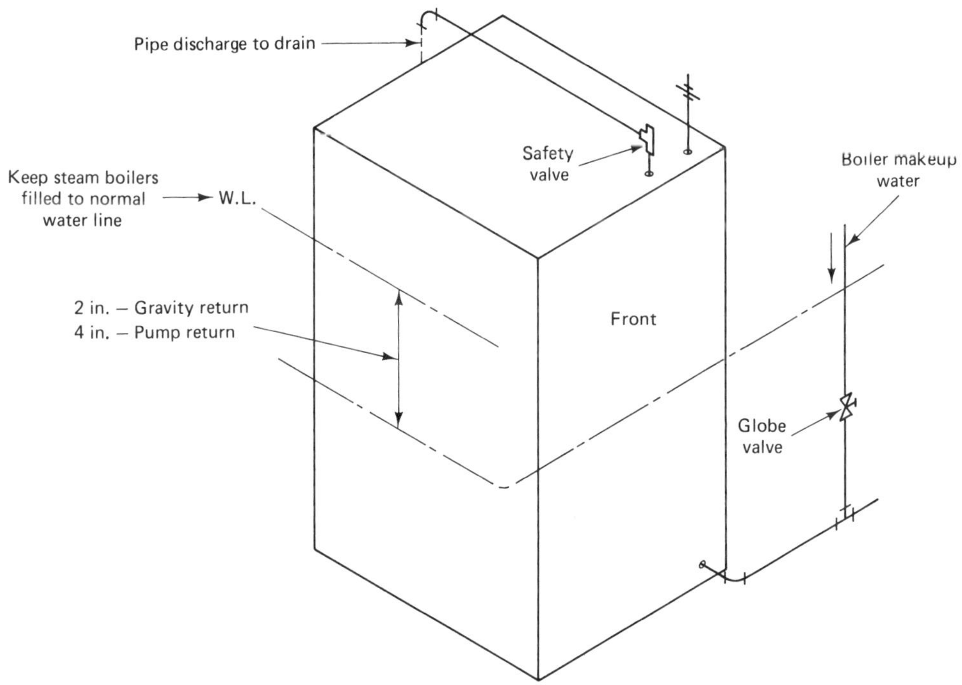

Figure 6.27 Manual feedwater piping to steam boiler. Boiler water level should be checked via the gauge glass periodically for safe operation.

Manual system. The water level in the boiler should be checked periodically by the homeowner via the glass gauge to ensure that safe operating water levels are present within the boiler and heating system. In a manual feedwater system, provision is made to connect a water line from a cold water supply pipe to a condensate return line. This supply line includes a gate valve that is opened to admit water into the boiler when required. This piping arrangement is illustrated in Figure 6.27. To add water to the system, the gate valve is opened while observing the level in the glass gauge. When the water level reaches the correct level, the valve is closed.

Automatic system. Water levels within the steam boiler can be maintained automatically by installing an automatic feedwater system. Typical of these is the unit shown in Figure 6.28. Automatic feedwater valves should be inspected and flushed periodically for proper trouble-free operation.

MAINTENANCE PROCEDURES

Maintenance procedures on steam boilers fall into several categories: maintenance of the burner unit, inspection of the venting system, inspection of all operating controls on the appliance, and periodic flushing of the system.

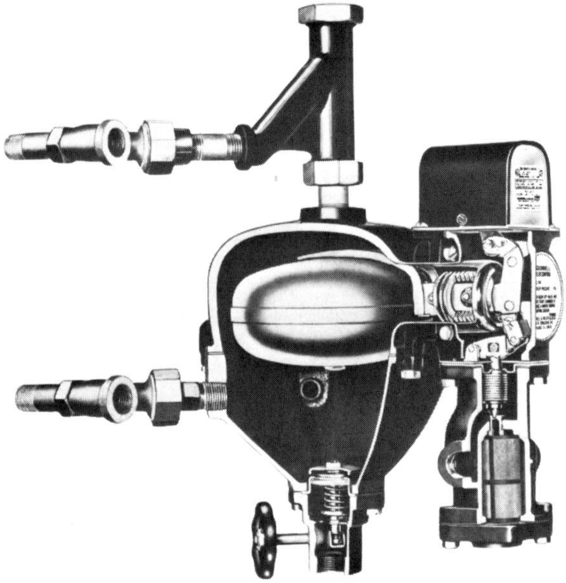

Figure 6.28 Automatic feedwater and low-level cutoff control. (Courtesy of ITT Fluid Handling Division.)

The procedures for inspection of venting systems, burner maintenance procedures, and control sequencing have been covered in other sections of this text and need not be gone into here. However, the procedures for flushing the system are unique to steam systems and should be paid close attention by the homeowner.

Flushing the system periodically ensures that rust, scale, and sediment from the heating system will not accumulate in the boiler, associated piping, or controls. Flushing is most critical in a system where a new boiler is installed into existing system piping. This often results in the breaking loose of scale and sediment that has collected over many years.

As is often the case, the major problems associated with scale take place in the wet condensate return lines when retrofit boilers are installed. In this instance, it is necessary to inspect all wet returns prior to completing the installation of the boiler. If necessary, lines that have excessive restrictions and sediment buildup should be replaced.

When installing new steam heating systems, the system should be initially cleaned for trouble-free operation. Oil, grease, and sediment that accumulate in a new system must be purged to prevent an unsteady water line within the boiler from developing. This can cause a carryover of water into the supply mains above the appliance. For best results, individual manufacturers' recommendations regarding cleaning procedures should be followed at the intervals specified.

7

Solid-Fuel and Multifuel Boilers

COMBUSTION CHARACTERISTICS OF SOLID FUELS

Efficient and safe solid-fuel appliance operation requires an understanding of the combustion characteristics of wood and coal. Since each of these fuels burns differently, boiler design can vary based on the type of fuel for which the unit is designed. Boilers designed to burn coal will usually burn wood quite well, assuming that the fuel box is large enough to accommodate the volume necessary for wood combustion. Boilers designed to burn wood primarily may not burn coal efficiently if the grate system and primary air inlets have not been engineered for coal combustion as well.

An examination of the combustion characteristics of each of these fuels will highlight some of the design features that are necessary for efficient solid-fuel combustion.

BURNING WOOD

Heat Value of Selected Cordwoods

Different species of wood vary in their heating content. In general, hardwoods are preferable to softwoods for use in almost all types of wood-burning boilers. Hardwoods are denser and contain fewer volatile materials than do softwoods. Since the density of wood is directly proportional to its heating value, hardwoods deliver much more heat per unit of volume than do most softwoods, since their density is much greater. The heating content of selected cordwoods is illustrated in Table 7.1.

Heating content can also vary from one cord of wood to another within the same species of wood. This variation is based on the different amounts of wood contained in a cord due to the volume of actual wood versus the amount of space present between the stacked logs. Thus the amount of wood present in each cord depends on how carefully it was stacked. A cord, the standard unit of measurement within the wood industry, is a stack measuring 4 ft × 4 ft × 8 ft.

The Btu content of wood is a function of its weight, or density, which is in part dependent on the moisture content of the log. Wood labeled as "dry" in reality has a moisture content of between 15 and 20%. Wet wood, on the other hand, can contain up to 100% moisture content. Naturally, the Btu content per cord of wood

TABLE 7.1 HEAT VALUE OF SELECTED CORDWOODS

Species	Millions of Btu per cord
Black locust	26.0
Shagbark hickory	25.5
White oak	24.5
Beech	21.0
Sugar maple	21.0
Red oak	21.0
Yellow birch	21.0
White ash	20.0
Red maple	19.0
Tamarack	19.0
Black cherry	18.5
Pitch pine	18.5
White birch	18.0
Silver maple	18.0
Norway pine	17.5
White elm	17.5
Grey birch	17.5
Hemlock	15.0
Red spruce	15.0
Balsam fir	13.5
Black willow	13.5
White pine	13.0
Basswood	12.5

Note: Btu values are approximate and based on a moisture content of 15 to 20%.

will vary depending on its moisture content and species, anywhere from approximately 6500 Btu/lb to almost 9500 Btu/lb.* Air-dried wood possesses a moisture content that is equal to its surroundings, usually within the limits listed above. High moisture content in fuelwood merely adds to its weight without increasing its heating value. High moisture content should be avoided since it lowers combustion chamber temperatures due to the large quantity of steam that must be driven off from within the wood before it will burn, as well as accelerating creosote deposits within the boiler and chimney flue.

Cost Analysis of Comparative Fuels

The explosive sales of solid-fuel heating equipment beginning with the oil shortages of the early 1970s have been well documented. Solid-fuel central heating boilers account for an increasingly larger percentage of all wood- and coal-burning appliances sold, as consumers update earlier purchases of radiant units or make first-time purchases. Solid fuels can sometimes offer significant cost advantages compared to fossil fuels for residential space and domestic hot water heating. To help determine what these cost advantages are, the use of Table 7.2 is suggested.

To illustrate this procedure, let us assume that a homeowner is presently using electricity for space heating at a cost per kilowatthour of 10 cents and is trying to decide whether to install a central wood-heating boiler in the home. Mixed hardwoods are available at $90 per cord. The current delivered price of electricity is $29.30 per million Btu. The cost of the same amount of energy supplied by the wood boiler would be $7.50 with the use of mixed hardwoods, a savings of approximately 75% on the cost of heating a home with wood versus electric space heating. Similar calculations can be performed for other types of fuels using the figures supplied in Table 7.2. It should be noted that these calculations do not include either the purchase of the appliances or their installation, which should be added to the respective fuel costs.

Also, the calculations do not include any value placed on the amount of time that must be devoted to the solid-fuel heating system. Since this time can often be considerable when cutting, splitting, and stacking wood, as well as increased chimney maintenance, is taken into account, it should be considered when making cost comparisons.

Combustion Characteristics

Wood combustion takes place in distinct phases or stages. When wood is first placed into the boiler, it contains between 15 and 20% moisture content by weight. Table 7.3 illustrates how moisture content affects the weight and heating value of selected fuel woods.

In order for wood to begin to burn, the trapped moisture within the wood

*A cord of dry oak, with a moisture content of 20%, contains approximately 8500 Btu/lb.

TABLE 7.2 CONVERSION FACTORS AND BTU EQUIVALENTS OF COMMON FUELS

Fuel type	Description
Natural gas	One therm = 100,000 Btu = 100 ft^2 Assumed efficiency of unit = 75% $/million Btu = 13.33 × $/therm
Fuel oil	138,000 Btu/gallon Assumed efficiency of unit = 65% $/million Btu = 11.15 × $/gallon
LP gas (propane)	93,000 Btu/gallon Assumed efficiency of unit = 75% $/million Btu = 14.34 × $/gallon
Electricity	3412 Btu/kilowatthour (kWh) Assumed efficiency of unit = 95% $/million Btu = 293 × ¢/kWh
Mixed hardwoods	24 million Btu/cord One cord = 128 ft^3/90 ft^3 wood Assumed efficiency of unit = 50% $/million Btu = $/cord ÷ 12
Mixed softwoods	15 million Btu/cord One cord = 128 ft^3/90 ft^3 wood Assumed efficiency of unit = 50% $/million Btu = $/cord ÷ 7.5
Coal	12,500 Btu/pound Assumed efficiency of unit = 60% $/million Btu = $/ton ÷ 15

Source: Adapted from *Heating with Coal* by John W. Bartok, Garden Way Publishing Co., Charlotte, Vt., 1980. Available from Storey Communications, Pownal, Vt., 05261.

must be driven off, either by the heat of the existing fire or by burning kindling placed in the combustion chamber to start the fire. In the first stage of the combustion process, the temperature of the wood rises, and the moisture content within is converted to steam. This steam escapes through the open cellular structure on the face and ends of the logs. The temperature in the firebox continues to rise until it reaches the ignition temperature of the wood, which is approximately 550°F.

When the actual burning of the wood and volatile gases driven off from within the wood begins, the second stage of combustion commences. This combustion stage is evidenced by the customary flaming characteristics of the wood fire. The flames in evidence are the result of the oxidation of the gases driven from within the logs. For efficient burning to take place, the wood should be stacked within the firebox so as to reflect heat back into the fire. Excessive space between the logs allows heat to escape to the cooler surfaces of the boiler and makes for a less efficient fire.

After all the gases have been driven off from within the wood, what remains is almost pure carbon in the form of charcoal, initiating the final phase of the com-

TABLE 7.3 EFFECT OF MOISTURE CONTENT ON SELECTED VARIETIES OF WOOD

	Moisture removed/cord			
	Weight per cord (lb)			
Species	Green	Air dried	In drying to zero moisture content (gal)	Approximate heat value: 60% efficiency furnace (Btu)
Ash	3840	3440	76	16,500,000
Aspen	3440	2160	171	10,400,000
Beech	4320	3760	116	17,300,000
Yellow birch	4560	3680	150	17,000,000
American elm	4320	2960	186	14,300,000
Shagbark hickory	5040	4240	131	20,400,000
Red maple	4000	3200	153	14,000,000
Sugar maple	4480	3680	132	17,400,000
Red oak	5120	3680	220	16,900,000
White oak	5040	3920	180	18,300,000
White pine	2880	2080	126	9,500,000

Source: Solid Fuel Furnaces and Boilers by John W. Bartok, Garden Way Publishing Co., Charlotte, Vt., 1982. Available from Storey Communications, Pownal, Vt., 05261.

bustion process. Temperatures within the bed of the fire during this stage can reach over 2000°F. The wood fire at this point is practically smokeless: The bed of hot coals glows with only short blue flames and radiates intense heat.

Primary and Secondary Air for Wood Combustion

Efficient wood burning depends on both primary and secondary air delivered to the combustion chamber. This airflow pattern is shown in Figure 7.1. Primary air furnishes the basic oxygen supply for the wood fire. This air pattern is directed over the wood pile from front to back, top to bottom. Secondary air is introduced into the upper section of the combustion chamber to aid in oxidizing the unburned gases, which help to reduce creosote deposits and add significant amounts of heat to the combustion process. Airtight solid-fuel boilers are designed to provide both primary and secondary air during all heating cycles of the boiler to raise water temperatures to a level sufficient for baseboard radiation to perform effectively. During the boiler "off" cycles both primary and secondary air is usually restricted, allowing the boiler in effect to "cycle" as do conventional fossil-fuel units.

Creosote Deposition

One problem associated with the burning of wood is the formation of creosote deposits on the interior surfaces of both chimney flues and the boiler. Creosote is black and shiny in appearance and tarlike in consistency. Complex in its chemical com-

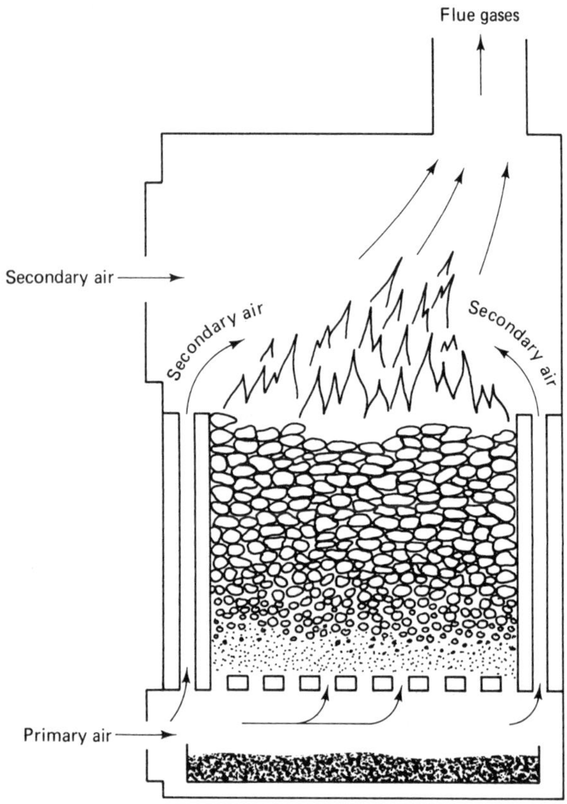

Figure 7.1 Primary and secondary airflow through a solid-fuel appliance. (Adapted from *Solid Fuel Furnaces and Boilers* by John W. Bartok, Garden Way Publishing Co., Charlotte, Vt., 1982. Available from Storey Communications, Pownal, Vt., 05261.)

position, creosote begins to accumulate when the chimney flue and internal boiler surface temperatures drop below approximately 240°F. Unburned gases from the combustion chamber condense on these surfaces and build up over a period of time. Although creosote deposits are common in all wood-burning appliances, they can be especially troublesome in solid-fuel boilers, since the firebox is water cooled. This results in internal boiler surface temperatures below the creosote condensation point during much of the operating cycle of the boiler. There are several techniques that are effective in helping to control creosote deposits:

1. Use of dry hardwoods whenever possible is recommended. Freshly cut wood often contains moisture content in excess of 50%, lowering combustion chamber temperatures when it is burned. Also, softwoods contain higher percentages of resinous materials than do hardwoods. These resins yield a great deal of smoke, a prime ingredient of creosote.
2. When outside temperatures are above 40°F, avoid burning the boiler under restricted primary air for extended periods of time. Small, intense fires burned with maximum primary air are more efficient than longer-burning ones fired

under restricted air. Also, the small intense fires will provide comfortable heating levels since building heat loss at higher outdoor temperatures is not excessive.

3. Set water temperatures high enough to maintain an interior boiler surface at temperatures that are high enough to minimize smoke condensation. These settings range between 180 and 200°F.
4. When outside temperatures are approximately 50 to 55°F, consider shutting down the solid-fuel boiler. At these outside temperatures, it is usually more efficient and less expensive to burn oil or gas for heating the home than burning either wood or coal.

Creosote Reduction Technology

The accumulation of creosote deposits within wood-burning appliances is a normal by-product of the wood combustion process. There are two technologies in common use to reduce creosote deposition: either treating the wood prior to burning it with a topical catalytic spray or by the use of a catalytic combustor located within the stove or flue pipe. The reduction of creosote can have three beneficial effects:

1. Increase available heat delivered by the appliance.
2. Reduce creosote deposition within the boiler and flue passages.
3. Reduce air pollution.

We will now examine both catalytic topical treatments and catalytic combustors used for reducing creosote formation.

Catalytic combustors. A catalytic combustor is composed of a heat-resistant ceramic honeycomb structure that is covered with a noble metal catalyst, such as platinum or paladium. A combustor of this type is illustrated in Figure 7.2. Catalytic combustors can be installed in wood-burning appliances as either retrofits or as part of a factory-equipped unit, as illustrated in Figure 7.3.

For catalytic burning to begin, the combustor must be heated to between 500 and 700°F to achieve a condition that is known as "light-off." As wood smoke comes into contact with the catalyst, it ignites at approximately 500°F, well below the 1000°F normally required for wood smoke to burn. During operation, the combustor becomes very hot, and although temperatures above 1800°F can damage the combustor, operating temperatures between 1200 and 1400°F are common. Note in Figure 7.3 that a flame guard is installed to prevent direct flame contact on the combustor surface during high-firing operation of the wood-burning appliance.

The catalytic combustors in both wood stoves and retrofit stovepipe applications are designed to operate in units that burn natural wood only. The burning of pressure-treated lumber, trash, or chemically laden wood can greatly reduce the life of the combustor. Although an occasional mistake is not likely to damage the combustor, extensive burning of material other than natural wood can result in

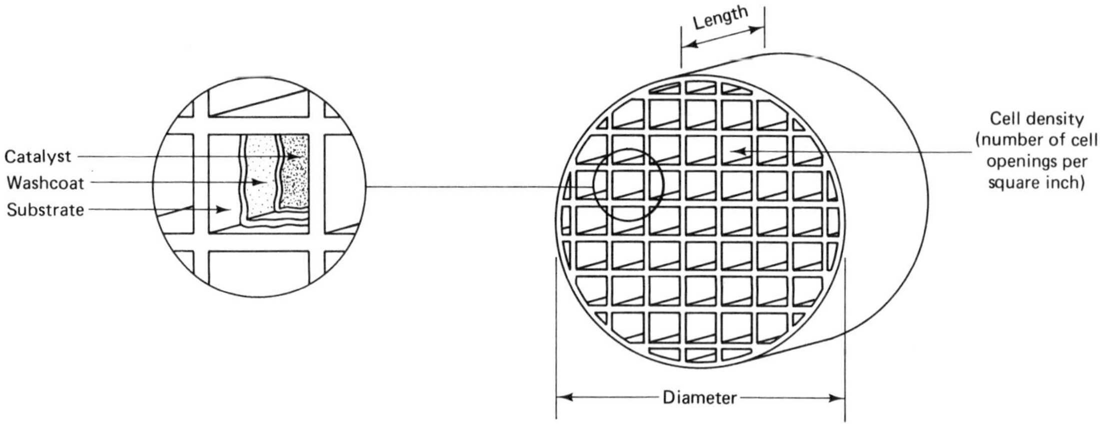

Figure 7.2 Catalytic combustor for wood-burning appliances. (Courtesy of Corning Glass Works.)

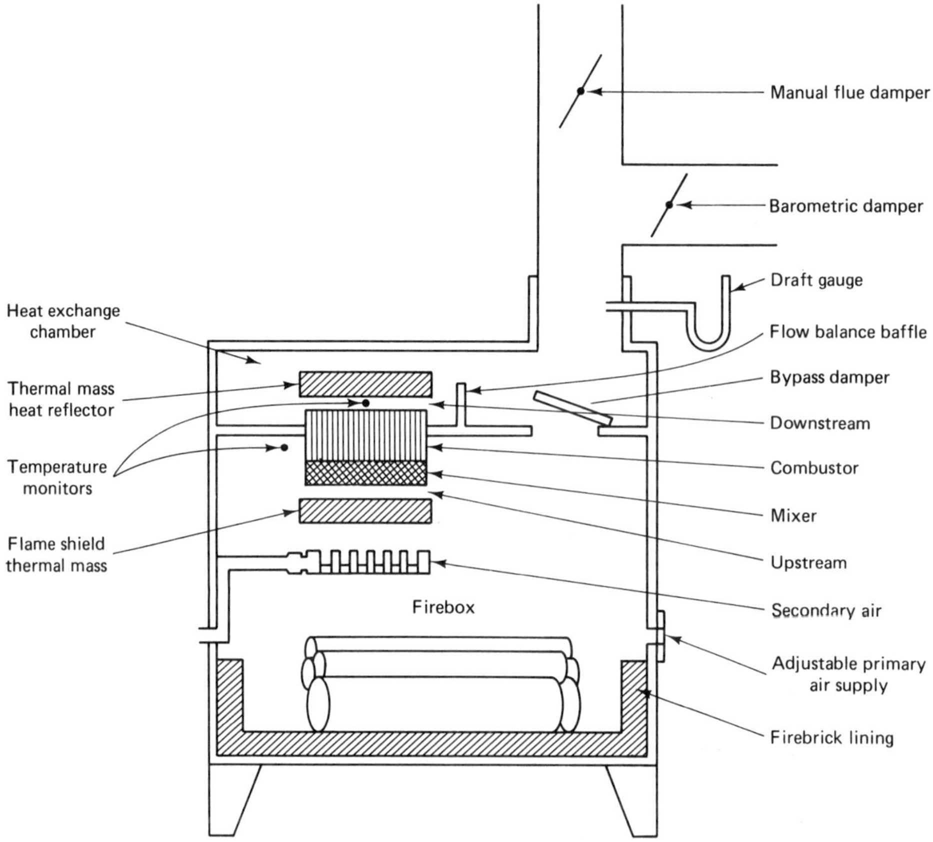

Figure 7.3 Cross section showing typical installation configuration of factory-equipped catalytic combustor wood-burning appliance. (Courtesy of Corning Glass Works.)

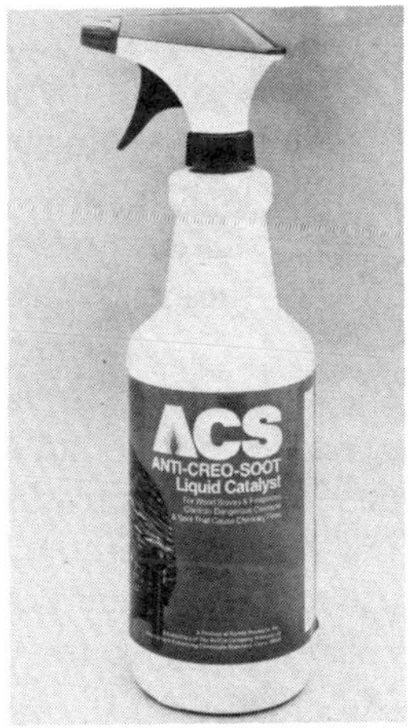

Figure 7.4 Catalytic topical chemical anticreosote spray. (Courtesy of Rolfite Co.)

either a deterioration of the catalyst or a plugging of the cells in the combustor, greatly reducing its operating efficiency and usefulness.

As the reliability and effectiveness of the catalytic combustor has increased, the number of retrofits and stoves equipped with these units has also increased. In addition, certain localities have begun to impose maximum air pollution emissions standards for solid-fuel appliances, a move that has fostered the use of catalytic combustion devices.

Topical chemical treatments. A second method to help prevent creosote buildup is through the use of special chemicals that are sprayed over the wood before it is burned in the appliance. One such chemical treatment is illustrated in Figure 7.4. These chemicals act as catalysts to enhance the oxidation of the wood, increasing the heat from the wood fire and helping to reduce some of the smoke emissions, a prime factor in the formation of creosote deposits.

While both topical treatments of the wood as well as the use of catalytic combustors can effectively reduce creosote deposition, the primary rule for efficient wood burning with a minimum of creosote formation is to burn only well-seasoned wood. Also, hardwoods are generally denser and contain fewer volatiles than do softwoods. Volatiles can be a major contributing factor to the formation of creosote and should be minimized when and where possible. These techniques, in conjunction with frequent inspection and cleaning of the chimney, will result in operating the solid-fuel appliance under maximum safety conditions, keeping the probability of chimney fires to a minimum.

BURNING COAL

Coal has been used for residential heating on an extensive basis in the United States for well over a century. Its decline in popularity, concurrent with the advent of relatively inexpensive fuel oil, natural gas, and electricity, was reversed in the early 1970s due primarily to the meteoric rise in the price of these fuels relative to coal.

Coal is classified according to its geological age. The major varieties of coal are listed in Table 7.4. Lignite, which is dark brown in color, is geologically the youngest of the coals, contains the least amount of energy per pound, and contains the highest moisture content. For these reasons, lignite is not often used as a fuel in solid-fuel boilers.

The bituminous varieties of coal are sometimes referred to as "soft coal." They contain higher heating values per pound than lignite and are used widely as fuel in industrial power plants. Where it is readily available, bituminous coal is also used as a fuel for residential heating. Although bituminous coals vary widely in physical characteristics, most contain a high percentage of sulfur, which can often be detected by the familiar "rotten egg" odor when it is being burned.

Anthracite is geologically the oldest of the coal varieties. It burns easily with relatively little smoke. Anthracite is the coal of choice for home heating since, as it is almost pure carbon, it yields a long-burning, clean fire that requires a minimum of attention. However, since the heating content, ash content, and fusion temperature of anthracite coals can often vary greatly, the purchaser should thoroughly research the characteristics of the prospective coal before it is delivered.

Coal Sizing

Each variety of coal is sold according to its mesh size, calculated according to the number of openings per inch in a wire mesh screen that it will fit through. Sizing coal is important since each solid-fuel boiler is designed to burn only a specific size of coal most effectively, based on grate configuration and size of the firebox. Although sizing techniques vary within the industry, the general classifications of coal are listed in Table 7.5.

Ash Content and Fusion Temperature

All coal contains a certain amount of fly ash. During combustion, this ash fuses into large chunks of noncombustible material commonly referred to as "clinkers." The tendency of the ash to combine is based on its fusion temperature. The higher the fusion temperature, the less likelihood there is for the fly ash within the coal to

TABLE 7.4 RANKING OF COAL BY GEOLOGICAL AGE

Coal variety	Age
Lignite (and sublignite)	Youngest
Bituminous (and subbituminous)	
Anthracite (and subanthracite)	Oldest

TABLE 7.5 ANTHRACITE COAL SPECIFICATIONS

Grade	Size (in.)
Pea	$\frac{1}{2}$–$\frac{3}{4}$
Nut	$\frac{3}{4}$–$1\frac{1}{2}$
Stove	$1\frac{1}{2}$–$2\frac{1}{2}$
Egg	$2\frac{1}{2}$–$3\frac{1}{2}$

Note: Sizes of coal are approximations.

fuse into larger chunks. Therefore, low ash content combined with high fusion temperatures are the most desirable characteristics for residential fuel coal.

Combustion Characteristics

The combustion characteristics of coal are in some ways similar to wood in that combustion takes place in distinct phases. For effective coal burning to take place, a wood fire is started within the combustion chamber using kindling in order to drive off moisture in the coal and raise the temperature within the firebox to a level sufficient to support combustion. The volatiles in the coal are then driven off and oxidized, and the remaining carbon in the coal is burned.

Because the ignition temperatures of coal are much higher than those of wood, a hot wood fire should first be established before attempting to light the coal. The ignition temperatures of common solid-fuel materials and derivatives are listed in Table 7.6.

After a well-established wood fire is present, coal should be added evenly to the firebox in a thin layer. Once the initial charge of coal has been ignited, the height

TABLE 7.6 IGNITION TEMPERATURE OF COMMON FUELS

Fuel	Ignition temperature (°F)
Paper	350
Peat	435
Wood	550
Fuel oil	560
Lignite	600
High-volatile bituminous coal	765
Low-volatile bituminous coal	870
Anthracite coal	925

Source: Solid Fuel Furnaces and Boilers by John W. Bartok, Garden Way Publishing Co., Charlotte, Vt., 1982. Available from Storey Communications, Pownal, Vt., 05261.

of the coal pile can be built up. Care should be exercised not to add too much coal before the initial charge is burning well. The combustion cycle of a coal-burning appliance is illustrated in Figure 7.5. Note that combustion of the coal bed takes place from the bottom up, with primary oxygen feeding the coal fire from below the grates. For best results, the coal bed should be a minimum of 4 to 8 in. thick. The majority of combustion occurs in the oxidation zone directly above the grates. Primary oxygen combines with the hot coal to produce CO_2 as the major by-product. The hottest part of the fire is located at the top of the oxidation zone. In the reduction zone, CO_2 is chemically reduced to carbon monoxide (CO). Carbon monoxide is colorless, odorless, and highly toxic.

From the reduction zone, the gases rise into the distillation zone where fresh coal is also added to the fire. Moisture and volatiles from the fresh charges of coal are distilled off and collect at the top of the zone. The introduction of secondary air into the distillation zone helps to complete the combustion of carbon monoxide and volatile gases.

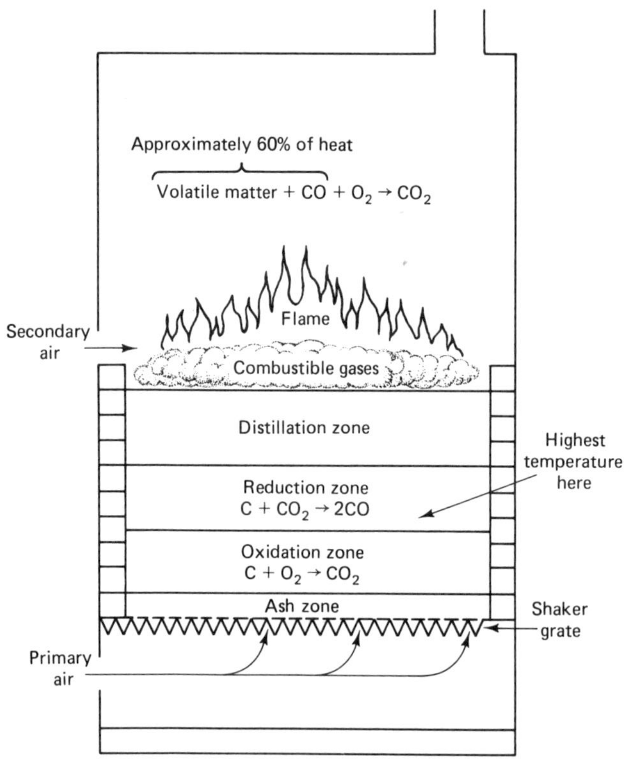

Figure 7.5 Coal combustion cycle. (Adapted from *Solid Fuel Furnaces and Boilers* by John W. Bartok, Garden Way Publishing Co., Charlotte, Vt., 1982. Available from Storey Communications, Pownal, Vt., 05261.)

SOLID-FUEL AND MULTIFUEL BOILER CONFIGURATIONS

The use of solid-fuel and multifuel central heating boilers has increased for two basic reasons: the rise in the price of conventional fossil fuels and utility-produced electricity relative to wood and coal, and the desire to move away from radiant stoves and kerosene heaters to the convenience of central heating systems. Since most homes already have a central heating system, an add-on or multifuel boiler allows the homeowner to enjoy the comforts of central heating while concentrating the majority of dirt and debris commonly associated with these systems to a basement or utility room.

Add-On Solid-Fuel Boilers

There are a variety of add-on solid-fuel boilers available. Although most of these units are domestically produced, a large number of imported appliances are also available. Prices range from a low of approximately $1100 to many thousands of dollars, depending on the type of boiler chosen. A typical add-on wood/coal boiler is illustrated in Figure 7.6. This type of unit is thermostatically operated and features a fan-induced draft system.

A cross section of the airflow pattern within a typical boiler is illustrated in Figure 7.7. In this type of design, characteristic of many units of the market, a draft fan is activated to build up the intensity of the solid-fuel fire within the boiler when the room thermostat calls for heat. When the boiler reaches operating temperature,

Figure 7.6 Add-on wood/coal boiler. (Courtesy of Royall Furnaces, Inc.)

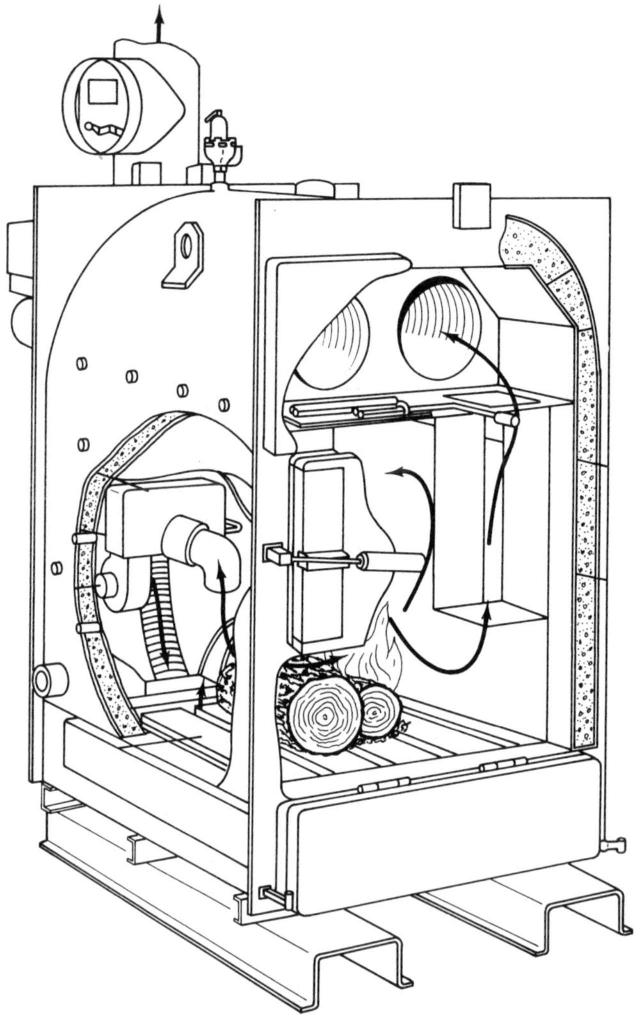

Figure 7.7 Combustion airflow pattern within solid-fuel boiler. (Courtesy of Rite-way-Dominion Mfg. Co., Inc.)

water begins to circulate throughout the heating system, raising the house temperature. When the room thermostat is satisfied, the thermostat relay turns off the draft-inducing fan, and the fire within the boiler dies down to an off-cycle standby level. This type of primary air induction system gives the boiler cycling characteristics similar to those of fossil-fuel units. For maximum combustion flexibility, the shaker grate system should be designed so as to enable the boiler to burn either wood or coal (Figure 7.8). The size of the coal to be used for best results depends on the design of the grate system. For each boiler, the manufacturer's literature should be consulted for recommended sizes of coal best suited for each application.

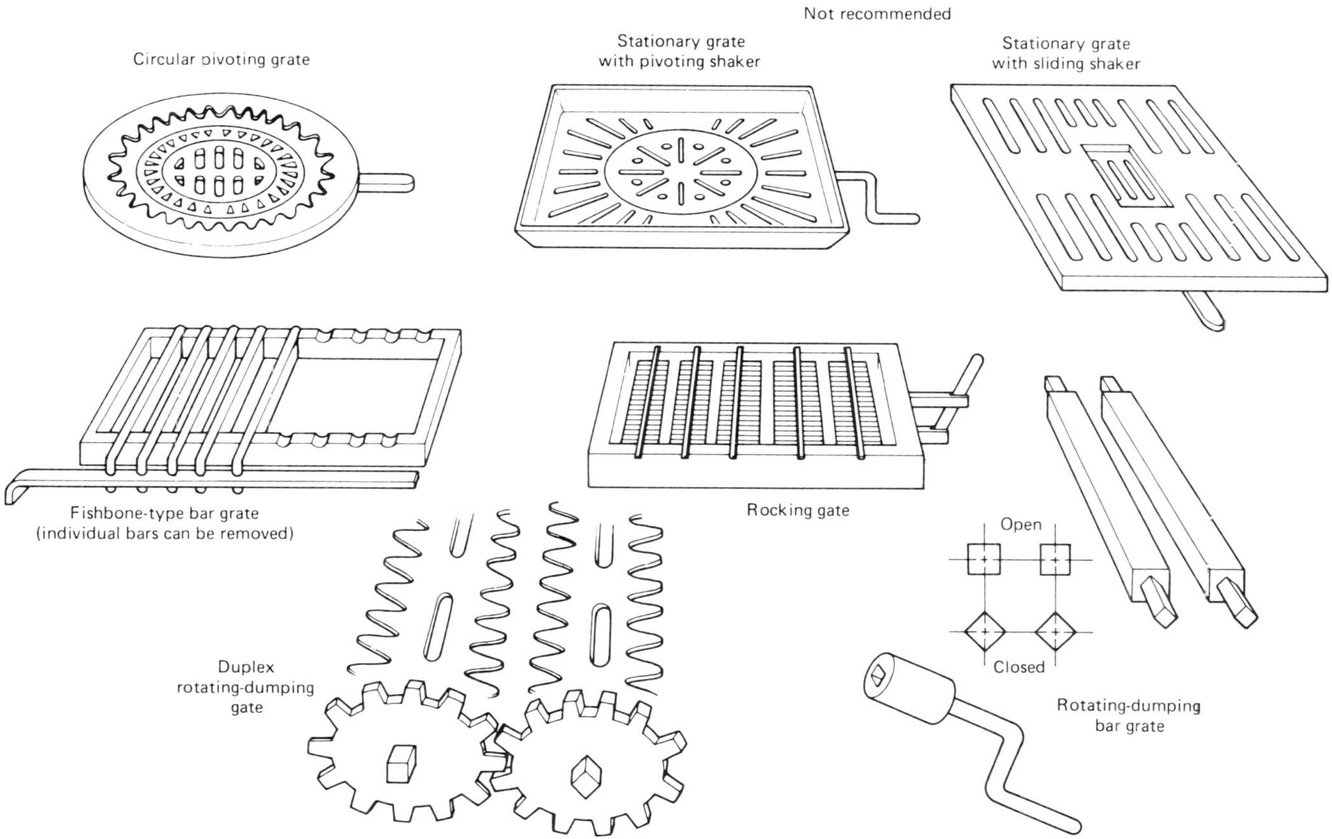

Figure 7.8 Variety of shaker grate configurations in solid-fuel appliances. (From *Solid Fuel Furnaces and Boilers* by John W. Bartok, Garden Way Publishing Co., Charlotte, Vt., 1982. Available from Storey Communications, Pownal, Vt., 05261.)

The boiler illustrated in Figure 7.9 features an immersion control that maintains water temperature at a preset level. In this design, water temperature remains fairly constant and the unit is on "standby" to allow for hot water transfer to the baseboard heating system whenever there is a call for heat.

Some add-on units employ a fire-tube design for maximizing heat transfer from the combustion chamber to the water. A series of tubes is incorporated in the boiler through which the flue gases must pass before exiting the stack pipe of the unit. In so doing, heat is transferred from these exiting gases to the water surrounding the firetubes. This configuration is illustrated in Figure 7.10.

Another type of boiler design that is available in both add-on and multifuel versions and that is somewhat different from conventional solid-fuel appliances is known as a wood gasification boiler (Figure 7.11). Gasification boilers operate differently from their conventional solid-fuel counterparts. In gasification units, an induced-draft fan supplies a large volume of air to the combustion chamber, which causes an intensely hot fire that heats up a huge mass of carefully configured and specially shaped refractory material. Once the boiler reaches operating temperature, the induced-draft fan stops, completely shutting down the fire. Because of the heat

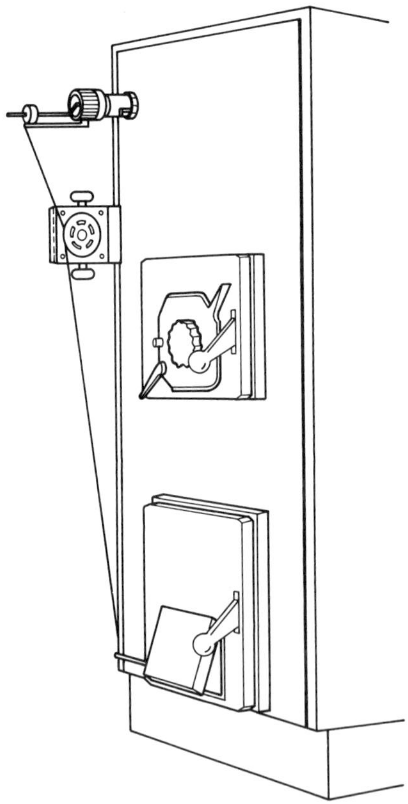

Figure 7.9 Immersion draft control to maintain water temperature levels within a solid-fuel boiler. (Courtesy of Ammark Corp.)

Figure 7.10 Firetube heat transfer boiler design. (Courtesy of Northland Corp.)

that is held in the refractory lining of the firebox, the fire can be reignited for a period of several hours whenever the boiler-water temperature drops below a predetermined level. At this level, the induced-draft fan again begins to operate, sending large amounts of air back into the combustion chamber. This reignites the fire due to the high temperature of the refractory, and the fire continues to burn until the operating temperature is again reached. This cycling continues as heat is required for both the baseboard heating system and tankless domestic hot water coil of the boiler.

Due to the intense heat developed in a gasification boiler, almost all the gases driven off from the wood are completely burned, resulting in combustion efficiencies in the neighborhood of 85%, with practically no accumulation of creosote or fly ash. Often, a firetube design is used within the combustion area to aid in complete burning of the wood gases and to improve heat transfer from the combustion chamber to the refractory lining and boiler water.

Draft Controls

There are several different methods utilized to regulate the draft over the solid-fuel fire to maintain boiler-water temperatures.

The first type, mentioned previously in our discussion of add-on solid-fuel boilers, is a fan-induced draft system which blows air over the fire chamber in a pattern determined by the baffling of the particular unit (Figure 7.12).

Another popular draft control is the immersion type. The immersion control is inserted into an immersion well in direct contact with boiler water. As water temperature changes within the boiler, a bimetallic temperature-sensing element expands or contracts, moving a lever that either raises or lowers an air intake door on the appliance, shown in Figure 7.13.

A third type of control for draft regulation is controlled by an electric motor

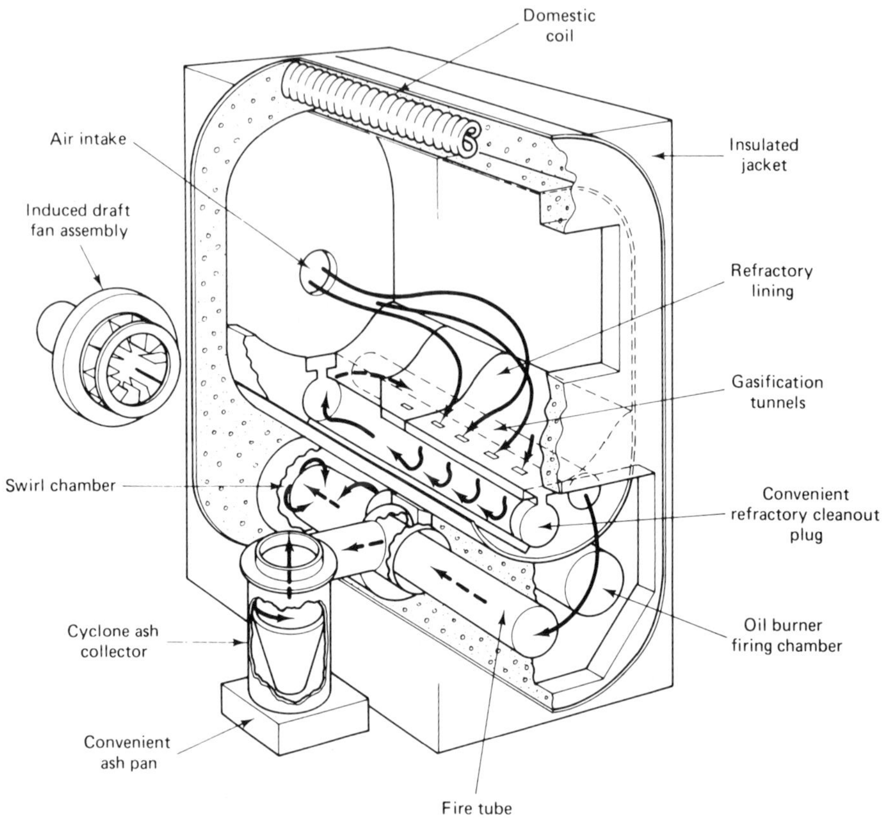

Figure 7.11 Wood gasification boiler. (Courtesy of Eshland Enterprises, Inc.)

Figure 7.12 Induced-draft fan. (Courtesy of Royall Furnaces, Inc.)

and is operated by an aquastat that monitors boiler-water temperature. Upon a drop in water temperature, the aquastat closes a relay that powers the electric motor to raise or open the draft control lever, which admits primary combustion air. When the boiler reaches operating temperature, the electrical circuit to the control is opened and the draft door will close automatically. One type of electrical draft control is illustrated in Figure 7.14.

Add-On Solid-Fuel Steam Boilers

Solid-fuel add-on boilers can also be equipped for use in steam heating systems, although their availability is more limited than that of hot water units. Modification for use in steam heating systems is done by reconfiguring the control package on

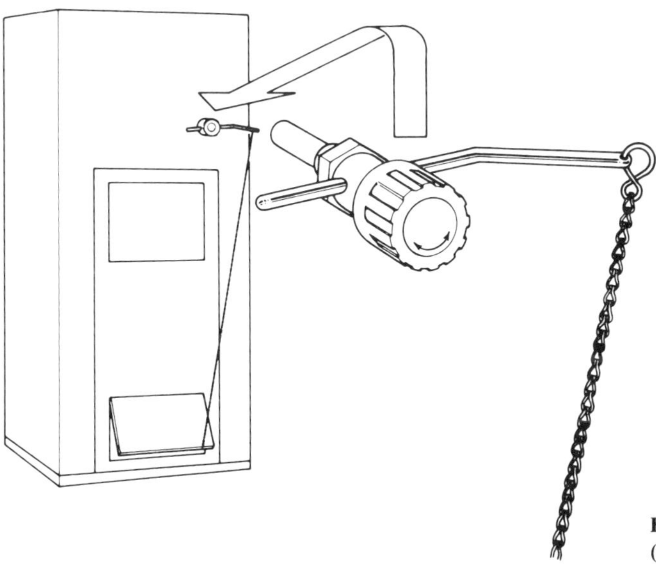

Figure 7.13 Immersion draft control. (Courtesy of Ammark Corp.)

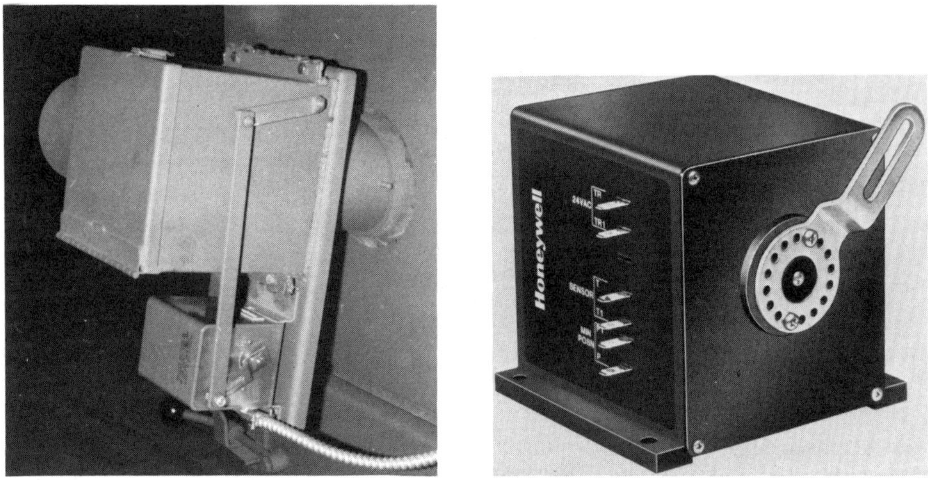

(a) (b)

Figure 7.14 Electrical draft control. [(a) courtesy of Eshland Enterprises, Inc.; (b) courtesy of Honeywell, Inc.]

the boiler with steam controls in place of the hot water operating controls. A typical add-on solid-fuel steam unit installation configuration is illustrated in Figure 7.15.

Multifuel Boilers

In both new and retrofit installations, it is sometimes desirable to install a multifuel boiler that has the capability to burn wood, coal, and either fuel oil or natural or LP gas, all with relatively high combustion efficiency. Multifuel units take up less space than do two separate units and generally vent into one chimney flue. While the cost of a multifuel unit is either usually the same as, or more expensive than, two separate units (one conventional boiler and one add-on boiler), many people prefer the convenience of utilizing only one appliance for both space heating and domestic hot water production.

Since the combustion characteristics of wood, coal, oil, and gas vary widely, two separate combustion chambers are generally employed in the design of multifuel boilers: one chamber for the wood and coal combustion, the other for the combustion of the fuel oil or gas. One such double-chamber arrangement is illustrated in Figure 7.16. Note that the two combustion chambers are engineered so that the oil combustion takes place updraft of the solid-fuel combustion. This design configuration helps prevent creosote from accumulating over the nozzle and electrode assembly of the oil burner or in the orifices of the gas jets. Both combustion chambers usually vent into one stack for effluent removal, although separate flues for each combustion chamber are common as well.

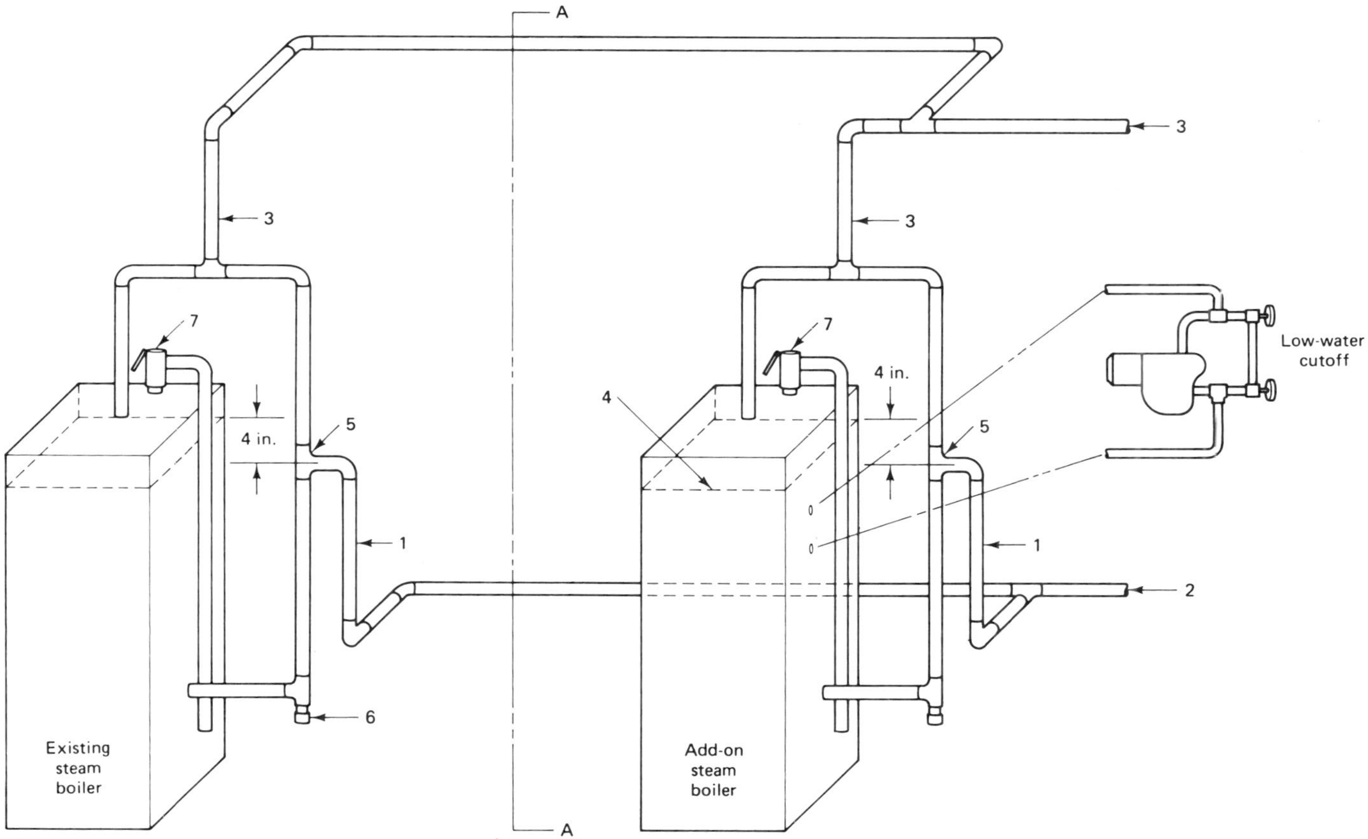

Maintain clearance of 24 in. from rear of boiler to wall or other obstruction.

Figure 7.15 Add-on solid-fuel steam boiler installation arrangement. 1, Hartford return connection same pipe size as return line; 2, return from system; 3, supply to system; 4, boiler-water level (see note); 5, the Hartford loop connection should be 4 in. below the water level; 6, boiler drain; 7, steam safety valve. *Note:* The water level of both boilers in the system, when two are used, must be the same. If one boiler has a lower water level than the other, raise the boiler with the lower level (on a concrete or steel base) so that the water level of the boilers are the same. (Courtesy of Northland Corp.)

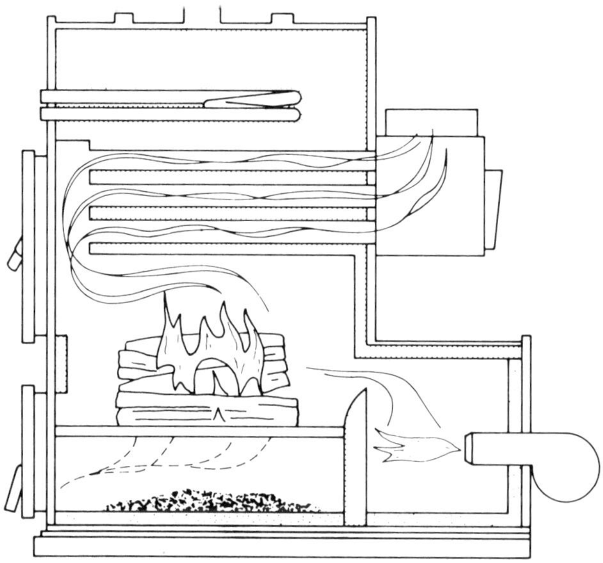

Figure 7.16 Double-combustion-chamber multifuel boiler. (Courtesy of Northland Corp.)

Control functions. The controls on a typical multifuel boiler are designed for two-stage heating: Wood or coal provides the first heating stage, or upper level of heated boiler water; oil or gas provides the second stage of heating should the water temperature drop to the second stage temperature set level.

The electrical wiring schematic for this type of control system is illustrated in Figure 7.17. The wiring diagram assumes installation of a mechanical damper on the solid-fuel boiler to control combustion air entering the appliance. This damper should be set to keep the boiler water at a temperature approximately 10 to 15°F higher than the oil burner aquastat relay. In this way, should the temperature of the water within the boiler drop to the oil burner set point, the contacts in the aquastat relay will close, firing the burner in the boiler to raise the water temperature up to the high-limit set point on the burner relay. This automatic backup feature is desirable, since there will be times that the heat loss of the house will exceed the ability of the solid-fuel unit to replace it or when the house will be unoccupied for extended periods.

Primary air in these units is controlled by fan induction, immersion, or electric draft controls. Ignition of the oil or gas unit on most multifuel units is controlled by a standard immersion-type aquastat. Also, an overheat aquastat is usually installed to prevent excessively high boiler-water temperatures (Figure 7.18).

The overheat aquastat automatically opens a zone valve or begins circulator operation to remove heated water from the boiler if the water temperature reaches the high-limit setting. The oil or gas unit is protected from high temperatures by the upper-limit control in the ignition aquastat.

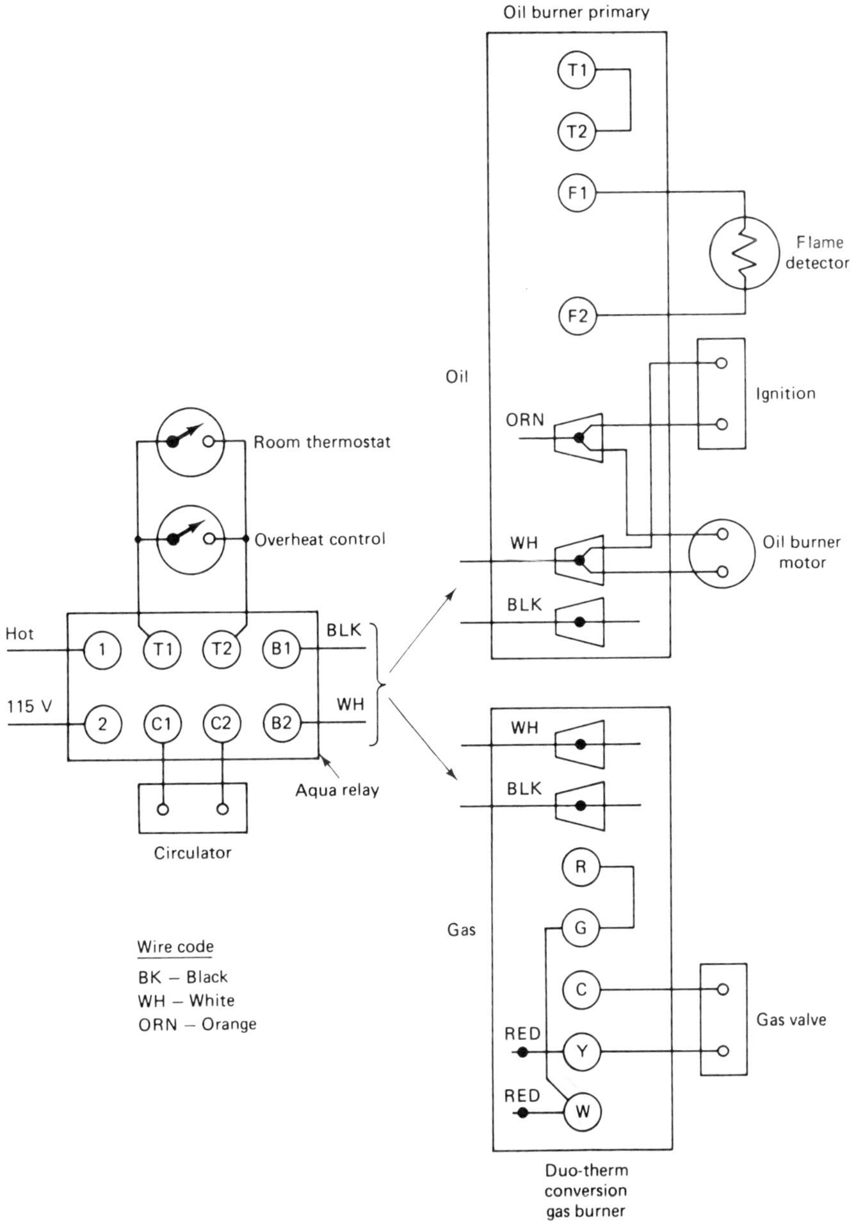

Figure 7.17 Typical wiring arrangement for two-stage control for multifuel boiler with mechanical damper. Both oil and gas wiring are shown. Use only the one that applies. (Courtesy of Northland Corp.)

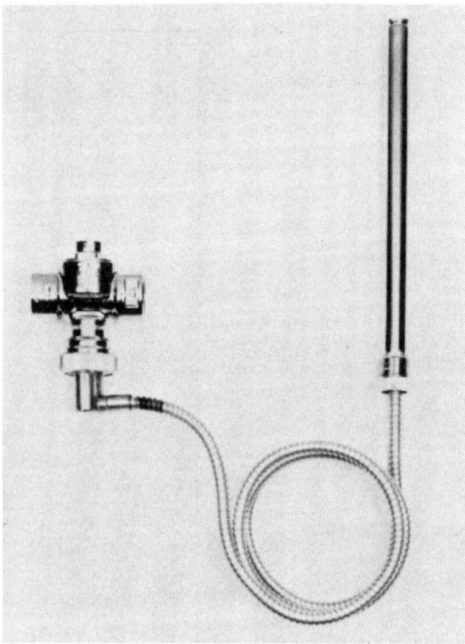

Figure 7.18 Overheat aquastat for solid-fuel or multifuel boiler. An overheat aquastat can be configured to open a zone valve, begin circulator operation, or introduce cold water into the system to keep temperatures within safe limits. (Courtesy of Ammark Corp.)

RATINGS OF ADD-ON AND MULTIFUEL APPLIANCES

Historically, regulations applied to the design and construction of solid-fuel add-on boilers have been far less stringent than those for conventional oil- or gas-fired units. In the rush to meet the demand for these units during the late 1970s and early 1980s, many were produced which were of questionable quality. Twenty-year warranties were offered by companies that went out of business within two or three years, leaving many consumers without access to replacement parts.

From these rather inauspicious beginnings, most solid-fuel boilers must now meet the same design and construction features as those of conventional units. The consumer should look at all testing procedures applied to a particular solid-fuel appliance to determine the validity of manufacturers' claims concerning the quality of their product line. In some instances, ASME labeling may be required for boiler certification, either by local building codes or by a homeowners' insurance policy.

The problem of quality has not been associated with multifuel units, however, since the standards associated with fuel oil and gas-fired appliances were carried over into the construction of multifuel boilers in order to maintain certification. Again, local building codes and insurance requirements should be researched to determine if any certification restrictions apply.

Due to the varying combustion characteristics of solid-fuel boilers, it is difficult to give them precise heat output ratings. Often, these ratings are based on the maximum number of Btu that can be delivered on an hourly basis. One modification of this procedure is to rate the boiler for the number of Btu per hour that it is capable of delivering for a specified period of time with sufficient fuel remaining in the firebox to ignite a subsequent fuel charge. For example, a boiler that is rated at 150,000 Btu/hr by the manufacturer may be able to maintain this heat output for only 2 hours before all the wood in the firebox has been consumed. If we assume that dry wood contains approximately 7500 Btu/lb, a 150,000-Btu/hr rating is misleading, since the unit will need to be refilled six to eight times each day with 20 lb of wood per charge. On the other hand, a boiler rated at 150,000 Btu/hr for 6 hours with sufficient wood for subsequent fuel ignition indicates a firebox of sufficient capacity to carry a 150,000-Btu/hr load for a reasonable period of time.

Rating techniques that specify the number of cubic feet or the square footage of a house that can be heated by a particular boiler should be viewed with skepticism. Residences differ greatly with respect to insulation characteristics and family patterns which dictate interior temperatures. Thus no single unit can be expected to provide equivalent comfort levels, as each home differs so greatly.

INSTALLATION METHODS AND PIPING PROCEDURES

There are several piping arrangements to choose from when installing solid-fuel and multifuel boilers, depending on whether or not the solid-fuel unit is an add-on or stand-alone boiler and how the homeowner wishes the system to operate. For example, add-on solid-fuel units can be installed whereby they furnish a continual supply of hot water to the backup fossil-fuel boiler. They can also be installed to furnish hot water for central heating only on a call for heat from the room thermostat. The basic piping configurations for add-on boilers are referred to as series, parallel, or gravity feed.

Series Piping

In series piping, the add-on unit is installed in a series circuit with the return lines from the baseboard heating loop and the backup fossil-fuel boiler. This piping arrangement is illustrated in Figure 7.19. The layout illustrates a single-zone heating system in which cool water returning from the baseboard heating loop is preheated by the solid-fuel boiler prior to entering the backup unit. A single circulator moves all the water through the heating system and both boilers. On a call for heat from the room thermostat, primary combustion air is admitted to the solid-fuel boiler, which raises the temperature of the water to the level required for space heating. When the low-limit aquastat setting has been reached, the system circulator comes on to move the water from the add-on unit to the backup boiler and from there through the heating circuit. Once the thermostat setting has been satisfied, the combustion air to the add-on boiler is restricted and the fire dies down to an off-cycle level, awaiting another call for heat.

Although Figure 7.19 illustrates a single-zone heating system, many situations are encountered where the add-on boiler is installed in a multizoned residential heat-

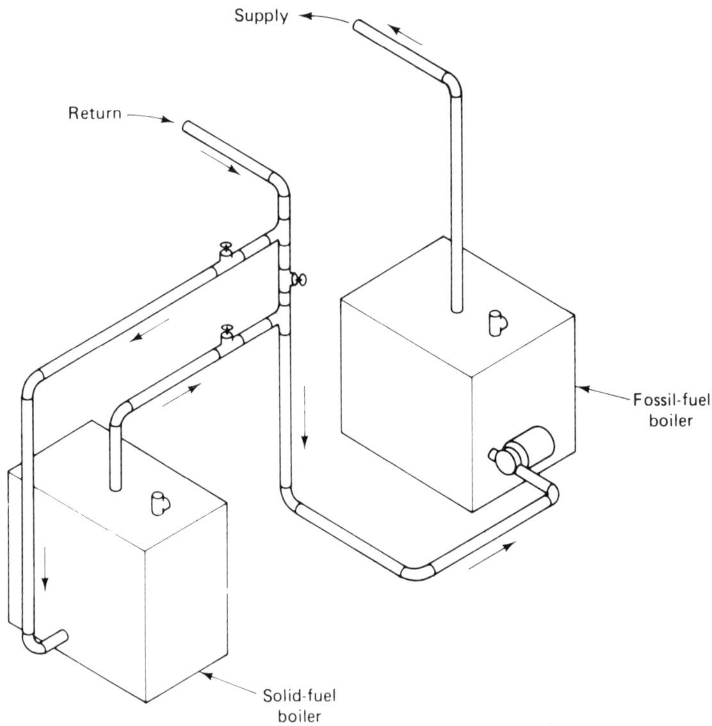

Figure 7.19 Add-on solid-fuel boiler installed in a series piping arrangement. The solid-fuel boiler preheats the return water in a series plumbing arrangement. (Courtesy of Martin L. Greenwald and Thomas K. McHugh, *Practical Solar Energy Technology,* © 1985, p. 142. Reprinted by permission of Prentice-Hall, Inc., Englewood Cliffs, NJ.)

ing system. While conceptually the same as the single-zone hookup, provision must be made to accommodate the additional heating zones if the add-on is to work properly. This is shown schematically in Figure 7.20. When using this configuration, one of the zones is selected for the series installation of the add-on boiler. This zone incorporates a bypass around the zone valve to allow free circulation of water within this zone. Operating in this manner, hot water circulates throughout the system upon either a call for heat from the zone thermostat or whenever the boiler-water temperature is above the low-limit aquastat setting on the add-on unit. The bypass around the zone valve is installed to allow for circulation of water even though the thermostat is not calling for heat, as long as the boiler-water temperature is above the low-limit setting. Thus heated water will circulate even though the zone valve is closed. This configuration is important since it allows the boiler to use the zone as a heat dump to prevent excessive water temperatures from developing. While the living area served by the zone may become warmer than usual since water will cir-

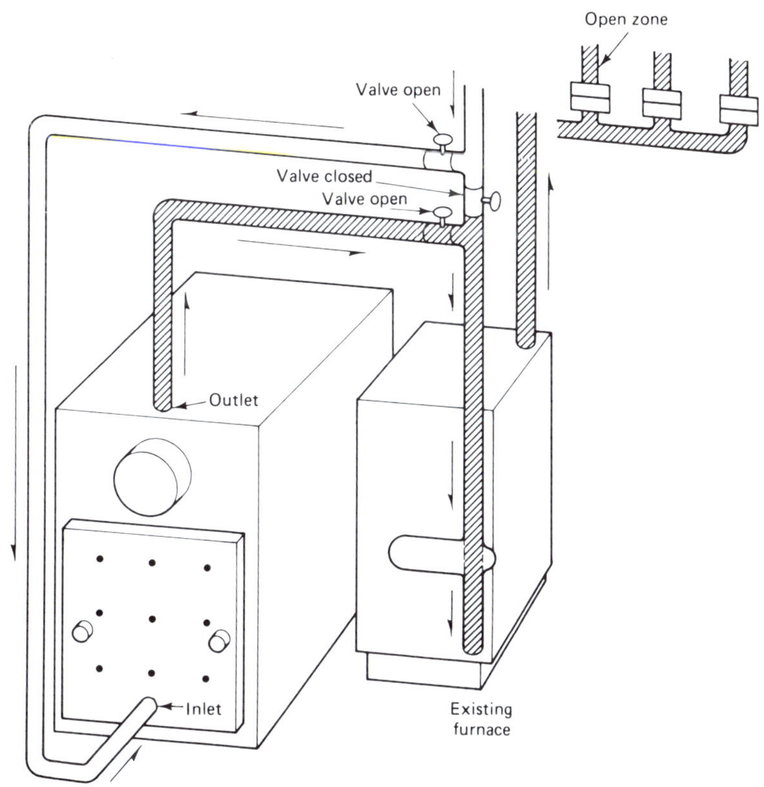

Figure 7.20 Multizone series hookup for add-on solid-fuel boiler. (Courtesy of Royall Furnaces, Inc.)

culate whenever the boiler-water temperature is above a predetermined set point, temperature control of the add-on unit is a highly desirable end result.

The series zone must be circulating hot water in order to get heat in zones other than the primary. This limitation in heating zones other than the primary series-installed zone limits domestic hot water production to times when water is circulating in the heating system.

The operating characteristics of multizone circulator systems is similar to the zone valve system. The add-on boiler controls the circulator for the zone into which it is installed. No bypass piping is necessary in this type of system since the add-on controls the operation of the zone circulator directly.

The zone that is chosen for an add-on boiler must be decided on very carefully. In most instances, the largest heating zone in the house or the zone that is used the most by the family is usually selected. These choices allow a maximum of heat distribution from the add-on system to the baseboard circuit, which is critical for the system to function properly.

Overheat control. Provision must be made in any solid-fuel heating system to control possible boiler overheating. Even though induced draft allows for cycling characteristics of the boiler and is usually designed well enough to prevent the add-on unit from overheating in most instances, this situation can sometimes occur. Even in systems in which draft control is accomplished by the use of immersion aquastats, overheating may occasionally take place. Overheating in solid-fuel appliances will usually occur for the following reasons:

1. Improperly set draft controls that allow excess primary air into the firebox and can occur on both induced-draft and immersion controls
2. Loss of electrical power, which results in no heated water circulating from the add-on boiler to the baseboard heating circuit
3. Low heating demand within the residence, which results in long standby periods in which excess heat may tend to build up in the add-on boiler
4. Improperly designed overheat protection, in which case excess heat, as it begins to build up, remains in the unit rather than being removed from it before the overheating situation becomes critical

The design of overheating protection in both single- and multiple-zone heating systems is essentially the same: The object is to move heated water throughout the heat distribution system whenever the boiler-water temperature reaches a preset high limit. Thus the system functions to remove excess heat from the boiler whether or not a particular zone is calling for space heating. This design is illustrated in Figure 7.21.

The overheat protection system is activated by the aquastat on the solid-fuel boiler. When the high-temperature overheat setting is reached, the aquastat starts a circulator which moves water from the add-on boiler through the baseboard heating circuit, dumping excess heat from the add-on boiler into the house. The boiler

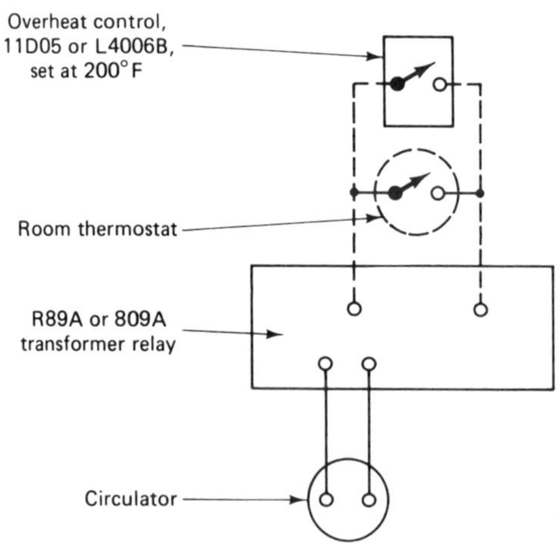

Figure 7.21 Overheat temperature control diagram. Solid line, 120 V; dashed line, 24 V.

will remain in this operating mode until the water temperature drops below the aquastat setting. Although this type of overheating control is highly effective in maintaining maximum temperatures in the boiler at safe operating limits, it can cause excessive temperatures in the overheat zone.

Aquastat and thermostat settings. For series piping to be cost-effective, control of boiler-water and house temperatures are important. Under normal conditions, the add-on boiler thermostat should be set 5 to 10°F higher than the backup unit. This arrangement allows true two-stage heating to take place: The add-on unit heats the residence prior to a call for heat by the backup system. For example, an add-on thermostat setting in this type of heating system should be set at approximately 72°F, while the backup boiler thermostat should be set between 60 and 65°F.

Proper aquastat settings in these installations are necessary to ensure that the add-on boiler furnishes the maximum amount of heat possible while the backup fossil-fuel boiler comes on only when absolutely necessary. Recall from our discussion in Chapter 2 relative to the operation of boiler aquastats that fuel ignition in a gas- or oil-fired unit will generally occur when the boiler is on standby, not calling for space heating, and the water temperature drops below the low-limit setting by a fixed differential, usually 10°F. This situation is typical in a residential hot water boiler equipped with a tankless coil.

In cold-start configurations in which boiler ignition occurs only when the thermostat is calling for heat, the burner will remain off regardless of internal water temperature until the thermostat activates the burner. Thus in a typical high-limit/low-limit aquastat system, the add-on solid fuel unit should be set so that when it is supplying hot water to the home heating system, its water temperature is hot enough to prevent the fossil-fuel backup boiler from coming on. Typical settings for these combination units are listed in Table 7.7.

To illustrate the use of Table 7.7, let us assume that a solid-fuel add-on boiler with an aquastat setting of 170 (low limit) and 190 (high limit) is interfaced with a backup oil-fired boiler which is set at 140 (low limit) and 160 (high limit). On a call for heat by the add-on solid-fuel thermostat, boiler-water temperature in the add-on unit will rise from standby temperature up to 190°F. When the water temperature reaches 170°F, the low-limit aquastat setting, the circulating pump will come on and heated water will flow from the add-on unit through the backup boiler, through the

TABLE 7.7 AQUASTAT SETTINGS FOR ADD-ON AND FOSSIL-FUEL BOILERS IN SERIES PIPING (°F)

Solid fuel		Fossil fuel	
Low	High	Low	High
180	200	150	170
170	190	140	160
160	180	130	150

baseboard radiation circuit, and back into the add-on unit. Since the backup fossil-fuel aquastat settings are 140 (low) and 160 (high), water from the solid-fuel add-on is hot enough to prevent the oil burner from coming on. Burner ignition will occur in this setup only if the water temperature from the add-on unit falls 10°F below the low-limit oil burner setting, or to 130°F, assuming that the backup thermostat is not calling for heat. While this type of control configuration will eliminate the backup boiler from operating during the vast majority of the heating season assuming an adequate supply of solid fuel, it is always ready to assume the space-heating functions of the home should the output of the solid-fuel boiler be insufficient to make up for the heat loss in the residence or should the fire in the add-on unit go out.

Parallel Piping

Parallel piping is a common configuration used to install an add-on solid-fuel unit into an existing heating system. Parallel piping differs significantly from the series design in several respects. A typical parallel piping arrangement is illustrated in Figure 7.22. In the parallel piping circuit, the add-on unit draws water from the bottom of the backup boiler, heats it, and then returns it to the top of the backup boiler in a closed loop. This parallel flow process takes place on a continual basis within the closed heating loop that connects the two boilers.

The add-on unit functions to maintain boiler-water temperature within both boilers at a predetermined level. A separate circulating pump is used to move the water between the two boilers. Two aquastats are used on the add-on solid-fuel unit. One controls the operation of the circulator to move water between the add-on and fossil-fuel unit; the other provides overheat protection.

The operating characteristics of parallel piping differ from those of the series installation. The series system responds to a residential thermostat to increase boiler-water temperature. This allows the boiler to cycle on and off during heating and nonheating cycles. In the parallel configuration, the add-on unit operates continuously to maintain boiler-water temperature. On a call for heat, the system circulator is energized to move heated water throughout the heating system. Since the parallel system is in the operating mode most of the time, this type of piping will generally consume somewhat more fuel than a system piped in the series, given equivalent heating loads.

Overheat protection. To prevent a buildup of excessive heat or pressure within the system, provision is made to move this excess heat into the heating system in a manner similar to that described for the series system. A typical overheat protective circuit is illustrated in Figure 7.23.

Gravity-Flow Piping

Hot water heating systems that do not use circulating pumps to move water throughout the heating circuit but rather rely on the relative density of hot and cold water

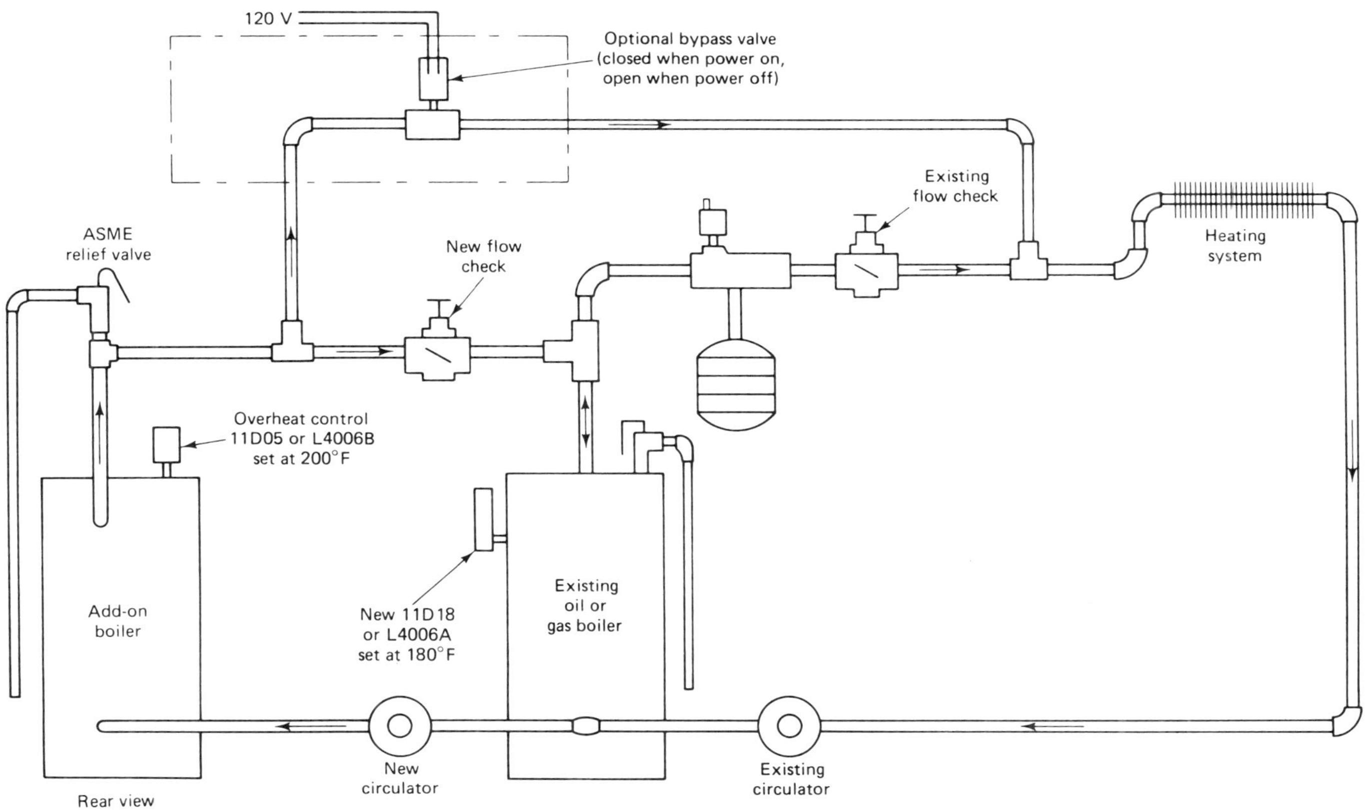

Figure 7.22 Parallel piping arrangement for add-on solid-fuel boiler. (Courtesy of Northland Corp.)

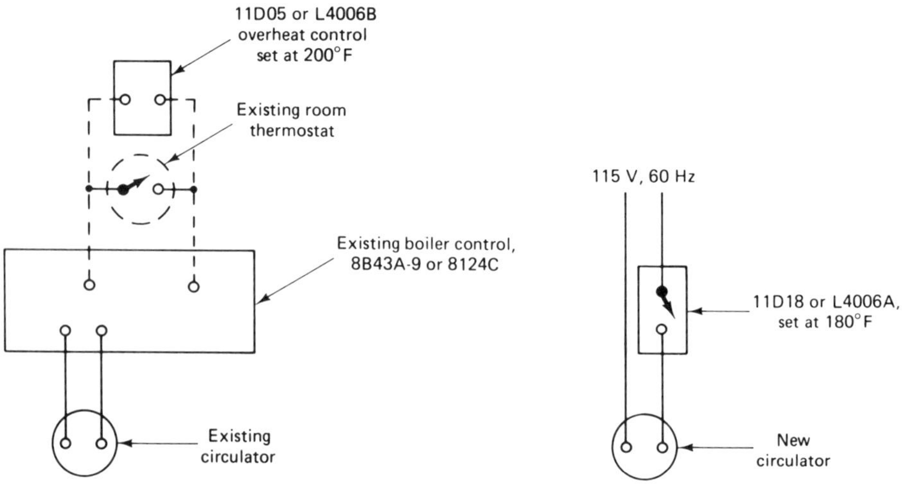

Figure 7.23 Overheat protection schematic in parallel add-on boiler installations. Solid line, 120 V; dashed line, 24 V. Suggested new settings for existing controls: high limit, 165°F; low limit, 160°F; circulator, 140°F.

for fluid movement are referred to as *gravity-flow systems.* The basic system configuration for this type of installation is described in detail in Chapter 1. Although gravity installations are not as common as either series or parallel configurations, they can still be effectively installed when a solid-fuel boiler is interfaced with a gravity-feed fossil-fuel unit. A typical piping layout for a gravity-feed system is illustrated in Figure 7.15.

In a gravity-flow system, the baseboard radiation is always located above the level of the boiler. Larger pipe sizes than those required for forced-circulation systems must be used to reduce frictional losses within the system.

Proper gravity-flow operation is more difficult to achieve than in either series or parallel configurations because overheating cannot be prevented as easily. In forced-circulation systems, the circulator is energized to remove excess heat from within the boiler, whereas gravity systems must rely on natural convection within the heating system to maintain safe water temperatures.

Gravity-flow systems control combustion air in a manner similar to boilers installed in other types of piping configurations, the most common being either fan-induced or immersion well aquastats. Circulation within the system takes place at a rate dependent on boiler-water temperature: The hotter the system water, the quicker will be the circulation; as the water within the system cools down, the rate of circulation slows down. Overheat dump zones can be installed in gravity systems that are controlled by aquastats that open a zone valve to allow for additional radiation to be placed in the heating circuit in the event that boiler overheating occurs.

CHIMNEYS

In any solid- or multifuel installation, the chimney is one of the most critical of all components; solid-fuel units place much more stress on chimneys than do their fossil-fuel counterparts. Thus chimney installations featuring both high-quality materials and installation techniques are mandatory for long-term safety and operational reliability.

Several decisions concerning a chimney need to be made when first installing a solid-fuel or multifuel boiler. All applicable building codes must be consulted to determine what, if any, prohibitions exist concerning chimney types, materials, and minimum sizes. For example, some localities may specify minimum flue diameters and building materials necessary for the chimney.

If a house already has a chimney, its suitability for use must be determined carefully and it may be used only if:

1. The chimney is not used by any other appliance.*
2. It is lined with high-temperature fireclay flue tile or code-approved lining material.
3. It is in good repair, with no loose mortar, fallen bricks, or joint sections.
4. It is the proper size and cross-sectional area to safely vent the maximum heat to be produced by the appliance.

If it is determined that a new chimney must be installed, the homeowner can decide to install one of two types of new chimneys, made from either masonry or factory-built stainless steel. Also, it must be decided whether the chimney will be installed inside or outside the house. Interior chimneys are preferable to exterior ones: They add heat to the inside of the house once they have reached operating temperature. Also, they remain warmer than exterior units during operation and tend to produce less creosote and other condensates.

Sizing the Chimney

Chimneys are sized according to the minimum diameter and height that are necessary to safely remove the maximum amount of heat that can be generated by a specific appliance. If the chimney is too small for a particular boiler, it will not possess a sufficient volume of moving air to vent heat safely from the appliance, which will cause the chimney to overheat. In general, the higher the chimney, the greater will be the chimney draft. The heat-carrying capacity of a masonry chimney as a function of the cross-sectional area of the flue is listed in Table 7.8.

*A separate flue is required for each appliance connected into a chimney. Under no circumstances should a solid-fuel or multifuel boiler ever be connected into a chimney that serves any other appliance, since hazardous conditions resulting from overheating, cross-over drafts, and creosote buildup can occur.

TABLE 7.8 HEAT CARRYING CAPACITY BASED ON CHIMNEY FLUE SIZE

Chimney flue size (nominal)	Capacity (Btu/hr)	Firing rate[a] (lb/hr)	
		Wood	Coal
6 in. round	97,000	12.1	8.1
7 in. round	131,000	16.3	10.9
8 in. × 8 in. square	150,000	18.75	12.5
8 in. round	172,000	21.5	14.3
9 in. round	218,000	27.3	18.2
8 in. × 12 in. rectangular	234,000	29.3	19.5
10 in. round	270,000	33.0	22.5
12 in. × 12 in. square	358,000	44.8	29.8

Source: Adapted from *Solid Fuel Furnaces and Boilers* by John W. Bartok, Garden Way Publishing Co., Charlotte, Vt., 1982. Available from Storey Communications, Pownal, Vt., 05261.

[a]Assume 8000 Btu/lb wood, 12,000 Btu/lb coal.

The cross-sectional area of the chimney should never be smaller than the cross-sectional area of the stovepipe or venting collar on the appliance. For example, a boiler with a 7-in.-diameter stovepipe collar requires a chimney with a minimum diameter of 7 in. Minimum requirements for chimney area and height are recommended by the appliance manufacturer for each of the units that they produce. These recommendations should be adhered to for proper operation of the unit.

Draft Measurements

The term "draft" applied to a chimney is a unit of measurement that is used to calculate the negative or evacuative pressure of chimney air currents. Draft is measured in inches of a raised column of water within a U-shaped tube called a *manometer.* A typical draft gauge or manometer, together with recommended procedures for performing draft measurement tests on chimneys, are described in detail in Chapter 4.

The draft requirements for wood and coal units differ from those requirements for conventionally fueled appliances. For example, oil-fired boilers usually require a draft of 0.02 in. of water column (w.c.) over the fire, while coal units burn most efficiently with a draft of between 0.05 and 0.06 in. w.c. Before venting any solid-fuel or multifuel boiler into a chimney, it is good practice to check chimney draft with a manometer to ensure that the draft falls within the manufacturer's recommended limits.

Curing Common Chimney Draft Problems

Most situations in which poor draft occur can usually be rectified once the cause(s) have been isolated. The first step in resolving a draft problem is measurement of the cross-sectional area of the flue. Chimneys with excessively large flues that vent solid-fuel boilers with a relatively small (8 in. or less) stovepipe often have poor draft. This occurs because the flue gases cool before they can be evacuated from the chimney due to the large volume of air present in the chimney. This problem is most often encountered in older homes in which one or two central chimneys vented several large fireplaces and wood stoves.

A physical inspection of the chimney should be made prior to any appliance installation. Figure 7.24 illustrates some of the problem area associated with poor draft and hazardous operating conditions of a typical masonry chimney.

Sometimes the chimney may continue to exhibit poor draft even after all efforts to isolate specific causes for the problem have failed. One procedure to improve poor draft involves the installation of a small draft-inducing fan into the appliance venting system. A small draft-induction fan is illustrated in Figure 7.25. These fans are available in a variety of sizes to fit most stove and venting pipes and will maintain a positive flow of air in the chimney for proper draft conditions.

At times, poor draft can be the result of continual downdraft air currents in the chimney due to atmospheric conditions or geographical configurations surrounding the chimney, such as trees, hills, and surrounding roof projections. Special chimney caps have been designed to help alleviate these problems and act as draft inducers while preventing downdrafts from entering the chimney flue. One such cap is illustrated in Figure 7.26.

Masonry Chimneys

Chimneys constructed from masonry materials are by far the most common. The cross section of a typical masonry chimney is shown in Figure 7.27. The placement of the flue liner should allow for a space between it and the surrounding concrete block to accommodate expansion and contraction of the flue liner during operation. This prevents the exterior block from cracking due to stress that would otherwise be caused by the flue liner being in solid contact with the surrounding block. During construction of the chimney, care must be taken to ensure that this space remains free of mortar, although it is sometimes filled with a high-temperature thermal insulating material to help maintain chimney temperatures and lessen the tendency for condensation of flue gases, which would produce excessive amounts of condensates.

Since rapidly fluctuating chimney temperatures and the corrosive action of the flue gases will accelerate the decay of brick and mortar, flue liners are mandatory for all masonry chimneys. It was common practice many years ago to construct masonry chimneys without internal flue liners. This practice often resulted in chimney fires that would destroy a house due to the spread of the fire from within the chimney to the adjoining wooden framing members, since there was no flue liner present to help contain the chimney fire. Flue liners are made from high-temperature fire clay, are highly resistant to corrosion under normal operating conditions, and greatly reduce the risk of chimney fire damage in a home when they are properly installed.

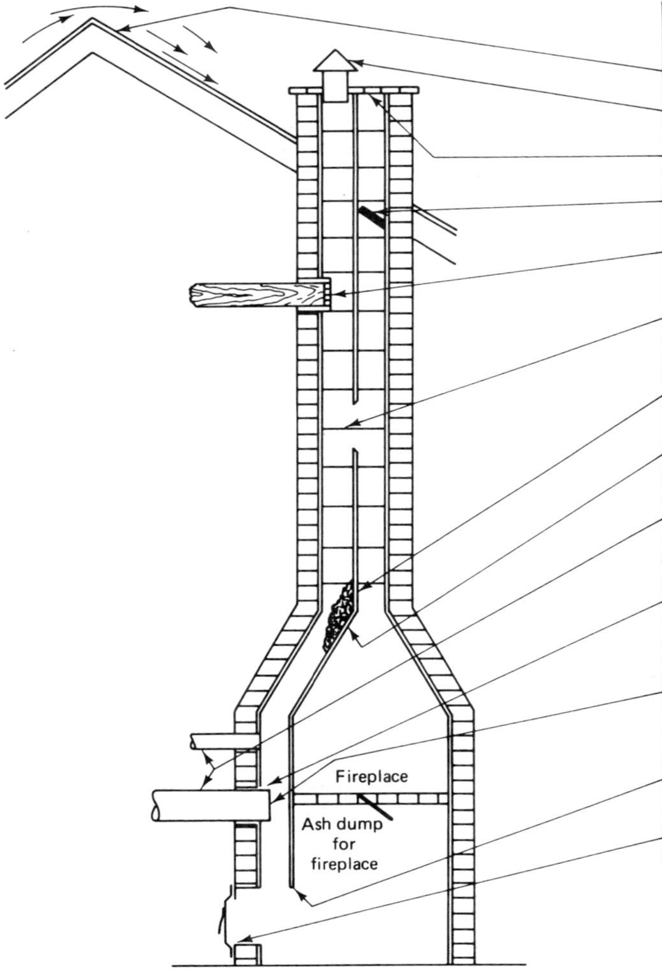

Common Chimney Troubles and Their Corrections

Troubles	Examination	Corrections
Top of chimney lower than surrounding objects	Observation	Extend chimney above all objects within 30 ft
Chimney cap or ventilator	Observation	Remove
Coping restricts opening	Observation	Make opening as large as inside of chimney
Obstruction in chimney	Can be found by light and mirror reflecting conditions in chimney	Use weight to break and dislodge
Joist projecting into chimney	Lowering a light on extension cord	Must be handled by a competent brick contractor
Break in chimney lining	Smoke test-build smudge fire blocking off other opening, watching for smoke to escape	Must be handled by a competent brick contractor
Collection of soot at narrow space in flue opening	Lower light on extension cord	Clean out with weighted brush or bag of loose gravel on end of line
Offset	Lower light on extension	Change to straight or to long offset
Two or more openings into same chimney	Found by inspection from basement	The least important opening must be closed, using some other chimney flue
Loose-seated pipe in flue opening	Smoke test	Leaks should be eliminated by cementing all pipe openings
Smoke pipe extends into chimney	Measurement of pipe from within or observation of pipe by means of a lowered light	Length of pipe must be reduced to allow end of pipe to be flush with inside of tile
Failure to extend the length of flue partition down to the floor	By inspection or smoke test	Extend partition to floor level
Loose-fitted clean-out door	Smoke test	Close all leaks with cement

Figure 7.24 Common chimney problems and corrective procedures. (Courtesy of R.W. Beckett Corp.)

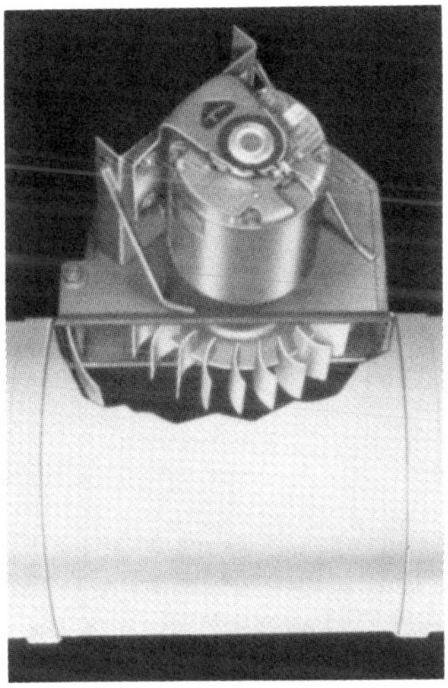

Figure 7.25 Draft induction fan. (Courtesy of Tjernlund Products, Inc.)

To support the weight of the masonry chimney and prevent sagging and stress cracks, concrete footings that go down approximately 18 in. below the frost line should be used. The footings should extend approximately 8 in. beyond the outside dimensions of the chimney on all sides, with a minimum thickness of 12 in.

Local building codes should be consulted for reference to specific guidelines that may apply to footing dimensions and other masonry chimney specifications.

Smoke pipe connections. One potential safety hazard in any chimney exists at the point where the connection from the solid-fuel boiler stovepipe meets the chimney. This location is sometimes referred to as the *chimney breeching*. The smoke pipe from the appliance should enter the chimney horizontally and should be flush with the interior lining of the chimney (Figure 7.28). Note that a thimble, or flue

Figure 7.26 Special chimney cap to help prevent downdrafts and improve chimney draft. (Courtesy of Improved Consumer Products, Inc.)

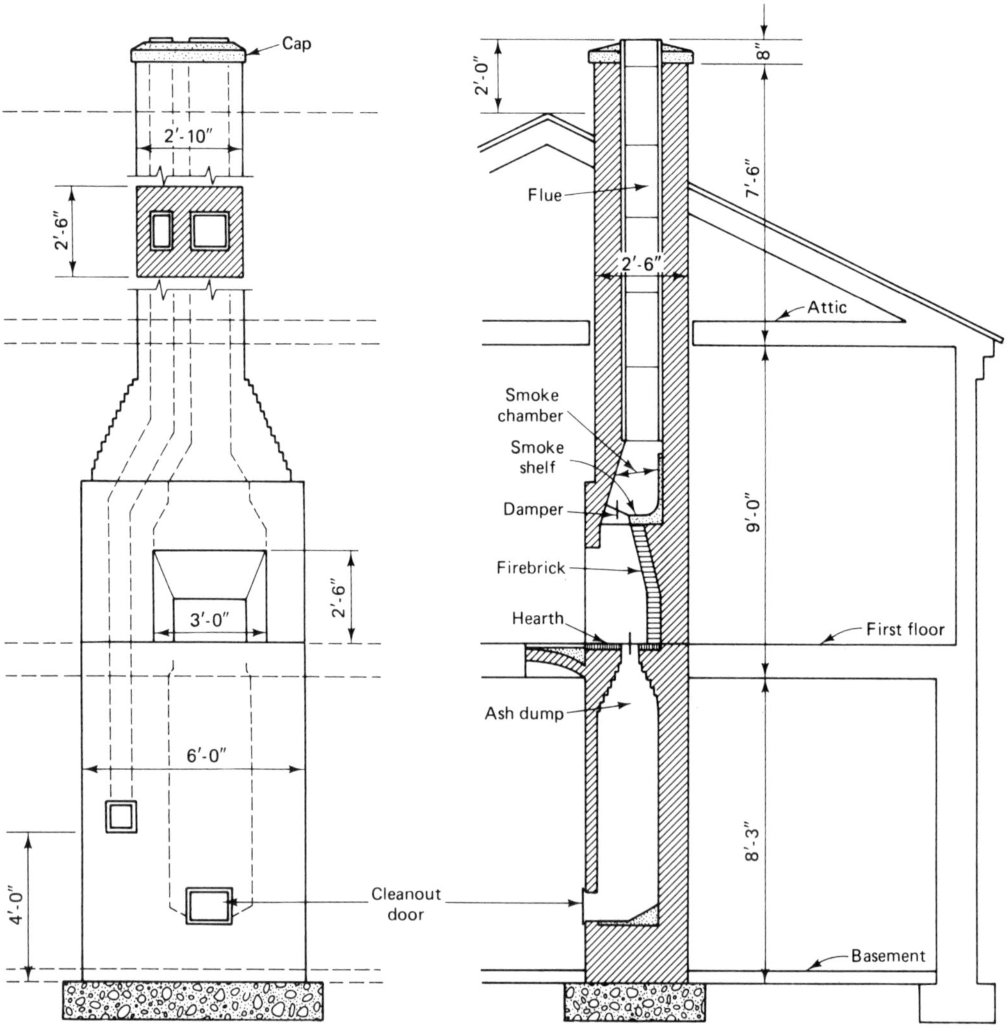

Figure 7.27 Masonry chimney construction for an entire chimney such as is commonly built to serve the house-heating unit and one fireplace. (Courtesy of U.S. Department of Agriculture, *Farmers Bulletin*, revised 1971.)

ring, is used to make this breeching connection. Thimbles are available in a variety of diameters to match the size of the chimney flue. Care must be taken to ensure that adequate clearances exist between the thimble, chimney, and all combustible surfaces. Also, a minimum distance of 36 in. should be maintained between the smoke pipe of the boiler and all combustible surfaces. All critical clearances should be checked for compatibility with local building codes. Smoke pipe should be a minimum of 24 gauge for long-term durability and safety. The junction of each pipe

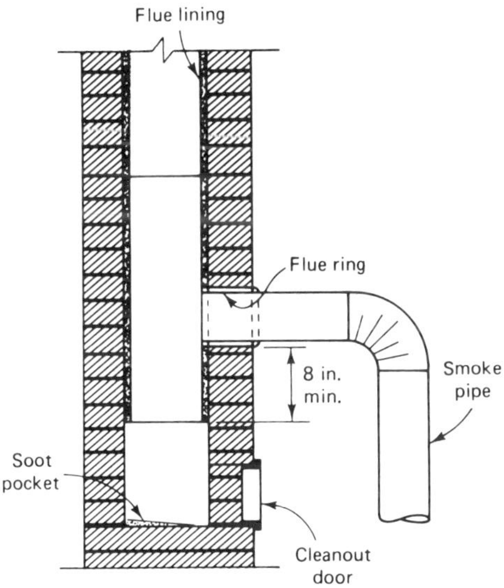

Figure 7.28 Chimney breeching connection. Note that the horizontal pipe is flush with the inside of the flue liner of the chimney. (Adapted from *Farmers Bulletin 1889,* U.S. Dept. of Agriculture, 1971.)

section should be secured with four screws and sealed on the inside of the pipe with high-temperature furnace or stove cement. The crimped ends of the pipe should face toward the solid-fuel appliance to facilitate creosote condensation to return back into the boiler, rather than to drip out of the pipe through an open crimped joint in the pipe (Figure 7.29).

Checking the chimney for leaks. Before any chimney is placed in service, it must be determined if it is free of imperfections that could result in unsafe operating conditions. A preoperating inspection of the chimney can prevent serious

Figure 7.29 Crimped end of stovepipe facing toward the solid-fuel unit. This arrangement facilitates creosote runoff back into the boiler, where it can be burned and prevents dripping of creosote on the floor.

problems from occurring later. Initial inspection should focus on all the joints in the chimney to determine if there is any loose or broken mortar, often indicated by dark stains on the exterior of the chimney. Also, the interior of the chimney should be inspected for fallen bricks, cracked or broken flue liners, or other obstructions that might interfere with proper operation.

Conducting a smoke test of the chimney should be done only if the initial inspection of the chimney indicates that it is in sound condition. A small smoky fire made up of grass or leaves should be started in a wire screen basket and placed in the bottom of the chimney, either inside the clean-out door or at the chimney breeching connection. All air intakes on the chimney as well as the top of the chimney should be sealed with wet burlap or canvas. The chimney should be inspected to see if any smoke escapes through the joints. If a finished wall of plaster or sheetrock butts up against a chimney wall, several test holes, approximately 1 to 2 in. in diameter, should be drilled through the finished wall to inspect the condition of the exterior of the chimney to determine if the finished walls are camouflaging decaying chimney joints. If problems are suspected, sections of a finished wall may have to be taken down in order to inspect the chimney more carefully.

Prefabricated Metal Chimneys

Factory-built chimneys that are manufactured from high-quality stainless steel alloys are available in a variety of configurations suitable for most residential applications. Several differences exist between prefabricated stainless steel chimneys and their masonry counterparts. On a comparative-height basis, stainless steel chimneys can sometimes be much less costly than masonry units, and their installation can usually be completed in one or two days. Both interior and exterior prefabricated chimneys are illustrated in Figure 7.30.

Interior chimneys are preferable to exterior units because interior chimneys are less prone to venting difficulties. The location of the chimney becomes important with prefabricated units that vent solid-fuel appliances. Excessively cold ambient temperatures can result in poor chimney draft as a result of rapid cooling of the vent gases, accompanied by high accumulations of creosote condensate in chimneys installed outside the home. In some instances, condensates such as sulfuric acid (from coal) and chlorides (from plastic trash), and high amounts of creosote, can accelerate chimney deterioration. Interior chimneys, on the other hand, will tend to have higher drafts and yield less creosote and other condensates. They will also operate safely for long periods because the flue gases tend to remain above the condensate dew point.

The minimum height and clearances of the chimney above the roof line are illustrated in Figure 7.31. Always consult local codes and building inspection departments for the applicable requirements.

Following these guidelines helps to minimize downdrafts from the roof and provides the minimum clearances necessary for chimney exhaust to be directed away from the roof of the home.

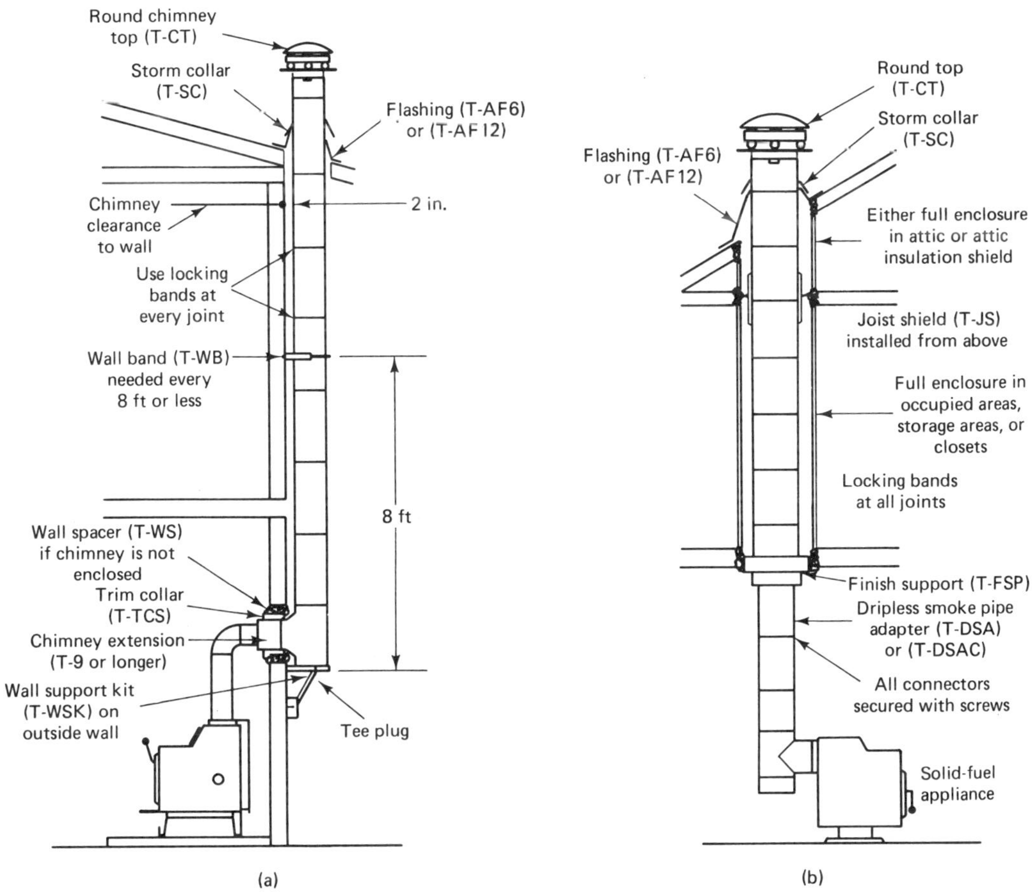

Figure 7.30 Interior and exterior prefabricated stainless steel chimneys: (a) basement heater with tee-supported outside chimney; (b) wood heater with chimney through an upstairs enclosure. (Courtesy of Selkirk Metalbestos Corp.)

Testing and certification procedures. Factory-built chimneys are subjected to tightly controlled testing procedures under the auspices of the Underwriters' Laboratories in both the United States and Canada. There are, at present, several different test procedures directed toward ensuring the operational reliability and safety of factory-built chimneys, depending on the type of appliance that will be vented into them. These test procedures are reviewed and revised periodically. Although specific tests vary, the usual procedure for testing factory-built chimneys involves exposing the chimney to specified flue gas temperatures for a measured period of time and then examining it to determine if the integrity of the unit has been maintained. Typical test temperatures are 1000°F for 4½ hours, then 1400°F for 1 hour, and then 1700°F for 10 minutes. The new (1984) HT standard requires a high temperature of 2100°F for 10 minutes repeated three times in succession.

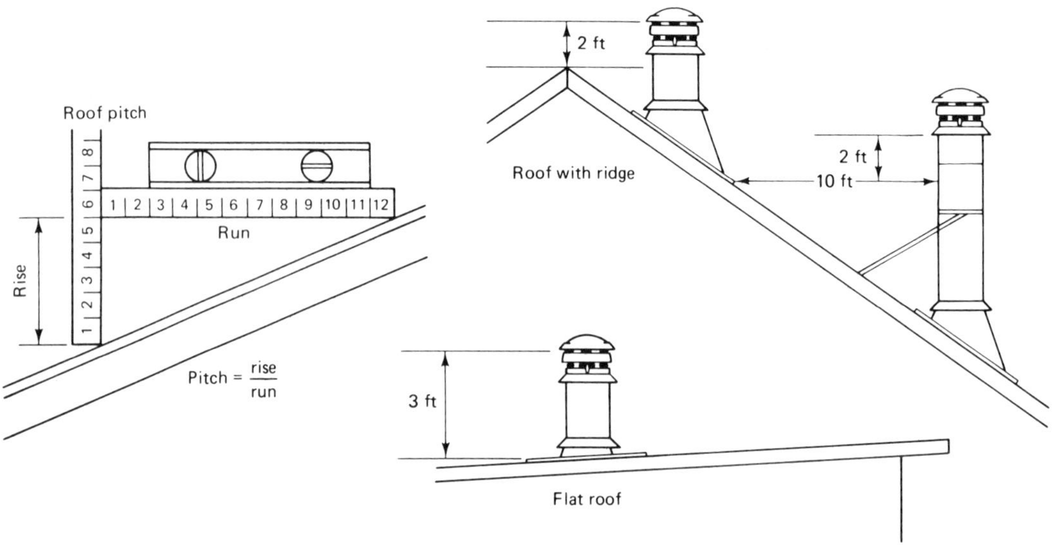

Figure 7.31 Minimum height and chimney clearance for safe operation. (Courtesy of Selkirk Metalbestos Corp.)

Due to the varying nature of these test procedures, questions regarding the UL listing of chimneys should be directed to the various Underwriters' Laboratories around the country or to the National Fire Protection Association in Quincy, Massachusetts.

Installation. For obvious safety reasons, it is extremely important that manufacturers' installation instructions be followed when installing any prefabricated chimney. Most of these units can be installed in a relatively short period. Based on the placement of the chimney, an assortment of specialized fittings are available to meet almost every type of installation prerequisite.

Although these units are well insulated and allow for relatively close distances to combustible surfaces during operation (usually 2 in.), local building codes should be consulted to ensure that the installation of the chimney will be in code compliance. Manufacturers' literature should be consulted to obtain the necessary support structures for the chimney as well as specific installation techniques recommended for safe installations. A finished prefabricated stainless steel chimney is represented in Figure 7.32.

Maintenance Procedures for Masonry and Factory-Built Chimneys

All chimneys, regardless of the quality of materials used in construction or the skill applied to their installation, are potential killers if not properly installed and main-

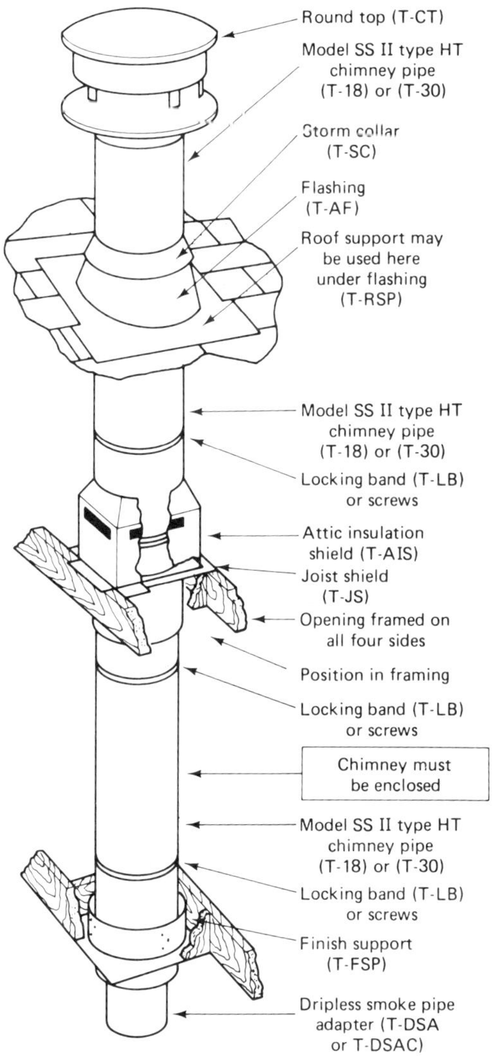

Figure 7.32 Installed factory-built stainless steel chimney. Minimum clearances to all combustible surfaces as required by the chimney manufacturer and local codes should be strictly followed. (Courtesy of Selkirk Metalbestos Corp.)

tained. Frequent ongoing maintenance procedures are mandatory when chimneys are used to vent solid-fuel and multifuel appliances, whereas gas- and oil-fired units usually require only an annual inspection and cleaning.

As a general rule, chimneys that vent solid-fuel and multifuel boilers should be inspected and cleaned on a regular basis. Heavy usage may require biweekly cleaning or inspection. At the end of each heating season in the springtime, the chimney should be thoroughly cleaned and oiled.

Creosote deposits can be particularly severe with wood-burning boilers, whether the chimney is located inside or outside the home. To prevent excessive

Figure 7.33 Chimney cleaning brushes with accessory rods. (Courtesy of Worcester Brush Co.)

creosote deposition, a prime cause of chimney fires, regular (biweekly or monthly) cleaning of the flue will keep creosote deposits at a safe level. Also, a failure to remove creosote can result in a significant loss of chimney draft as the cross-sectional area inside the chimney is reduced. Chimney brushes, available in a variety of flue sizes, are used for this purpose (Figure 7.33).

When cleaning chimneys that vent coal-burning appliances, steel bristle chimney brushes should not be used since they can generate sparks within the flue liner as the brush is moved up and down the chimney. Sparks that are created within the chimney in this way can spontaneously ignite with the coal dust that is produced, resulting in explosions of considerable magnitude. Also, a semiannual inspection of the chimney should be undertaken to help detect early signs of chloride deterioration due to the salts usually found in coal, thus preventing hazardous operating conditions. Following these procedures, most chimneys can be expected to provide years of safe, trouble-free service.

TANKLESS COILS

There are numerous ways in which tankless coils can be configured in heating systems that feature solid-fuel add-on and multifuel boilers. The specific piping arrangement for the coil is dependent on the basic layout of the heating system and the components in the system. The most common tankless coil arrangements used with solid-fuel appliances are:

1. Backup fossil-fuel boiler with coil—add-on no coil
2. Backup fossil-fuel boiler no coil—add-on with coil
3. Backup fossil-fuel boiler with coil—add-on with coil
4. Backup fossil-fuel boiler with coil and aqua booster—add-on solid-fuel boiler no coil
5. Solar domestic hot water system with tankless coil

Whether or not to include a tankless coil on a solid-fuel add-on or multifuel boiler depends on the answers to the following questions:

1. Does the present heating system contain a tankless coil within the fossil-fuel boiler?
2. Does the homeowner wish to turn off the fossil-fuel heating system during the summer months to conserve fuel, thus requiring an additional means for producing domestic hot water during the nonheating season?
3. Does the residence have a solar domestic hot water heating system, or is the installation of one planned in the near future?
4. Will the homeowner be relying on the add-on solid-fuel boiler for the majority of the residential space-heating requirements, or will the add-on boiler merely furnish supplemental heat on an occasional basis?
5. Which type of piping configuration will be used to install the add-on boiler (series, parallel, or gravity)?
6. What are the chemical characteristics of the residential water supply system?

To illustrate how the answers to these questions can affect the operation of a tankless coil and thus affect decisions concerning their purchase and installation, consider the following residential heating situation. A homeowner has decided not to rely on the tankless coil in the fossil-fuel boiler for domestic hot water production during the nonheating season. In this instance, a separate hot water heater will be required to produce domestic hot water when the boiler is not in operation. Also, depending on the type of domestic hot water heater chosen for this application, it might be more cost-effective to use the water heater year round rather than relying on the tankless coil for hot water, even during the winter heating months.

Chemical Characteristics of the Home Water Supply

The chemical characteristics of the residential water supply might indicate excessive amounts of dissolved minerals, which could cause continual coil clogging. Mineral and chemical content of the water can be determined by the use of an inexpensive test kit designed for this purpose, shown in Figure 7.34.

Figure 7.34 Water testing kit. Most test kits of this type are stocked to test water hardness, iron content, and pH values. (Courtesy of Hach Co.)

Often, companies that sell and install water softeners will test a sample of water for the homeowner at no charge. A sample bottle of water is drawn from a tap within the home and tested. A detailed printout highlighting all the pertinent chemical characteristics of the water supply is sent to the customer together with suggestions on how best to treat impurities and dissolved minerals. A typical printout of this type is shown in Figure 7.35.

Piping Configurations

The piping configuration of an add-on boiler should be taken into account when considering the installation of an optional tankless coil within the unit. Series add-on boilers may circulate hot water throughout the heating system and backup boiler only in response to a call for heat by a room thermostat, limiting hot water production to heating cycle "on" times only. Thus, if domestic hot water demand is fairly constant throughout the day and the backup fossil-fuel boiler contains the only tankless coil in the system for domestic hot water production, an add-on unit piped in series will not be able to furnish as much hot water as its parallel piped counterpart. To gain a clearer understanding of how the tankless coil can operate within the heating system, let us examine each of the piping configurations in more detail.

OWNERS NAME & ADDRESS George McNally

DEALERS NAME & ADDRESS Shawanga Valley Energy Systems

WHOLESALERS NAME & ADDRESS Ral Pump - Box 547, Middletown, NY 10940

INFORMATION TO BE ANSWERED IN FULL:

Source of Water: City ___ Well X Other ___ Pump Capacity ___ Pressure Tank Size
Water Pressure 40 lbs. Family consist of 6 people. No. Bathrooms 2 Type of piping: Copper ___
Iron ___ Other ___ Automatic Washer: Yes X No ___

VISUAL EVIDENCE CUSTOMER COMPLAINT:

1. ___ Blue discoloration of fixtures.
2. X Red or brown discoloration of fixtures.
3. ___ Pipes are eaten away or corroded.
4. ___ Pipes clogged or filled with deposits.
5. ___ Rotten odor.
6. X Soap will not lather well.
7. ___ Glassware dries cloudy.
8. ___ Water becomes cloudy.
9. ___ Water filled with dirt or silt.
10. ___ Bad odor.
11. ___ Other ___

ANALYSIS:

TOTAL HARDNESS: 15 Grains per gallon - Total hardness is the mineral matter in water that forms scale in piping, and causes excessive soap consumption, damages plumbing, burns out heater coils, and rots clothing. Water over 4 grains hard should be softened.

IRON: .5 Parts per million (ppm) - Iron in water stains plumbing enamel, discolors clothing, imparts bad taste and odor in food, and quickly rusts out water heaters and piping. Iron removal filter required where iron content is excessive.

pH 7.8 pH is the index for determining the alkalinity or acidity of the water. Acid waters attack metals and cause corrosion in copper tubing and copper plumbing parts. Water above 7.0 is alkaline - below 7.0 is acid.

OTHER: Note -- This analysis assumes the water suitable for drinking: if purity is questionable, contact your State Health Department for a *Bacteriological Analysis.*

EQUIPMENT RECOMMENDED, BASED ON INFORMATION SHOWN ABOVE. This recommendation lists the various units which can be used to treat this water. All recommendations assume that "hot" and "cold" water, except lawn hydrants, is to be treated and estimates 50 gallons of treated water per person per day. The amount is considered ample, including using automatic washing machine and dishwater. If only hot water is to be treated, units having approximately 3/4 the capacity of units recommended may be selected.

CUSTOMER REFERENCE: Manual ___ Semi-Automatic ___ Electric Automatic ___

RECOMMENDATION:

L56–32 Softener.

Figure 7.35 Water analysis printout. (Courtesy of Lancaster Pump, a Division of C-B Tool Co.)

Backup boiler with coil—add-on no coil. When the fossil-fuel boiler contains the only tankless coil for production of domestic hot water in a dual-boiler heating system, it is often left to the solid-fuel unit to furnish the circulating hot water necessary to heat the tankless coil. The piping arrangement for the residential water system is illustrated in Figure 7.36. This type of piping is common to many residences in which the fossil-fuel boiler furnishes all of the domestic hot water and must operate all year long. When a solid-fuel unit is added to the heating system, domestic hot water production is dependent on the burning of either wood or coal, which furnishes the hot water within the heating system necessary for indirect water heating.

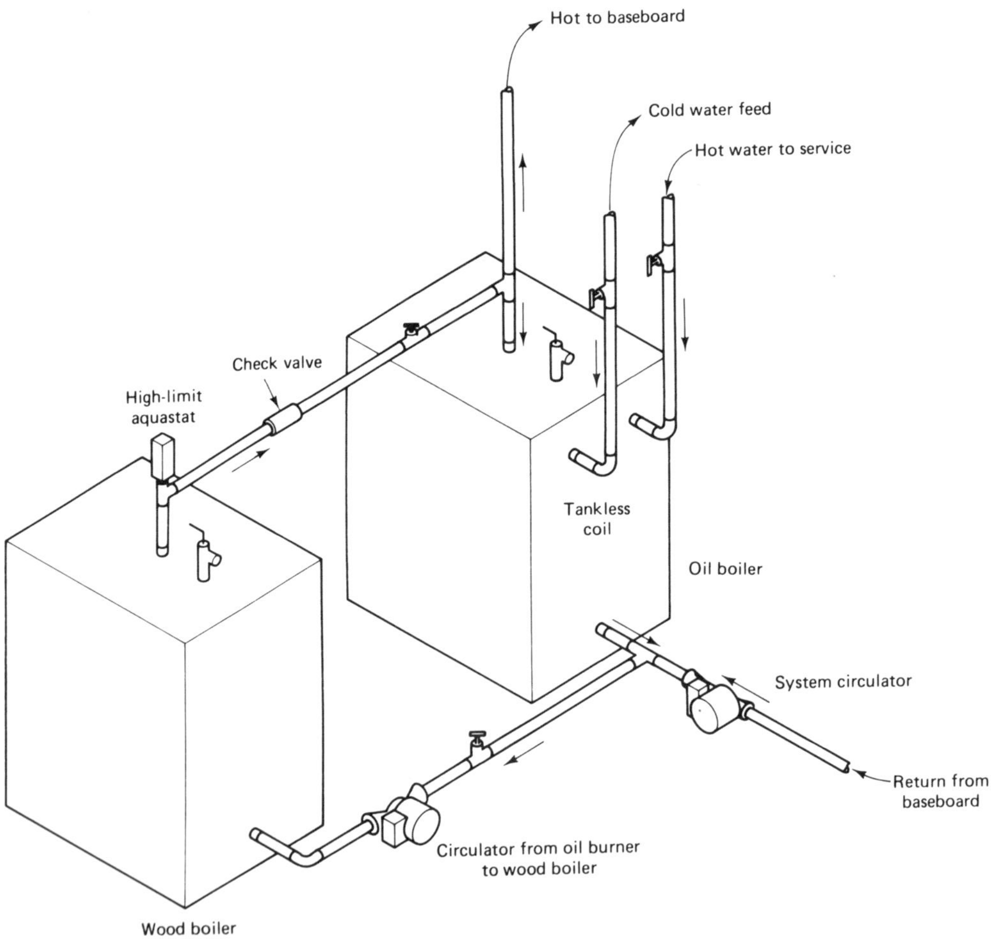

Figure 7.36 Piping arrangement of tankless coil in backup fossil-fuel boiler, add-on solid-fuel unit contains no coil but is piped in parallel to keep hot water in the backup boiler so that the coil is ready to provide domestic hot water at all times.

Backup boiler no coil—add-on with coil. Many dual-boiler heating systems feature a tankless coil in the add-on unit only. In this type of installation, the add-on is purchased with a tankless coil to replace a stand-alone domestic hot water heater. During the heating season, the add-on unit will furnish the majority of domestic hot water: The stand-alone water heater is used during the nonheating season only. The piping arrangement for the hot water house supply system is illustrated in Figure 7.37. Note that both the water heater and the tankless coil in the boiler are equipped with a three-valve bypass circuit. In this way either the coil or the water heater can operate independently. It is common practice to use the water heater as a storage tank during the heating season, with the boiler preheating the cold water and directing it into the storage tank. The storage tank enables the homeowner to overcome the limited storage capacity of the coil. It should be remembered that tankless coils, whether installed on add-on solid-fuel boilers or on conventionally fired appliances, are rated on an intemittent rather than a continual draw. Thus a coil that is rated at 3 gal/min will furnish, on the average, 42 gal of hot water per hour, based on a boiler-water temperature of 200°F.

Backup boiler with coil—add-on with coil. Add-on solid-fuel boilers equipped with tankless coils are often installed into heating systems in which the backup boiler already has a tankless coil in place. In this arrangement, the two coils are best plumbed in series, with the add-on coil feeding preheated water into the backup boiler, as illustrated in Figure 7.38. Efficient installation procedures would suggest that a bypass be installed on both the feed and delivery lines of each coil to facilitate maintenance or repairs to either unit.

Backup boiler with coil and aqua booster—add-on no coil. The limited storage capacity of most tankless coils, and their hot water production being based on an intermittent rather than a continuous draw, sometimes necessitates the installation of a system that minimizes these restrictions. Typical of these is the external heat exchanger and installation shown in Figure 7.39. This setup is a modification of the standard tankless coil and uses a separate storage tank with an heat exchanger external to the boiler for domestic hot water heating and storage. Such a system offers greater flexibility than a coil alone in that it provides upward of 40 gal of water storage: Coil capacity is generally under 2 gal. Also, the buildup of lime and other mineral deposits that can clog the coil, a common problem with many water systems, is eliminated with this arrangement since the external heat exchanger loop contains relatively inert water that is drawn from within the residential heating system. The booster acts as a separate heating zone, activating a circulating pump to move heated water through the coil to heat the water within when the storage tank water falls to a specified low-limit aquastat setting.

Solar domestic hot water system with tankless coil. An increasing number of solar domestic hot water heating systems are being installed in conjunction with coil-equipped hot water boilers. The combination of solar with a coil maxi-

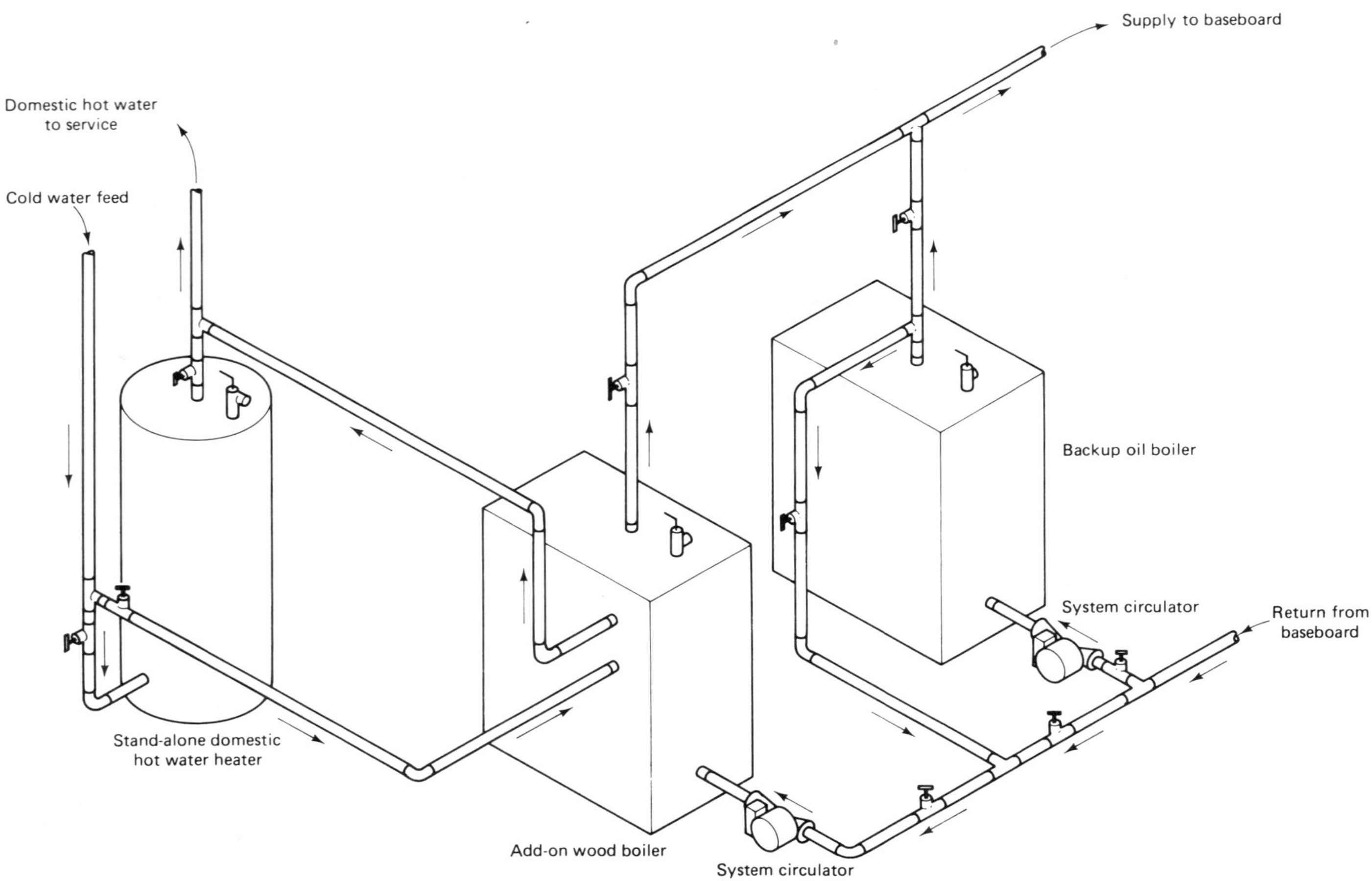

Figure 7.37 Add-on wood boiler with tankless coil supplies domestic hot water to the residence during the winter. A valved bypass is included so that the stand-alone water heater can provide domestic hot water during the nonheating season. The circulator and valving arrangement on the two boilers allows for either a series or parallel piping arrangement between the add-on and backup boiler for maximum flexibility.

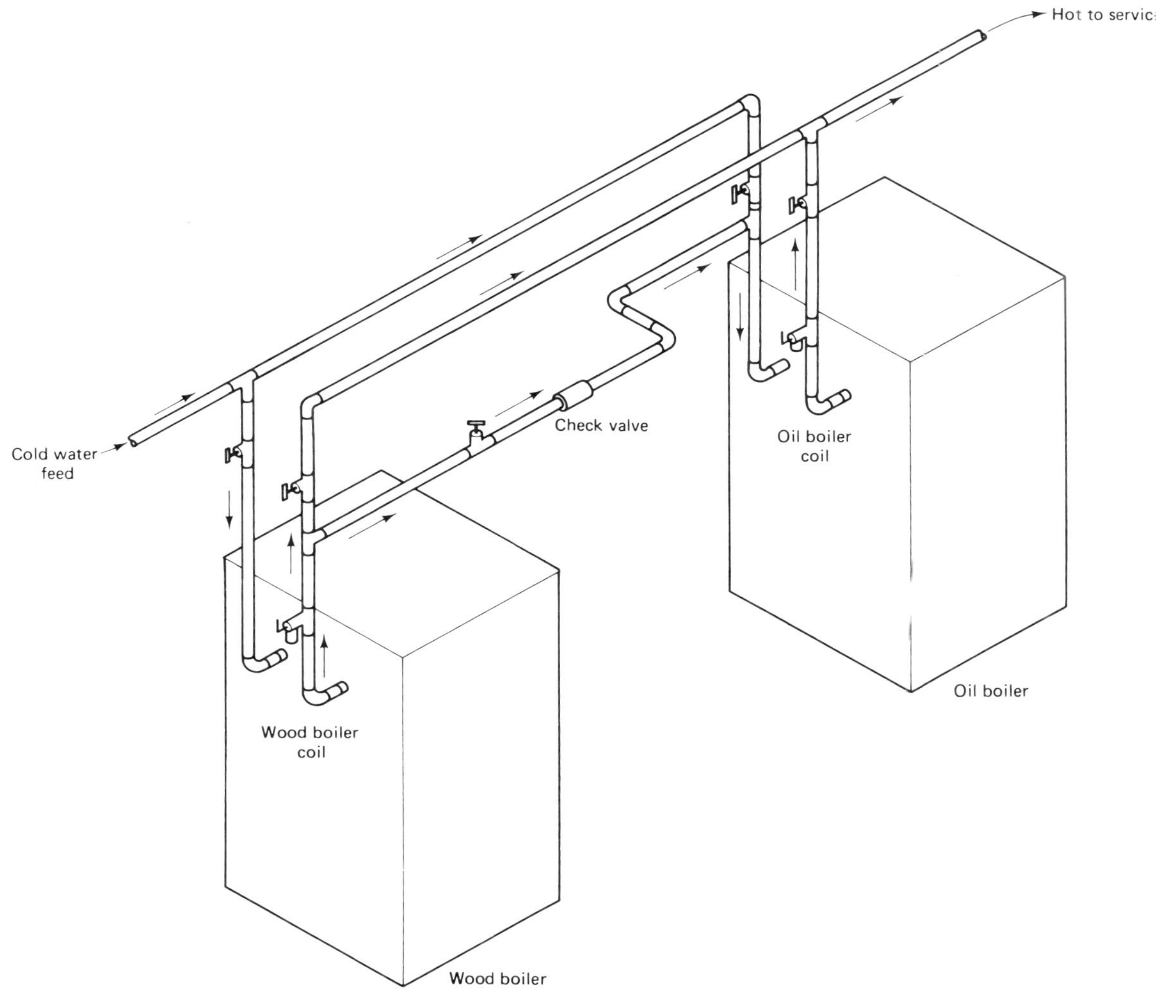

Figure 7.38 Piping arrangement of tankless coil installed in both add-on solid-fuel and backup fossil-fuel boilers. Valving is arranged to enable bypassing either tankless coil.

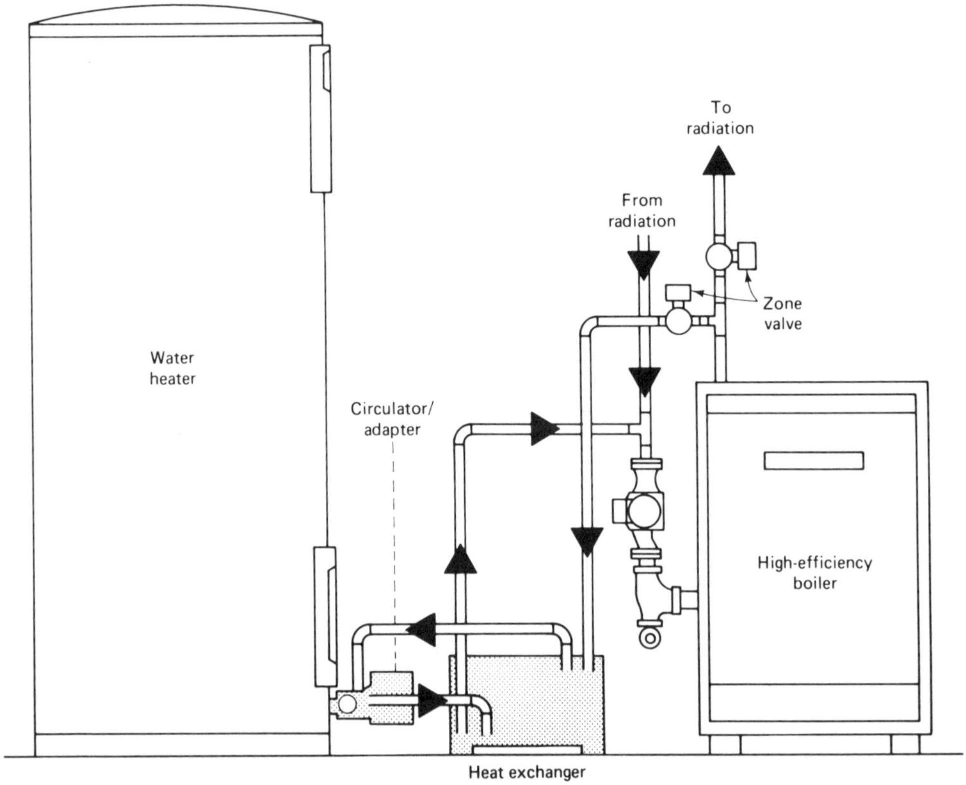

Figure 7.39 External domestic hot water heat exchanger and recommended installation. (Courtesy of Weil-McLain, a Marley Co.)

mizes the efficiency of both the solar and residential heating system. One major advantage of this type of installation is that the hot water boiler can be turned off during the nonheating season, resulting in substantial fuel savings. The illustrations that follow highlight some interface variations possible with solar and tankless coil-equipped solid-fuel boilers. Similar plumbing circuits can be configured with fossil-fuel and multifuel boilers as well.

Coil Preheating Arrangement

Figure 7.40 illustrates a solar domestic hot water system arranged to preheat the cold water supply prior to entering the tankless coil. Note that a bypass on the tankless coil is installed to allow the solar water heater to feed the hot water service line directly during the nonheating season. Most solar tanks are equipped with built-in backup electric- or gas-fired heating units. Thus the backup system will maintain hot water at the desired temperature within the solar storage tank in the absence of sun, or whenever hot water demand within the house is in excess of the recovery ability of the solar system.

When the boiler is located in damp basement areas, it is sometimes desirable to direct the solar-heated water through the boiler prior to its entering the hot water service line, even during the summer months. This procedure maintains some heat within the boiler, which helps to prevent corrosive condensation from taking place

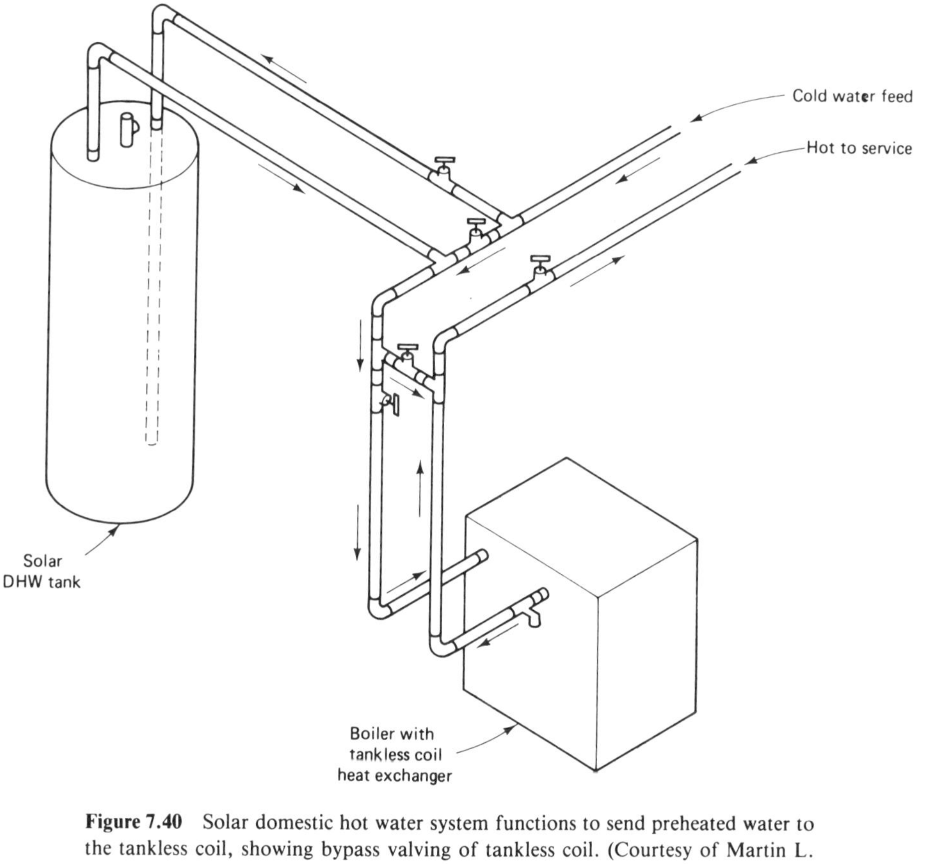

Figure 7.40 Solar domestic hot water system functions to send preheated water to the tankless coil, showing bypass valving of tankless coil. (Courtesy of Martin L. Greenwald and Thomas K. McHugh, *Practical Solar Energy Technology,* © 1985, p. 94. Reprinted by permission of Prentice-Hall, Inc., Englewood Cliffs, NJ.)

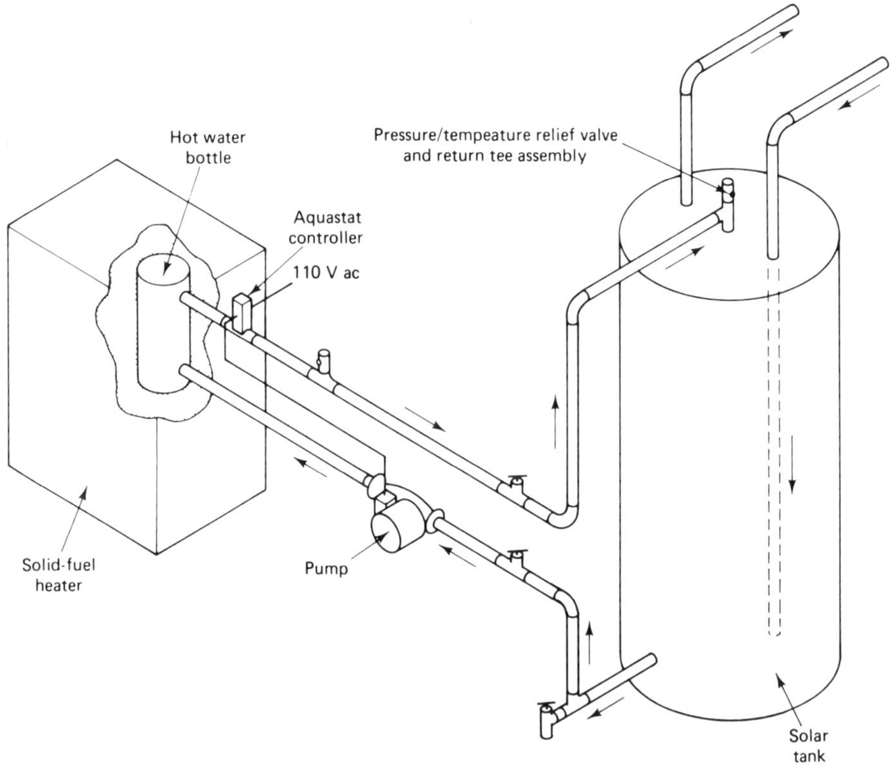

Figure 7.41 Parallel piping arrangement between solar tank and solid-fuel boiler. Water within the solar storage tank is heated by the coil in the boiler to enhance overall system output during the winter months.

on the interior and exterior boiler surfaces. Although the output temperature of the solar-heated water will be somewhat lower when the hot water is routed in this manner, the heated water within the boiler is advantageous in reducing corrosion.

Parallel configuration. A modification of the solar preheating piping arrangement is one in which the solar storage tank is connected in parallel to either a tankless coil or hot water bottle within the solid-fuel unit (Figure 7.41). Equipped with a stainless steel low-head circulator and switching aquastat, water from the solar tank can be heated by the tankless coil within the boiler and returned via the circulating pump to the top of the solar storage tank. This system is a combination of the aquabooster in Figure 7.40 and a standard solar hot water preheating system. The solar system operates independently of the boiler coil, with the added benefit of domestic hot water production capability isolated from limited sunlight availability during the winter months. During the summer months the boiler can be turned off; backup for the domestic hot water system is provided by the solar storage tank conventional heating elements.

8

Maintaining and Upgrading the Existing Heating System

SYSTEM IMPROVEMENT WITHOUT MAJOR SYSTEM MODIFICATION

It is common to encounter many residential heating situations in which excessive fuel consumption can be corrected without completely overhauling the existing system. Rather, tune-up and calibration procedures on specific system components can often bring about significant fuel savings. Taken individually, savings realized as the result of one small modification may be small; with conscientious effort, however, many small savings can be substantial. In this section we examine several areas and components of a typical heating system that are often overlooked but which can adversely affect overall system performance, efficiency, and fuel consumption if not properly calibrated or maintained.

Thermostats

Room thermostats must be properly calibrated if the heating system is to operate properly and efficiently. Ideally, the thermostat should activate the heating system when the room temperature drops to between 2 and 4°F below the thermostat set point. The heating system should run only long enough to bring room temperatures back up to the thermostat set point. If the thermostat is calibrated incorrectly, the heating system will cycle longer than is necessary, using excessive fuel without providing even interior comfort levels.

Also, the ability to selectively raise and lower the thermostat set point based on the living patterns of the family can produce significant fuel savings.

There are two basic service procedures applicable to most heating thermostats: calibration of the thermostat and proper setting of the heat anticipator circuit. Each of these procedures is examined below.

Calibration. Although most thermostats are factory calibrated, these settings can deteriorate over a period of time due to aging of the components. If the accuracy of the thermostat is suspect, the following calibration procedure is recommended.

1. Place a thermometer of known accuracy next to the thermometer of the thermostat cover.
2. Allow 10 minutes for the thermometer reading to stabilize, and observe the readings on each.
3. If both thermometers read the same, discontinue calibration procedure.
4. If the reading on the thermostat thermometer is different from that on the calibration thermometer, insert a screwdriver into the thermostat adjusting screw slot and turn until both thermometers read the same.
5. Check the thermometer calibration after one week to ensure that the accuracy of the settings has been maintained. If the thermostat reading is found to be inaccurate after this time, the unit is probably defective and should be replaced.

Heat anticipator settings. The operation of the thermostat heat anticipator is examined in detail in Chapter 3; however, periodic inspection or adjustment of the anticipator may be necessary if components are changed or if present settings are suspect.

Adjustment of the heat anticipator regulates the flow of electrical current through a resistor within the thermostat which is wound in series with the bimetallic thermostat temperature sensor. When the anticipator is properly matched to the current draw on the boiler operating control, the heat anticipator can accurately "predict," or anticipate, the lag incorporated into the heating system that is required to bring the room up to the thermostat set temperature after the heating system has turned off by virtue of heat still contained within the baseboard or radiation. Incorrect heat anticipator settings will cause the boiler heating cycle to be either too short or too long, and questionable heating system performance can result.

To set the anticipator properly, find the current draw for the particular boiler operating control or aquastat. These electrical ratings are usually located on the electrical schematic of the control device adhered to the inside of the control box cover (Figure 8.1). Once this information has been located, the heat anticipator setting is adjusted to match the current draw (Figure 8.2). Setting and calibrating the thermostat in this manner helps to ensure that the heating system will call for

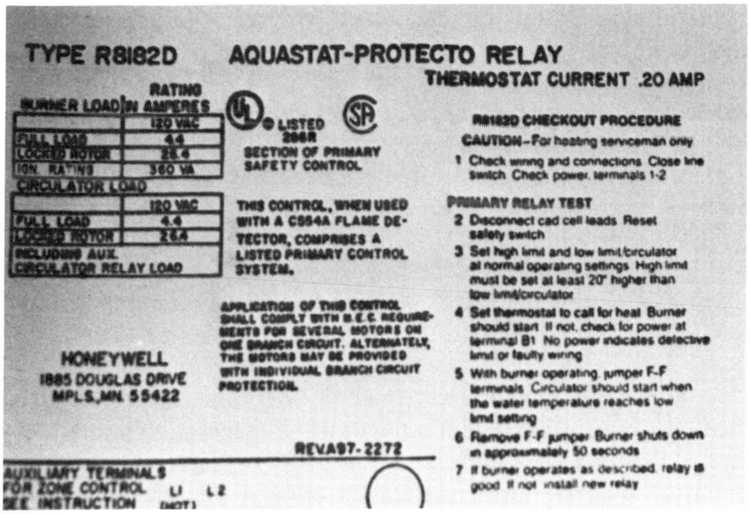

Figure 8.1 Current draw listing for setting thermostat heat anticipator. (Courtesy of Honeywell, Inc.)

Figure 8.2 Location of heat anticipator setting on room thermostat. (Courtesy of Honeywell, Inc.)

heat within accurately predetermined temperature ranges and that appropriate cycling times will achieve desired comfort levels.

Maintenance Procedures for Oil-Fired Appliances

System maintenance of oil-fired appliances results in minor modifications to the heating system by way of accessorizing or upgrading existing system components. Applicable procedures within this section include: (1) servicing and upgrading the fuel delivery system, (2) inspecting the burner mechanism and replacing the nozzle, (3) inspecting the air intake mechanism and air shutters, (4) inspecting and maintaining the draft system, and (5) testing the burner efficiency.

Fuel delivery system. Fuel oil filters should be changed at least once each heating season. If dirty fuel or continual condensation in the fuel system is a problem, a second filter should be installed: If two filters are appropriate, one should be installed at the outlet of the fuel oil tank, the other just prior to the intake on the fuel pump of the burner.

In underground tank installations, one filter may be installed where the lines from the oil tank enter the foundation wall of the home, and the other at the oil burner pump as described above.

All oil delivery piping should be inspected annually for leaking connections. All feed lines should be mechanically protected from accidental damage. If oil lines have been run on the surface of the floor, concrete or mortar thresholds should be poured over the lines to prevent them from being crushed or dented.

Oil tanks should be inspected periodically for water condensation that can build up at the bottom of the tank. This inspection is performed by wiping a special paste on the end of a dipstick that changes color when it comes into contact with water. The stick is then dipped into the tank and withdrawn to observe if any color changes have taken place in the paste (see Figure 4.12).

Burner mechanism and nozzle. Annual diassembly of the burner ignition system components and nozzle replacement is recommended for proper oil burner performance.

When diassembling the burner, the electrical circuit to the burner must be disconnected. As the unit is diassembled, particular attention should be paid to the surface of the cad cell flame sensor and the ignition electrodes.

The cad cell should always be wiped clean with a soft cloth whenever the burner is disassembled for inspection and component replacement. The electrodes should be checked for soot buildup and point deterioration. When inspecting the point assembly, refer to the manufacturer's specification for the proper point gap and height adjustment above the nozzle centerline and readjust if necessary (Figure 8.3).

The fuel oil nozzle should be changed once each heating season: The old nozzle should be replaced with a new one of the same rating, spray angle, and spray pat-

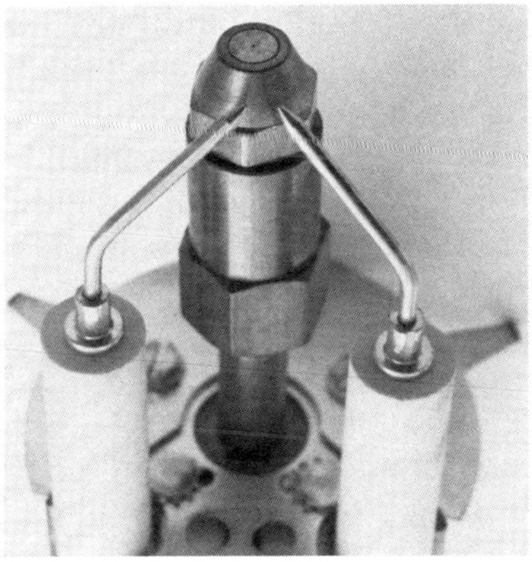

Figure 8.3 The electrode assembly should be gapped according to the manufacturer's specifications. Both electrode gap and position of the nozzle centerline are critical for proper functioning of the burner.

tern. Boiler specifications should be consulted when performing this maintenance procedure to ensure accuracy in selection of the replacement part.

Air intake mechanism and air shutters. Clean combustion air must be supplied to the burner in sufficient quantities for efficient combustion to occur. Also, a minimum of air transfer should take place during heating "off" cycles to keep heat losses caused by air movement through the appliance and into the chimney as small as possible. To keep off-cycle heat losses to a minimum, centrifugal air restrictors can be used (see Chapter 4, "Off-Cycle Air Restrictors").

For the proper amounts of combustion air to be available, the following conditions must be met:

1. Sufficient combustion air must enter the boiler room, preferably from outside the building.
2. The air intake shutters and blower wheel must be operating properly and without undo restriction caused by a buildup of dirt and lint.
3. Provision must be made to limit off-cycle air from entering the venting system of the boiler.

To determine if sufficient combustion air is present for the boiler, a survey should be made of the boiler room and adjoining spaces. The results should be compared with the requirements listed below to determine if adequate combustion air is being supplied to the unit.

Combustion Air Specifications

1. When a boiler is installed in a building of unusually tight construction, a direct opening to an outside wall, properly protected, should be provided with a total surface area equivalent to 1 in.^{2} for each 5000 Btu of boiler capacity.

2. When a boiler is installed in a confined area, provision for outside air must be made by either furnishing two openings in a wall to an adjoining room that has adequate air from the outdoors or by providing two openings to the outdoors or crawl spaces that are in unrestricted contact with the outdoors. When ducting into adjoining rooms communicating with unrestricted outdoor air, one opening should be provided within 12 in. of the ceiling, the other within 12 in. of the floor. Each opening should be sized to provide 1 $in.^2$ for each 4000 Btu of boiler input.

For other situations designed to provide adequate outdoor combustion air, the reader should consult the specifications in Chapter 5 relating to ventilation and combustion air requirements for gas-fired boilers, and manufacturers' literature regarding installation practices for specific appliances.

To ensure that airflow into the burner is free of unwanted dirt or dust buildup, the intake shutter on the burner should be vacuumed periodically (Figure 8.4).

Draft system. Since many chimneys supply draft in excess of that required by most residential oil burners, the chimney draft should be checked and regulated at least once each heating season.

If the venting system does not contain a barometric draft damper, one should be installed or replaced as the situation dictates. The placement of the draft damper should be as suggested in Figure 8.5.

The damper, which opens and closes in response to changing chimney draft, is adjusted to maintain an over-the-fire draft of between 0.02 to 0.03 in. of water column in most residential oil-fired heating equipment. The damper is adjusted by the position of a counterweight on the damper plate that can be set to maintain the desired draft setting (Figure 8.6).

On occasion, a particular chimney may not be capable of producing the necessary draft required for proper operation of the boiler. This condition is evidenced by excessive oil fumes and smoke in the vicinity of the burner when it is operating. In this instance, a specialist should be consulted to determine if local geographical

Figure 8.4 Location of air intake shutter. The shutter should be cleaned periodically to prevent a buildup of dirt and dust.

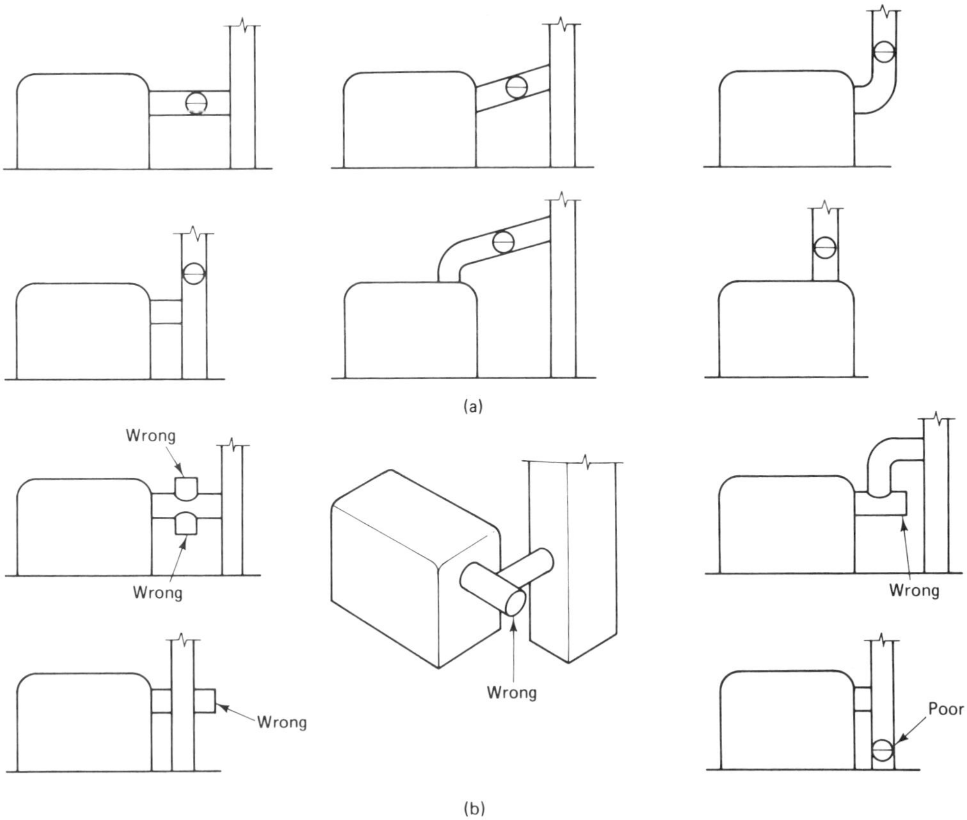

Figure 8.5 (a) Proper and (b) improper locations for barometric draft dampers on oil-fired appliances.

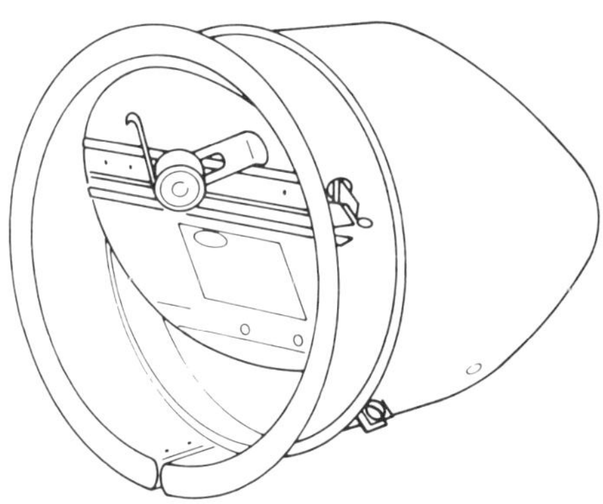

Figure 8.6 Barometric draft damper adjustments are performed by changing positions of the damper counterweight until the correct draft in the unit is achieved. (Courtesy of the Field Controls Company.)

configurations or some aspect of the construction of the chimney have caused the problem, and how best to correct it.

At times, a draft-inducing fan can be installed within the chimney flue or boiler stack pipe to bring the draft up to required levels for proper operation. An induced-draft chimney fan is illustrated in Figure 8.7. For other chimney problems, see Chapter 7, relating to masonry and factory-built chimneys.

Combustion efficiency. After all components of the system have been checked and/or upgraded, the burner should be tested to determine the combustion efficiency of the unit. The specific procedures involved for efficiency testing are given in Chapter 4; however, the overall limits for optimum performance of most residential oil-fired heating equipment powered by high-speed flame retention burners are listed in Table 8.1.

Efficiency testing is easy to interpret when the basics of combustion physics are applied to burner technology. Since air contains approximately 20% oxygen by volume, complete combustion of the carbon with the fuel oil should yield a CO_2 level of 21% in the flue gas when no excess combustion air is being supplied to the burner. Since a level of 0% excess air will not allow for proper mixing of all of the fuel oil with the incoming combustion air, excess levels of air are required. In general, excess air should be between 50 and 100%, with acceptable levels of CO_2 measured in the flue gases of between 10.5 and 12.5%, respectively. It should be emphasized that too much excess air will cool the fire and should be avoided. The overall objective in the efficiency testing procedure is to obtain the highest CO_2 levels possible with smoke levels below No. 1. The reader is referred to Chapter 4 for specific procedures involved in the performance of various efficiency tests.

Figure 8.7 Chimney fan for draft induction. (Courtesy of Tjernlund Products, Inc.)

TABLE 8.1 PERFORMANCE PARAMETERS FOR HIGH-SPEED OIL BURNERS

Smoke level	Percent carbon dioxide (CO_2)	Stack temperature (°F)
0–1[a]	10–12.5	500–600

[a]Maximum allowable level.

Maintenance Procedures for Gas-Fired Appliances

Although normal maintenance and component upgrading of gas-fired appliances is generally simpler than for oil-fired units, the same basic procedures apply.

Air intake. Ensure that adequate combustion air is entering the boiler room. Consult Chapter 5 for details covering the provisions to be present for adequate incoming and ventilation air.

All air shutters on the unit should be checked for dirt and dust accumulation. Shutters on both the main burner and pilot burner should be inspected. This procedure is especially important in atmospheric combustion units, since all entering air is moved under the influence of chimney draft without the assistance of induced-draft fans.

Pilot flame. The pilot flame should be checked periodically for proper position and flame configuration. In a typical thermocouple pilot flame arrangement, the flame should be positioned so that it comes into contact with the hot junction of the thermocouple (Figure 8.8). In electronic ignition systems, the ignition electrode should be positioned for smooth lighting of the pilot flame. The flame-sensing probe should be in the path of the central part of the pilot flame (Figure 8.9).

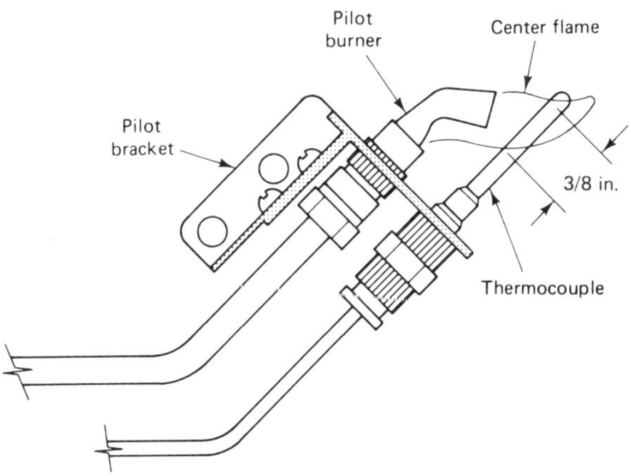

Figure 8.8 Typical pilot flame. (Courtesy of Burnham Corp.)

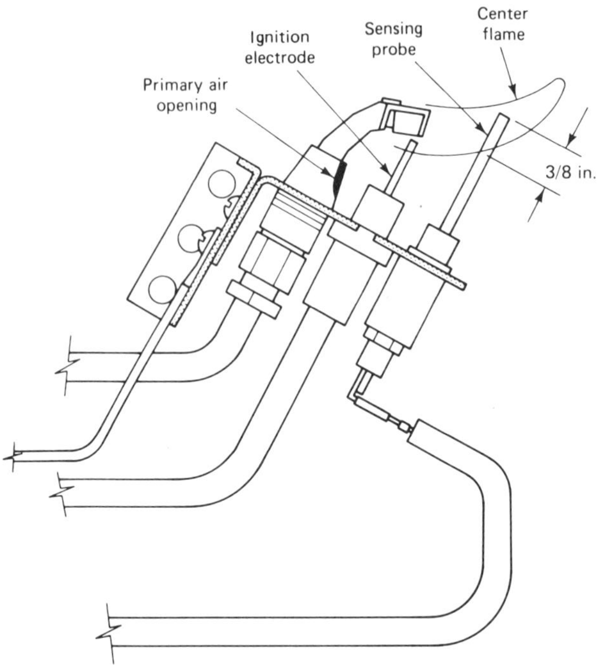

Figure 8.9 Flame position of electronic ignition pilot. (Courtesy of Burnham Corp.)

Main burner flame. The main burner flame should always be inspected when the boiler is initially activated at the beginning of the heating season, and at least semiannually thereafter. The main burners should be set to deliver the maximum amount of heat that is obtained by adjustment of the primary air shutter. Proper flame adjustment is illustrated in Figure 8.10.

The main burner flame should have a clearly defined inner cone with no yellow tipping. Most burner flames will occasionally have orange or yellow streaks in the outer mantle caused by dust particles. This streaking should not be confused with the yellow tipping caused by a lack of adequate primary air. Also, the orifices should be inspected annually and properly lined up with the burner intake manifold.

Vent system. The vent system of the boiler should be inspected annually to ensure that no deterioration has taken place in the piping configuration and that the integrity of all joints, supporting straps, and chimney connections are intact. For specifications relating to gas-fired venting system installation and maintenance procedures, the reader is directed to Chapter 5.

Combustion efficiency. Combustion efficiency measurements for gas-fired boilers are similar to those taken for oil-fired units; however, different readings should be expected for CO_2, stack temperature, and draft levels. For these tests to be accurate, the following information must be known:

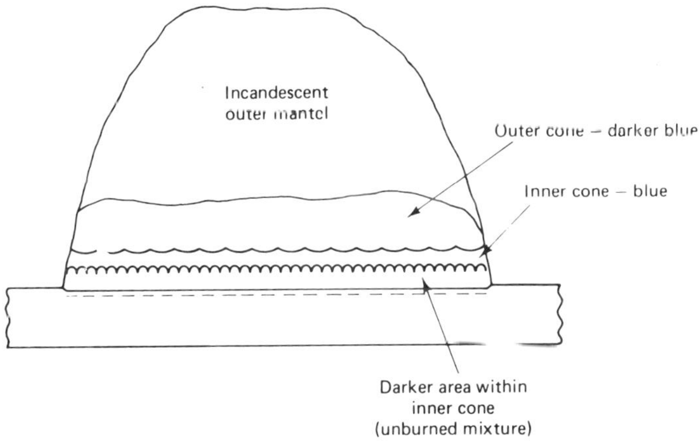

Figure 8.10 Primary air shutter and main burner flame adjustment. (Courtesy of Burnham Corp.)

1. Btu rating of the fuel gas
2. Gas pressure within the feed and residential piping
3. Accuracy of the meter with regard to delivered Btu input to the appliance

Chapter 5 should be consulted for specific information regarding the foregoing three considerations.

Aquastat Adjustments

Since aquastats regulate the water temperature within the heating system, their proper adjustment is critical for efficient heating system operation. The most important settings on the aquastat are those that regulate the high- and low-water temperature limits of the heating system. For best results, these settings should be seasonally adjusted. This procedure helps to keep the boiler-water temperature synchronized to the temperature conditions of the changing seasons. For example, during the coldest winter months in many parts of the country, the boiler should be set to produce water temperatures between 180 and 200°F for space heating. However, during the swing seasons, the fall and spring months, for example, 150°F boiler-water temperature might suffice. Often these settings will be arrived at by experimentation. What is most important, however, is that building heat loss will vary over the course of the year depending on the outdoor temperature. The boiler should be adjusted accordingly to provide the most comfortable heat depending on seasonal requirements. Since the heat loss of the home will be perhaps half as much during times when the outdoor temperatures average between 40 and 50°F, as opposed to the coldest winter months, when temperatures average −15° and 25°F, it makes sense to operate the boiler at lower water temperatures when practical. Significant fuel savings can result from this procedure since fewer Btu are required to provide for reduced temperature levels. For a more detailed explanation of the function of aquastats, the reader is directed to Chapter 4.

Continuous Circulator Operation

Continuous circulation is employed with cold-start boilers, whereby the thermostat controls the burner mechanism only; the circulator runs continuously. In this type of system, circulating water in the heating system tends to be warmer during the colder months of the heating season and somewhat cooler during the warmer heating months and swing seasons in response to the frequency of boiler cycling.

Continuous circulator operation causes fewer temperature changes within the heat distribution units, resulting in more-even interior temperatures, as the baseboard is continuously tuned into interior house temperatures. Also, rapid expansion and contraction of the piping and baseboard circuit is practically eliminated in this mode of operation.

Continuous circulation operation is not suitable for heating systems in which domestic hot water is produced via a tankless coil in the boiler, since this would cause overheating of the house. A typical electrical circuit illustrating continuous circulator operation in the cold-start mode is shown in Figure 8.11.

Balancing the Heating System

Proper system balancing involves specific fluid flow rates through the heating system that have been calculated to provide the most effective and evenly balanced heating available from a particular system. This is accomplished through an adequate temperature drop and heat transfer from the circulating water to the interior of the home.

Most residential baseboard radiation is rated for heat output based on a flow rate of either 1 gal/min (500 lb/hr) or 4 gal/min (2000 lb/hr). At the specified flow rate, the temperature drop across the baseboard circuit is designed to be approximately 20°F. This figure will vary, however, from one heating system to another depending on the length of the circuit, the number of fittings and valves in the system, and the size and length of the heating distribution units.

The problem most often encountered in residential heating systems is one in which the flow rate throughout the baseboard circuit is too high, resulting in a very small temperature drop. Sometimes this temperature drop can be as low as 3 to 6°F.

The results of this type of flow rate are that relatively little heat is transferred from the boiler to the home. The water, moving at excessive velocities, returns to the boiler with much of the heat that should have been transferred to the baseboard.

This problem can be rectified by reducing the flow rate. If the system has balancing or ball valves installed in the piping circuit, they should be adjusted so that the temperature drop across the circuit is increased to approximately 10 to 15°F. If the system does not contain balancing valves, they should be installed.

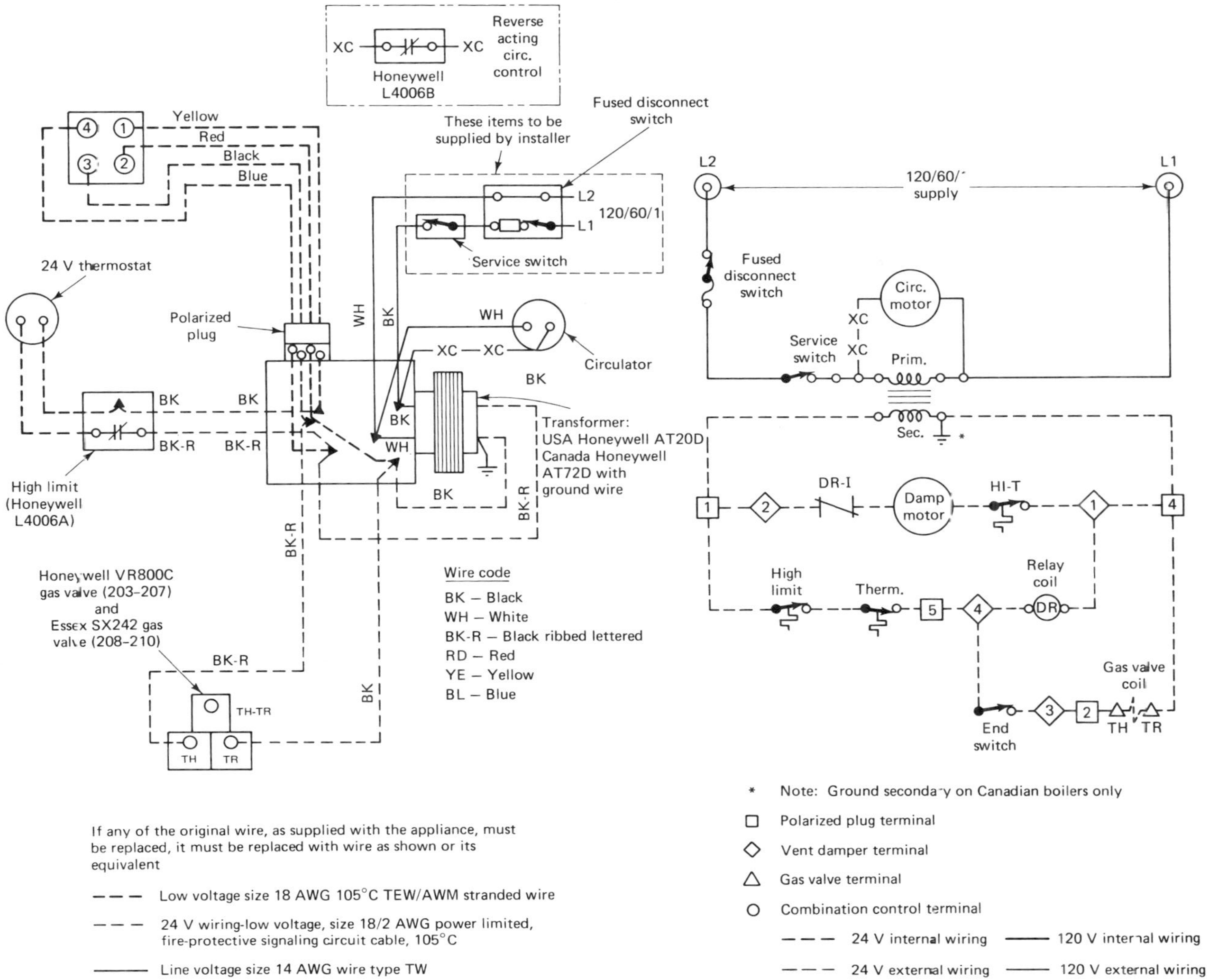

Figure 8.11 Cold-start configuration with continuous circulator operation, showing 24-V continuous or gravity circulation with vent damper. (Courtesy of Burnham Corp.)

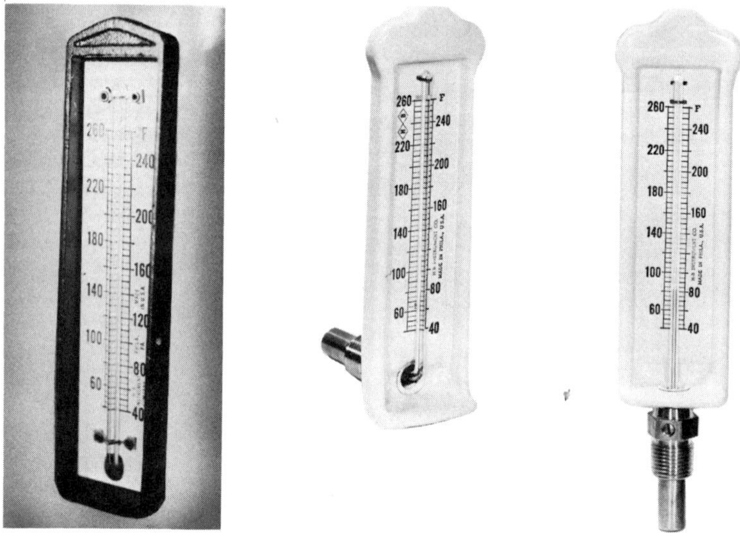

Figure 8.12 Thermometer inserted into baseboard return loop. (Courtesy of H. B. Instrument Co.)

In some instances, an excessively high-temperature drop may be encountered across the baseboard circuit. This situation is generally the result of an undersized circulating pump or an extremely long baseboard circuit. This situation can usually be corrected by the installation of a circulating pump that is properly sized to the heating circuit, or by breaking up the long series baseboard loop into smaller, separate circuits.

To measure the temperatures across the loop, surface probe digital thermometers may be used. Sometimes it is beneficial to install a small immersion well mercury thermometer into each return loop of the heating system (Figure 8.12). Also, there are devices available that can be installed into the heating system circuit to measure the fluid flow rates within the piping. Typical of these instruments is the unit illustrated in Figure 8.13. Once the system has been balanced in this fashion, a more even and efficient heating system will result.

UPGRADING MAJOR SYSTEM COMPONENTS TO IMPROVE OPERATION

At times it may become necessary to consider major revisions of components within the heating system to improve overall performance and fuel consumption. These revisions involve almost every major aspect of the heating system. This section deals with the most likely revisions to be undertaken by the homeowner.

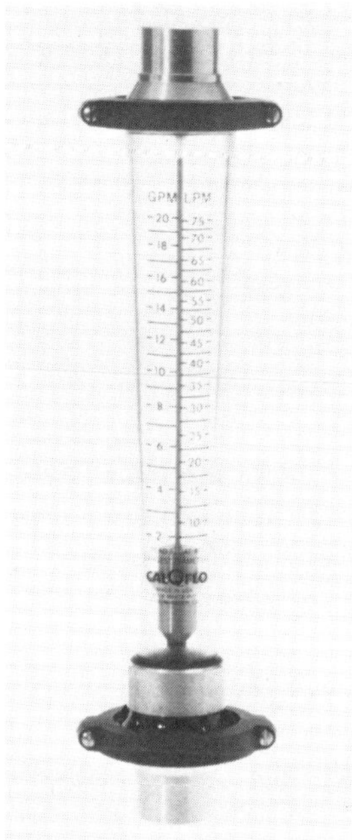

Figure 8.13 Series flow meter for measuring flow rate through the heating system. (Courtesy of Blue White Industries.)

Thermostats

The standard room thermostat used for residential heating is a simple on/off switch that controls the burner mechanism in response to a drop or rise in room temperature. While setback and two-stage thermostats have been available for some time, their use was not particularly widespread in residential applications until spiraling fuel costs forced homeowners to consider using these devices. Also, the burgeoning field of microcomputers and solid-state electronics has given rise to sophisticated multifunction thermostats that allow very precise system operation heretofore unavailable. Room thermostats provide a reasonable place to begin upgrading the various components of the residential heating system.

Setback. Adapting interior house temperatures to the living patterns of the family enables the homeowner to achieve significant savings of fuel by reducing or elevating the room heating when necessary. Sometimes, setback periods may occur several times a day; therefore, thermostats that offer multiple setback capability can be highly cost effective. For example, consider the living patterns of a family that rises at 6:00 A.M. in order to leave for work at 8:00 A.M. They return home at 6:00 P.M. and retire at 11:00 P.M. The thermostat settings for this particular situation could be programmed as shown in Table 8.2. In this arrangement, the heating system would be providing high-temperature heating for only 8 hours each day rather than 24 hours. Note that the setback periods end $\frac{1}{2}$ hour before the heat is required

TABLE 8.2 TYPICAL SETTINGS FOR AN AUTOMATIC SETBACK THERMOSTAT

High-temperature set point	Low-temperature set point
5:30 A.M.	8:00 A.M.
5:30 P.M.	11:00 P.M.

to be brought up to normal levels, allowing ample time for the system to pick up. A multiple setback thermostat for this purpose is illustrated in Figure 8.14.

The savings available to the homeowner by virtue of single- and multiple-setback thermostats can be substantial. Table 8.3 illustrates the savings that are available for both heating and cooling setback schemes in a variety of geographical areas within the United States.

Two-stage. A two-stage thermostat can be incorporated into a residential heating system to produce system operating temperatures below those temperature limits achievable with conventional single-stage thermostats. The organization of the control functions in a typical two-zone gas-fired hydronic system are shown in Figure 8.15.

Note that each of the two zones in the heating system is equipped with a separate two-stage thermostat. In this arrangement, the first stage of the two-stage thermostat controls the operation of the zone valve and circulator: When the zone calls for heat, the zone valve will open to allow circulation of water in the heating loop. The second stage of the two-stage thermostat controls the burner. The operating sequence of the system is as follows: When the first stage calls for heat, the zone valve opens and the zone circulator is activated to move water throughout the heating loop. The circulator is controlled by a reverse-acting aquastat that prevents the circulator from operating should the water temperature fall below the low-limit set point. Whatever heat is in the water will be transferred to the baseboard radia-

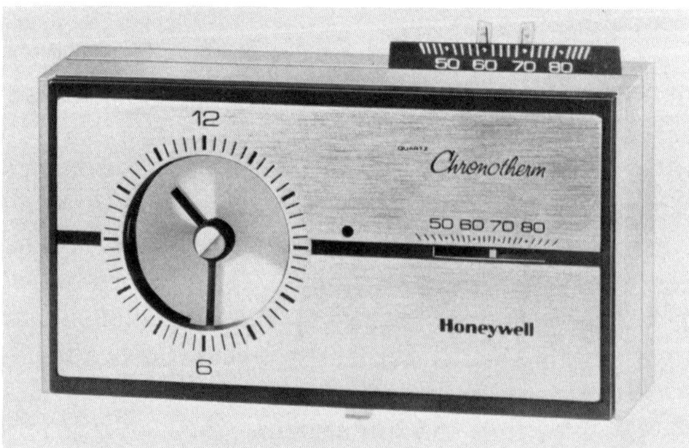

Figure 8.14 Multiple setback thermostat. (Courtesy of Honeywell, Inc.)

TABLE 8.3 PERCENT SAVINGS USING TOTAL ENERGY CONSERVATION SYSTEM

	Single heating setback (70° to 60° 8 hrs/day)	Double heating setback (70° to 60° 8 hr twice/day)	Single cooling setup (75° to 80° 11 hr/day)	Double cooling setup (75° to 80° 9 hr/day 7 hr/ night)
Billings, MT	25	34	12	19
Boise, ID	25	34	11	18
Boston, MA	26	35	16	23
Buffalo, NY	25	34	17	25
Charleston, SC	43	51	16	22
Cheyenne, WY	25	34	15	20
Chicago, IL	25	34	16	23
Cincinnati, OH	26	35	15	22
Cleveland, OH	26	35	16	23
Colorado Springs, CO	25	34	14	19
Columbus, OH	26	35	15	22
Denver, CO	26	35	13	20
Des Moines, IA	25	33	15	22
Detroit, MI	25	34	16	25
Indianapolis, IN	26	35	15	22
Kansas City, MO	26	35	13	19
Louisville, KY	26	36	14	21
Madison, WI	25	38	16	22
Memphis, TN	30	40	14	20
Milwaukee, WI	25	33	16	22
Minneapolis, MN	26	33	15	23
New York, NY	26	35	16	23
Oklahoma City, OK	28	38	14	19
Omaha, NB	25	33	15	22
Philadelphia, PA	26	36	16	23
Pittsburgh, PA	26	35	16	23
Portland, OR	31	41	14	23
Salt Lake City, UT	26	35	13	19
Seattle, WA	31	41	19	26
Sioux Falls, SD	25	35	14	21
Springfield, MA	25	34	16	23
St. Louis, MO	26	35	14	21
Syracuse, NY	25	34	16	24
Washington, DC	27	37	16	23

Source: Honeywell, Inc., 1982.

Note: Savings figures apply to Honeywell system only.

tion and interior spaces of the residence. Should the room-temperature drop to the second-stage setting, usually between 3 and 5°F below the first-stage set point, the burner will ignite. This raises the boiler-water temperature to the high-limit setting. As the room temperatures are raised above the second-stage ignition level, the second stage deactivates, turning off the burner. Heat transfer continues via the cir-

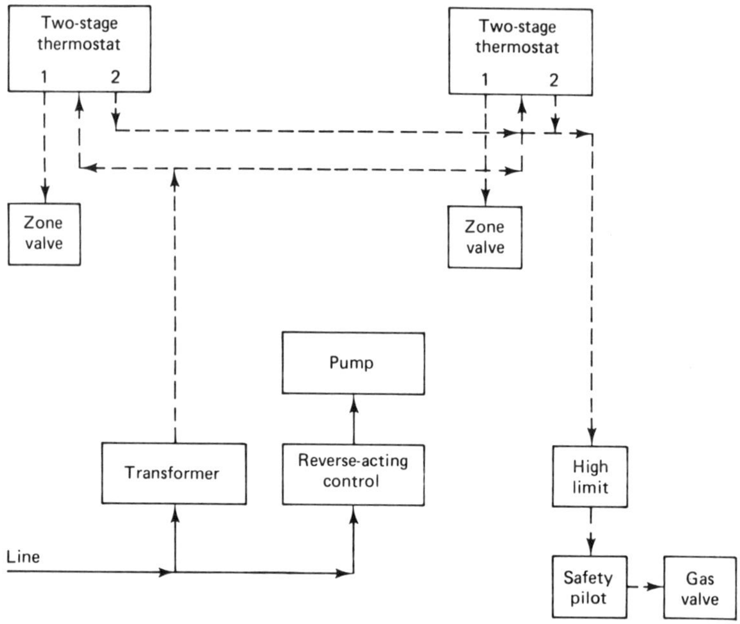

Figure 8.15 Control functions in a two-stage, two-zone, gas-fired hydronic system. (Courtesy of the Hydronics Institute.)

culator operation until the first-stage setting of the thermostat has been satisfied. Sometimes a second high-limit control may be required for added overheat protection.

This control arrangement helps to synchronize boiler-water-temperature to outdoor-temperature conditions by activating the burner only after all available heat has been transferred from the heating system to the home. Thus during periods of reduced heating demand, when the outdoor temperature is in the mid-50s, lower-temperature water will adequately heat the home without the necessity of operating the burner.

Indoor/Outdoor Temperature Modulation

Very efficient heating occurs when the boiler-water temperature is synchronized to outdoor temperatures rather than operating at the same temperature throughout the year, regardless of the heating load placed on the home by outdoor weather conditions. Thus, whereas 180°F water may be required to heat a house when the outdoor temperature is 0°F, the same house can easily and efficiently be heated by 140°F water when the outside temperature is 55°F.

To accomplish the task of correlating boiler-water temperature to outdoor conditions, small computers are used as illustrated in Figure 8.16.

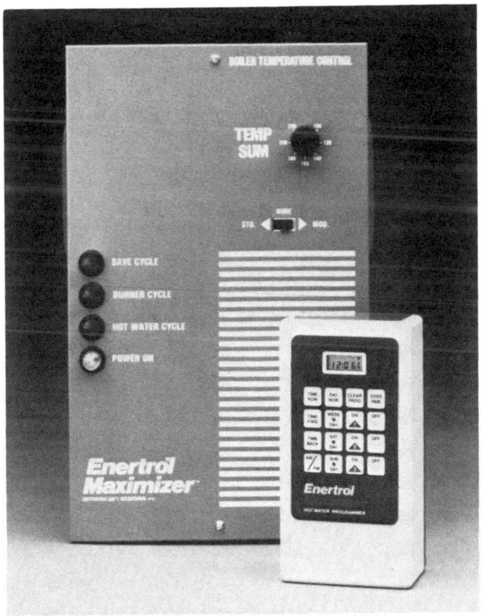

Figure 8.16 Indoor/outdoor temperature control for central heating system. By sensing outdoor temperature and heating system output, the precise amount of energy required to satisfy the room thermostat is computed, eliminating excess fuel consumption. (Courtesy of American Stabilis.)

The sequence of operation of this device is simple: As outdoor temperature increases, boiler-water temperature will be controlled to decrease; as the outdoor temperature decreases, the control will automatically increase the boiler-water temperature in synchronization to the extent of ambient temperature rise. A variety of these outdoor reset controls are available and are adaptable for use in hot air heating systems as well as hydronic applications.

Most of these devices employ two separate sensing elements: One element measures the boiler-water temperature, the other measures outdoor temperature.

These units will more than pay for themselves in one heating season due to reduced fuel consumption as a result of automatic temperature modulation.

Heating System Design Modifications

Most improvements that can be undertaken in redesigning a hydronic heating system that is not functioning up to par focus on modifications of the existing baseboard circuitry—usually breaking down the heating system piping circuit into separate zones. Other major modifications of an existing heating system often involve the changeover from a steam heating system to a forced-circulation baseboard hot water system, and the replacement of old, obsolete, inefficient boilers with newer, high-efficiency units.

In undertaking modifications of the baseboard circuit, it is often possible to reconfigure the feed and return lines in the vicinity of the boiler, keeping the changeover relatively simple. Sometimes, however, new baseboard needs to be added, and certain essential piping may be inaccessible. In these instances the changeover may result in major alterations to the building structure and temporary dislocation of residents.

When making a conversion from a steam to a hot water baseboard heating system, the changeover may be limited to a conversion of boiler trim and operating controls. Often, however, the changeover from steam to hot water involves replacement of all existing radiators or piping with new radiation, as well as feed and return lines necessary for the new heat distribution units.

Simple boiler replacements are perhaps the easiest of all major modifications to undertake. The old unit is removed and the new one is installed in its place, replacing old trim and control features with newer efficient components.

Each of these modifications will be examined to determine the scope of the project as well as the most efficient ways to undertake these changes.

Zoning. Incorporating separate heating zones into a hydronic system where none existed previously is an excellent way to maximize efficiency. Zoning allows only those areas within the residence that are in use or where heat is required to operate independently of those where heat is not needed. Zoned hydronic systems offer a great deal of flexibility that is hard to duplicate with other types of central heating.

Let us assume that a small home presently has the single-zone heating system illustrated in Figure 8.17. The layout of this home is such that the major family living areas, such as the kitchen, dining room, and living rooms, are separated from the bedroom areas, which are located on the opposite side of the home. Although this example might seem somewhat simplistic, most homes can usually be divided into two areas of "major" and "minor" occupancy over a 24-hour period. The home in Figure 8.17 can easily be divided into two separate heating zones: one serving the bedroom areas, the other serving the remainder of the house. To accomplish this zoning, changes have to be made to the piping circuitry at points A, B, and C. Once these changes have been made, the system configuration should now appear as illustrated in Figure 8.18.

The piping illustrated in Figure 8.18 represents feed and return lines that have been added to the system to change the system to a two-zone configuration. Note that although the system uses separate circulators for each heating zone, zone valves will produce equally effective results if they are used instead of separate zone circulators. Chapters 1 and 2 should be consulted for specifics concerning zoning configurations for both circulators and zone valves.

The heating system in Figure 8.18 is also flexible in that future expansion of the system can be undertaken with little disruption to the existing system. For example, let us assume that a room is to be added onto the house that adjoins the living room, which includes a garage on the lower level and a bedroom above, next to the living room. If the heating capacity of the boiler is sufficient to carry the added heating load (generally between 5000 and 15,000 Btu per room, depending on construction techniques), either a separate zone could be added to the present

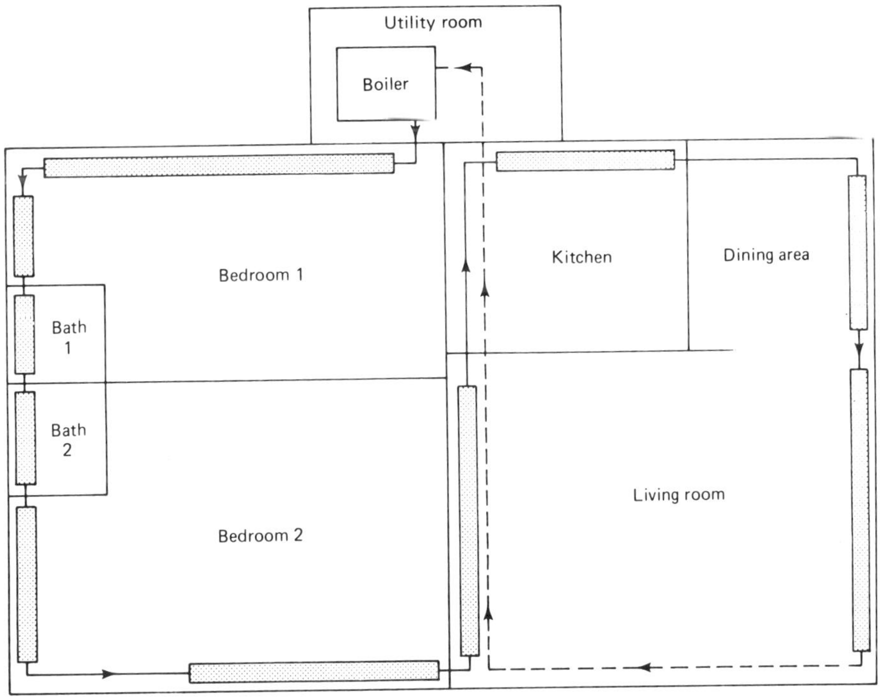

Figure 8.17 Single-zone residential heating system (not to scale). Solid lines, feed lines; dashed lines, return lines; shaded areas, baseboard radiation (representational only, not sized).

system or the heating zone that presently serves the living room could be extended to heat the addition.

If, during system modifications, existing heating zones are extended, it should be remembered that the heating capacity of each $\frac{3}{4}$-in. circuit should be limited to 40,000 Btu. If the expansion of the new baseboard is to exceed 40,000 Btu, larger feed and return branch lines will have to be incorporated into the system. In this instance, consult the pipe sizing requirements listed in Chapter 2 for limitations of differently sized circuits and baseboard configurations.

Conversion of steam to hydronic baseboard heating. Due to the radically increased costs of fuel that occurred during the 1970s, it is not uncommon to find that many older steam heating systems are being converted to more efficient baseboard forced-convection heating systems. While a simple conversion from steam to forced hot water heating using the same boiler may save between 10 and 20% on fuel costs, an additional updating of the older boiler to a new high-efficiency unit can often increase fuel savings to over 40%. To understand these conversions, the present steam system should be classified in terms of its design characteristics. While Chapter 6 examines steam heating in detail, for this analysis, steam systems can be

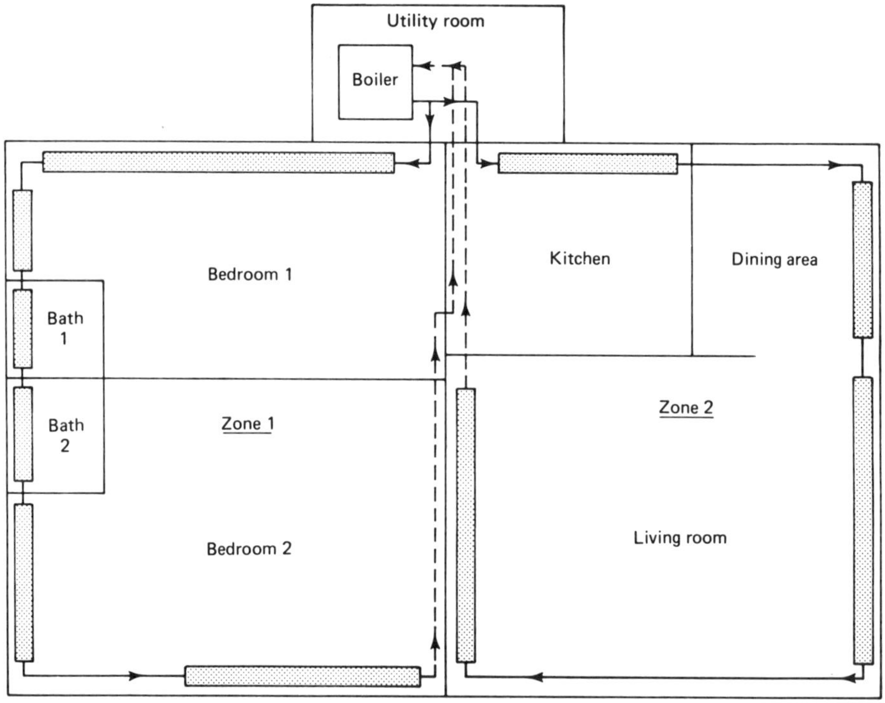

Figure 8.18 Reconfigured two-zone heating system. Solid lines, feed lines; dashed lines, return lines; shaded areas, baseboard radiation. No zone controls are illustrated.

classified as either one- or two-pipe circuits. Each of these systems is illustrated in Figure 8.19.

In a one-pipe system, both the steam and condensate flow in the same pipe (Figure 8.19a). In the two-pipe system, separate return lines are provided for condensate return (Figure 8.19b). Since one-pipe systems predominate in residential steam heating, all references in this section refer to the one-pipe system. For other systems, the reader is referred to Chapter 6.

The conversion of the one-pipe steam system to forced-circulation hot water baseboard is accomplished in three steps:

1. Removal and replacement of boiler controls and trim
2. Modifications of main piping
3. Piping of radiators, or addition of new baseboard heating units

Each of these stages is examined next in greater detail.

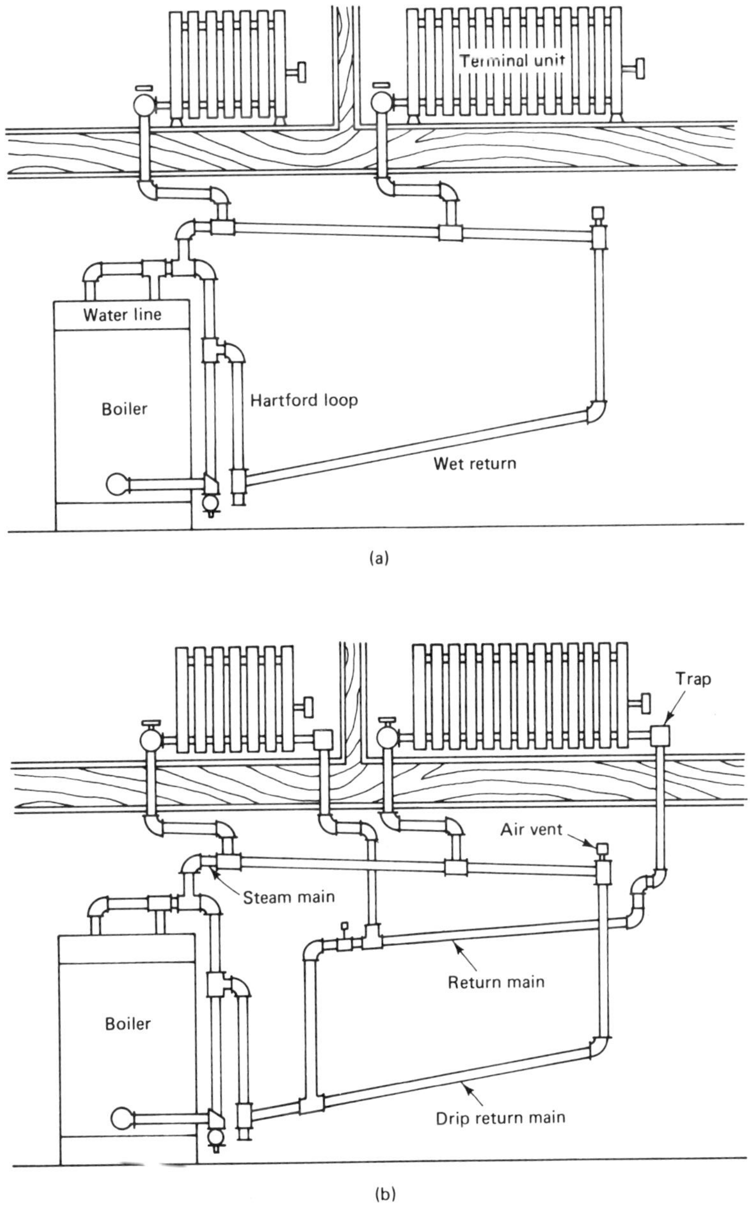

Figure 8.19 Steam heating systems: (a) one-pipe; (b) two-pipe. (Reprinted by permission of Small Homes Council, Building Research Council, University of Illinois, Technical Note 13.)

Stage 1: Removal and Replacement of Boiler Piping and Trim. Figure 8.20 illustrates the piping configuration of a typical steam boiler. When converting from a steam to a forced-circulation hot water system, all steam piping around the boiler must be removed, including all old operating controls. All steam risers, manifolds, traps, and the Hartford loop must be discarded. All controls, including the safety valve, gauge glass system, low-water cutoff, and steam gauges must also be removed. During this procedure, it is critical to make sure that the boiler and burner, whether it be gas or oil, is in reliable, serviceable condition. Particular care should be taken when removing any piping that is directly connected to the boiler tappings. Old piping, embrittled by years of use, can sometimes crack at the weakest point: the threaded connections.

After all the steam piping and trim has been removed, the hot water control system and piping in the vicinity of the boiler is added as illustrated in Figure 8.21. Also, a new thermostat, compatible with the boiler control system, should be installed at this time.

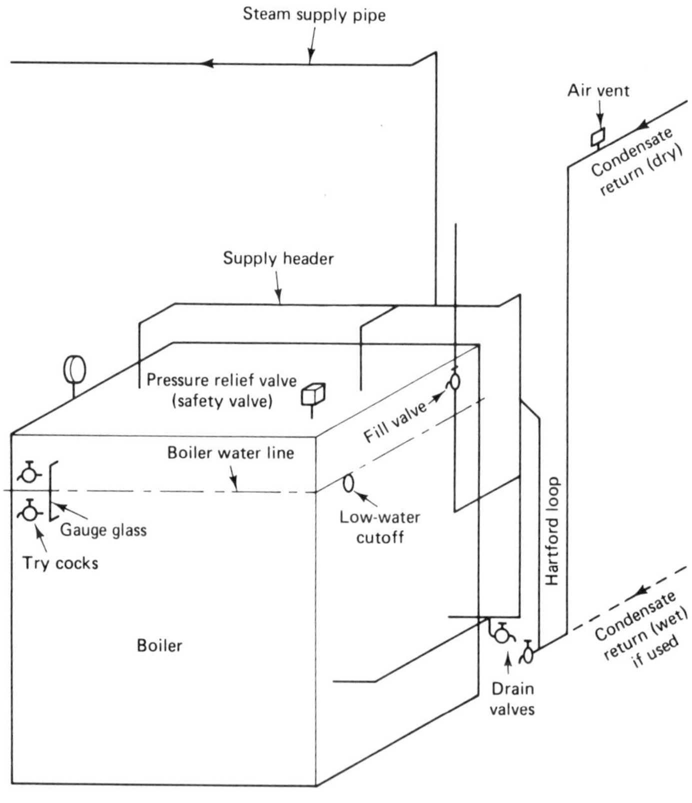

Figure 8.20 Typical piping configuration in the vicinity of a steam boiler. (Courtesy of the Hydronics Institute.)

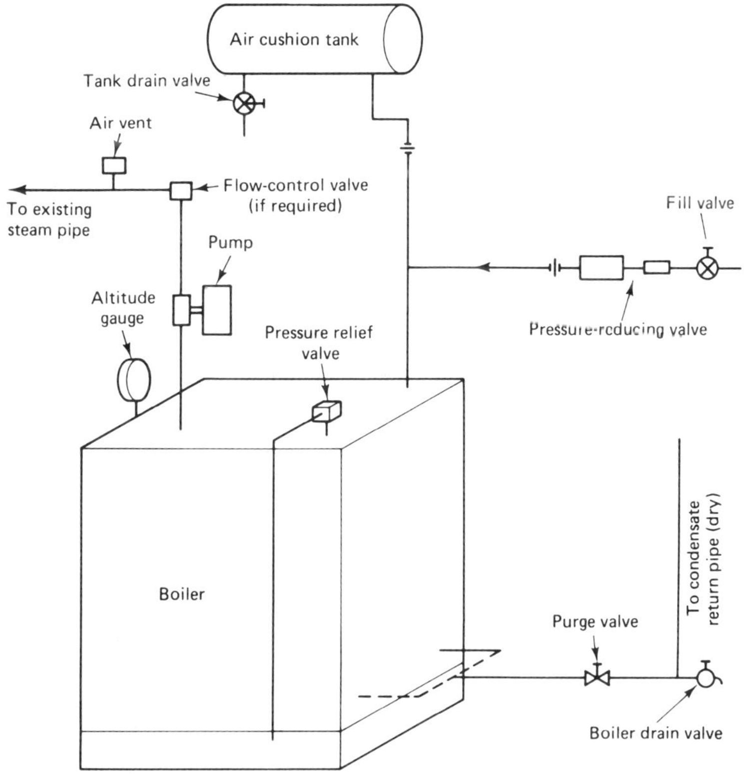

Figure 8.21 Hot water trim required for installation on a steam boiler when revamping the system. (Reprinted by permission of Small Homes Council, Building Research Council, University of Illinois, Technical Note 13.)

The expansion tank should be sized to accommodate the volume of water contained in the total heating system, including boiler contents and the volume of water to be housed in the piping and heat distribution units. In addition to zoning provisions, the piping configuration should include purging arrangements for each zone installed upstream of the circulator. If the steam boiler contained two or more condensate return lines, the open tappings should be plugged at this stage of the conversion.

Stage 2: Modifications of Main Piping. When converting a one-pipe steam system to a forced-circulation baseboard hot water system, all return piping from the radiators must be installed from scratch, since none existed before.

Ordinarily, the radiators will have two tappings, one on either end of the unit. However, the end of the radiator opposite the feed pipe will be plugged. These plugs will have to be removed, bushed down to proper size, and plumbed into the return line to the boiler. The return line for the radiators is illustrated in Figure 8.22.

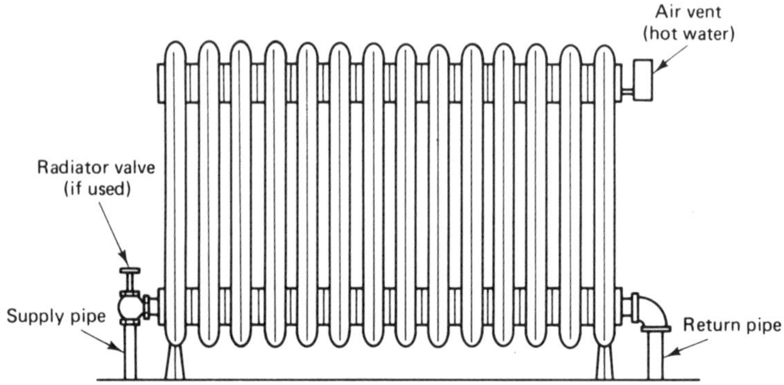

Figure 8.22 Steam radiator with return line installed in tapped plug hole. Note the location of the air vent for use in hydronic systems. (Reprinted by permission of Small Homes Council, Building Research Council, University of Illinois, Technical Note 13.)

Stage 3: Piping of Radiators, Addition of Baseboard. Existing steam radiators will need to be examined carefully during all system conversions. At the very least, old steam radiators will need to be reconfigured for the hot water system; in some instances they will need to be replaced altogether.

Most steam radiators have an air vent installed on the lower side opposite the steam feed line (Figure 8.23). These vents need to be removed and the holes plugged. Hot water air vents should be installed at the top of the radiator located at the nipple connection. If no hole exists in the radiator at this location, one should be drilled and tapped to accept the air vent using a $\frac{1}{8}$-in. NPT tap (Figure 8.24).

Steam radiator replacement. Old steam radiators should be replaced if they do not have bottom nipple connections that are tapped for installation of a return line, or if they only have a bottom set of nipple connections and none at the top, as shown in Figure 8.25.

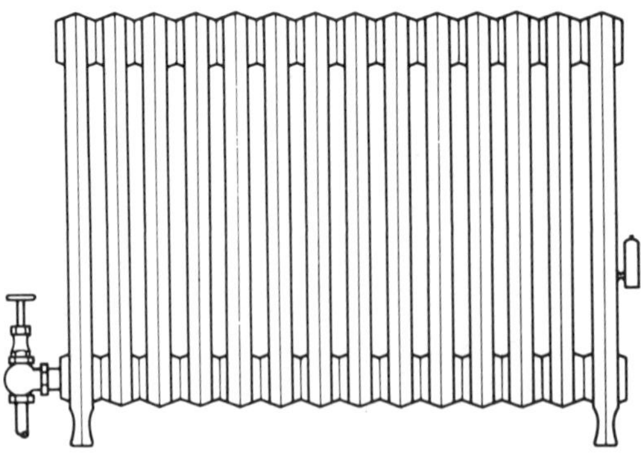

Figure 8.23 Venting of radiator in one-pipe steam system. A, Air vent; use lower opening in radiator. B, Radiator valve.

Figure 8.24 Placement of air vent on steam radiator used in hot water baseboard heating system. (Courtesy of Flair Mfg. Corp.)

Often, old-style radiators will be replaced by new baseboard radiation to improve the appearance of the room or to increase the heating capacity of the system in a particular room. When this is to be done, a design heat loss of the room needs to be performed. New radiation should be sized according to the room heat loss for the proper amount of delivered heat based on the operating temperature of the water in the heating system, usually between 180 and 200°F.

All new pipe runs should be sized according to the guidelines offered in Chapter 2. While most of the original piping in the system will be adequate for most modifications, all existing pipe sizes should be checked for suitability in the new system.

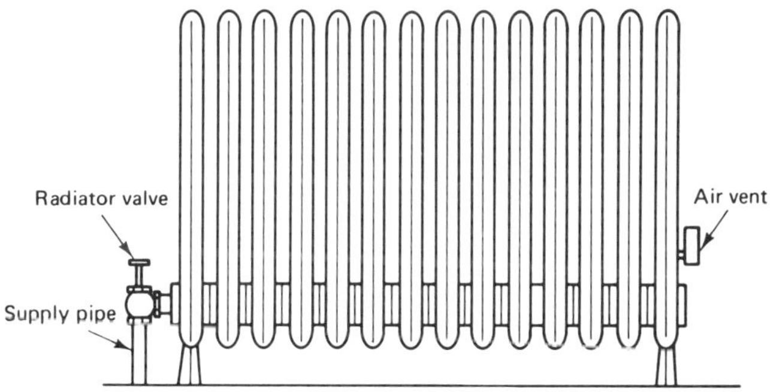

Figure 8.25 Old-style steam radiator (no top push nipples) that must be replaced when converting from steam to hot water baseboard heating. (Reprinted by permission of Small Homes Council, Building Research Council, University of Illinois, Technical Note 13.)

Boiler replacement. While many factors are involved in the decision to replace an older boiler with a new unit, none has had a greater effect on this decision than the introduction of new high-efficiency boilers. Incorporating the latest solid-state electronic circuitry and improved burner technology and boiler design, a simple boiler replacement in a steam heating system can offer fuel savings that are often in excess of 25%. For example, a new induced-draft gas-fired steam boiler with intermittent pilot ignition, solid-state circuitry, and improved boiler design and insulation can offer savings in excess of 20 to 30% over older units. Typical of these is the appliance shown in Figure 8.26. In this instance the replacement cost of a new steam boiler must be weighed against the cost of replacing and converting the entire heating system from steam to baseboard hot water. Careful cost comparisons will need to be performed before an informed decision can be made.

Also, many localities have energy conservation tax incentives that may help pay for the new boiler in the form of income tax credits or refunds.

In general terms, if the heating appliance is more than 20 years old, the savings can be of sufficient consequence to warrant replacement of the boiler, or at the very least the firing mechanism. Also, many steam boilers presently operating with oil or gas were converted from coal operation many years ago. In this instance, the savings to be gained from a boiler replacement are substantial, since the old coal-fired steam boilers had little or no internal baffling, limiting efficient transfer of heat.

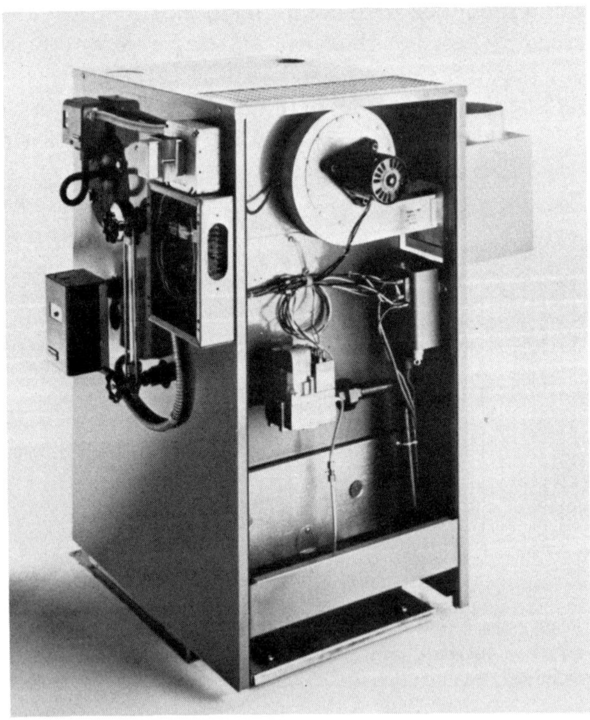

Figure 8.26 High-efficiency steam boiler. (Courtesy of Burnham Corp.)

Updated Fuel Ignition Systems

If it has been determined that an existing boiler, although 15 or 20 years old, is still in serviceable condition, it will almost always be highly cost effective to replace the existing burner with a newer, high-efficiency unit. Most gas-fired boilers, while not able to benefit from induced-draft technology conversions due to inherent deficiencies in the design of the boiler itself, can benefit from the installation of retrofit intermittent electronic spark ignition. Most oil-fired units can be equipped with new high-speed flame retention burners to replace older, low-speed units.

Gas retrofits. A typical retrofit installation for a gas-fired boiler featuring electronic ignition is shown in Figure 8.27. In this system, while the original pilot burner is kept in order to maintain the proper lighting of the boiler's main burners, most of the other controls are replaced and rewired as illustrated in Figure 8.28.

Oil retrofits. High-speed flame retention burners were discussed in detail in Chapter 4. As a major system modification and/or updating, burner replacement is perhaps one of the most frequently undertaken system updates. The same prerequisites applied to gas units are applicable to burner modifications; the boiler must be in serviceable condition for the update to be cost-effective.

The selection of a flame retention burner in a retrofit situation or update application involves several factors that must be taken into account. The size and shape of the combustion should be measured carefully in order to size properly the burner blower tube and nozzle firing angle. Also, the condition of the existing target

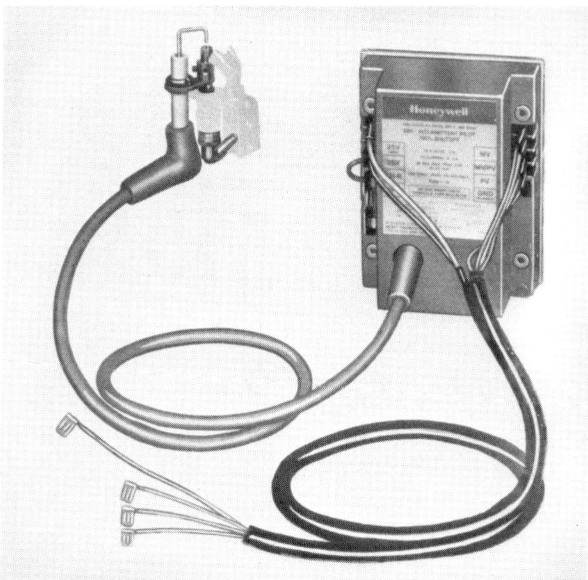

Figure 8.27 Retrofit solid-state ignition system kit for gas appliances. These kits contain all necessary controls to replace a standing pilot with an intermittent ignition system. In most instances, the original pilot burner on the gas appliance is used to ensure proper lighting of the main burners. (Courtesy of Honeywell, Inc.)

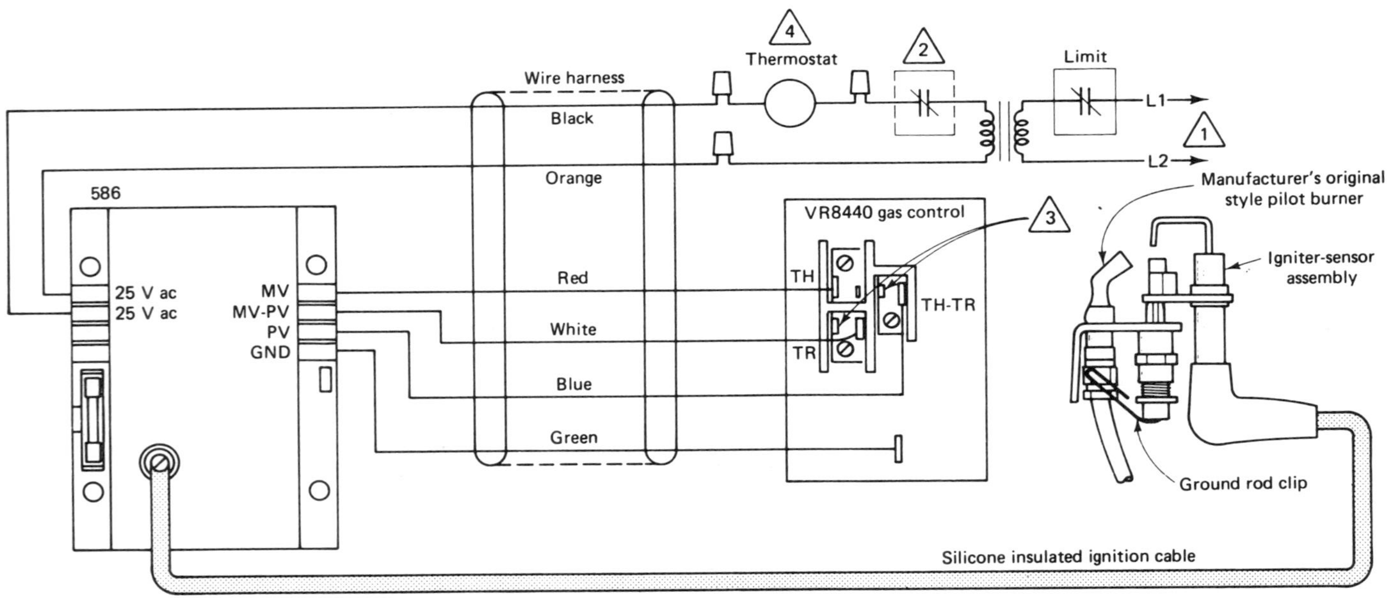

Figure 8.28 Wiring of retrofit electronic ignition system. 1, Power supply: disconnect means and overload protection provided as required; 2, alternate limit controller hookup; 3, factory-installed wiring, do not change; 4, set heat anticipator to 0.8 A. (Courtesy of Honeywell, Inc.)

wall must be determined in order to decide if new target walls are necessary or if the combustion chamber must be rebuilt.

After all these variables have been determined, the Btu capacity of the boiler, together with a sizing chart provided by the oil manufacturer, are used to determine the final specifications and sizing information for proper burner selection (Figure 8.29). In addition to installing a new burner, new switching controls (aquastats, and

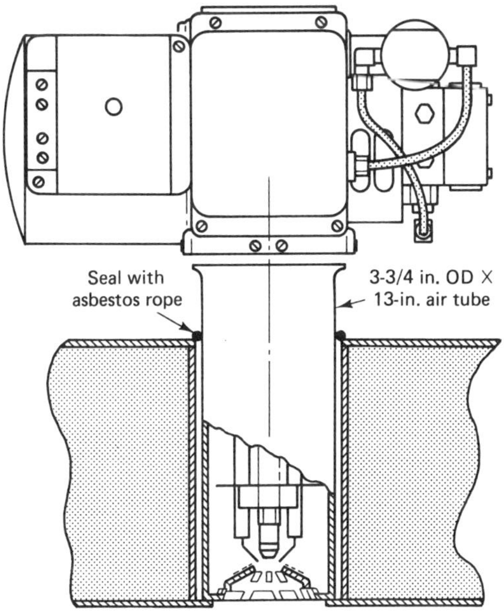

The following GE boiler models to be converted:

Models	Air tube combination	Suggested nozzle size[a]
LA 3	AFD 130 DN	1.00 (gal/hr)
LA 6	AFD 130 DO	1.75 (gal/hr)
LA 7	AFD 130 DP	2.00 (gal/hr)
LA 20	AFD 130 DR	0.75 (gal/hr)
LA 22	AFD 130 DN	1.00 (gal/hr)
LA 25	AFD 130 DN	1.00 (gal/hr)
LA 32	AFD 130 DO	1.25 (gal/hr)
LA 34	AFD 130 DP	2.00 (gal/hr)
LA 35	AFD 130 DO	1.25 (gal/hr)
LA 42	AFD 130 DP	2.25 (gal/hr)
LA 44	AFD 130 DP	2.75 (gal/hr)
LA 45	AFD 130 DP	2.00 (gal/hr)
LA 60	AFD 130 DP	2.50 (gal/hr)

[a] Some of these suggested rates represent a "down rating" of firing from original GE specifications. Conversion experience, however, has shown that stack temperatures have been lowered and effective heating has not been altered. Use 60° solid nozzles. (If a longer flame is required, use 45° solid nozzles.)

Figure 8.29 Conversion chart illustrating the conversion of a boiler series to updated high-speed flame retention burners. Charts of this type are typical and are usually available for a variety of boilers. In each case, the manufacturer of both the boiler and oil burner should be consulted prior to any conversion. (Courtesy of R.W. Beckett Corp.)

Figure 8.30 Retrofit off-cycle stack damper kits. (Courtesy of Flair Mfg. Corp.)

pyrostats, if applicable) should also be installed at this time. This type of update should provide many years of additional service from the unit, with accrued savings in fuel consumption that will range from 15 to 25% in most instances.

Automatic Vent Dampers

Separate automatic vent dampers, as described earlier in this chapter, together with their use in conventional oil- and gas-fired units (Chapters 4 and 5), should be installed in conjunction with a new boiler ignition system in order to maximize efficiency by minimizing system heat loss during off-cycle periods (Figure 8.30).

Flue Gas Heat Reclaimers

Another device that is sometimes installed as a retrofit component is a flue gas heat reclaimer. This device operates as a simple air-to-air heat exchanger, capturing waste heat from the flue gases as they exit the boiler. When the heat exchanger reaches a

Figure 8.31 Flue gas heat reclaimer. (Courtesy of Miracle Heat, Inc.)

preset temperature, a fan automatically comes on to blow hot reclaimed air into the interior of the residence. A typical device of this type is illustrated in Figure 8.31.

The effectiveness of flue gas heat reclaimers varies from one boiler installation to the next. For example, on modern condensing boilers that feature very low flue gas temperatures, these devices are of little value since flue temperatures exiting the boiler are very low. On the other hand, in less-efficient units in which stack temperatures in excess of 500 to 600°F are not uncommon, these devices can prove to be effective in reclaiming heat that would otherwise be lost up the chimney.

It should be noted that in instances where chimney draft is minimal and problematic, the use of these devices is not recommended. Since they effectively cool the flue gases by removing heat from the stack, their continual use tends to lower chimney draft. In situations where draft is not a problem, however, flue gas reclaimers can safely be incorporated into the appliance venting system.

Appendices

APPENDIX A

PIPE SIZING TABLE FOR TWO-PIPE REVERSE-RETURN SYSTEM

Available head (ft of water)	Total length of circuit (as measured on piping layout)															
4	35	45	50	60	65	70	75	80	90	100	110	130	150	180	220	290
5	45	60	65	70	80	90	95	100	120	130	140	160	190	230	290	360
6	55	70	80	90	100	110	120	130	140	160	180	200	240	290	350	450
7	65	90	100	110	120	130	140	150	170	190	210	240	290	340	420	540
8	75	100	110	130	140	150	160	180	200	220	250	290	330	400	490	620
9	85	110	130	150	160	170	190	200	230	250	290	330	380	450	560	710
10	100	130	140	170	180	190	210	230	260	290	320	370	430	510	620	790
11	110	140	160	190	200	220	240	260	290	320	360	410	480	570	690	880
12	120	160	180	200	220	240	260	290	320	350	400	450	530	620	760	960
14	150	190	210	250	260	290	310	340	380	420	470	540	620	730	900	1120

Pipe sizes: trunk, circuit, or branch (in.)	Capacity of trunk, circuit, or branches (gal/min)															
$\frac{3}{8}$[a]	0.9	0.8	0.7	0.7	0.6	0.6	0.6	0.6	0.5	0.5	0.5	0.5	0.4	0.4	0.3	0.3
$\frac{1}{2}$	2.3	2.0	1.9	1.8	1.7	1.7	1.6	1.5	1.5	1.4	1.3	1.2	1.2	1.1	0.9	0.8
$\frac{3}{4}$	5.0	4.3	4.1	3.8	3.7	3.6	3.4	3.2	3.1	2.9	2.8	2.6	2.4	2.2	2.0	1.8
1	9.6	8.3	7.7	7.3	7.0	6.8	6.5	6.3	5.9	5.7	5.5	5.0	4.6	4.3	3.8	3.4
$1\frac{1}{4}$	—	18	17	16	15	15	14	14	13	12	11	11	9.7	9.0	8.3	7.3
$1\frac{1}{2}$	—	27	25	24	23	22	21	20	19	18	17	16	15	13	12	11
2	—	—	—	—	—	42	40	39	38	36	34	32	29	27	24	21
$2\frac{1}{2}$	—	—	—	—	—	—	—	—	60	57	54	52	47	44	38	32

Source: The Hydronics Institute.

To use this table:

1. *Circuit or branch size.* Enter the left-hand column at the available head pressure of the pump selected. Read to the right to the figure closest to the total length of the circuit. Read down this column to the lower portion of the table to the gal/min capacity equal to or greater than the gal/min required for the circuit or branch. Read to the left-hand column to determine the pipe size required. Repeat the last step for each portion of the supply and return piping in the circuit.
2. *Trunk size.* Follow the same procedure using the head of the pump selected, the length of the longest circuit, and the gal/min being carried by the trunk. If multiple pumps are installed, size the trunk based on the lowest pump head, the length of the longest circuit, and the gal/min being carried by the pipe.

[a] $\frac{3}{8}$-in. copper tubing only.

APPENDIX B

PIPE SIZING TABLE FOR ONE-PIPE SYSTEM

Available head (ft of water)	Total length of circuit [measured length of circuit (not including branch piping) plus an allowance of 12 ft for each heat distributing unit connected to circuit]															
4	35	45	50	60	65	70	75	80	90	100	110	130	150	180	220	290
5	45	60	65	70	80	90	95	100	120	130	140	160	190	230	290	360
6	55	70	80	90	100	110	120	130	140	160	180	200	240	290	350	450
7	65	90	100	110	120	130	140	150	170	190	210	240	290	340	420	540
8	75	100	110	130	140	150	160	180	200	220	250	290	330	400	490	620
9	85	110	130	150	160	170	190	200	230	250	290	330	380	450	560	710
10	100	130	140	170	180	190	210	230	260	290	320	370	430	510	620	790
11	110	140	160	190	200	220	240	260	290	320	360	410	480	570	690	880
12	120	160	180	200	220	240	260	290	320	350	400	450	530	620	760	960
14	150	190	210	250	260	290	310	340	380	420	470	540	620	730	900	1120

Pipe or tubing size (in.)		Capacity of trunk or circuit and branches (gal/min) [pipe or copper tubing with one-pipe fittings (except $\frac{3}{8}$-in. branches[a])]															
Trunk or circuit	Branch																
$\frac{3}{4}$		5.0	4.3	4.1	3.8	3.7	3.6	3.4	3.2	3.1	2.9	2.8	2.6	2.4	2.2	2.0	1.8
	$\frac{3}{8}$[a]	0.9	0.8	0.8	0.7	0.7	0.7	0.7	0.7	0.7	0.4	0.4	0.4	0.4	0.4	0.4	0.3
	$\frac{1}{2}$	1.8	1.5	1.5	1.2	1.2	1.2	1.2	1.2	1.2	0.8	0.8	0.8	0.8	0.8	0.8	0.6
1		9.6	8.3	7.7	7.3	7.0	6.8	6.5	6.3	5.9	5.7	5.5	5.0	4.6	4.3	3.8	3.4
	$\frac{3}{8}$[a]	0.7	0.7	0.6	0.6	0.6	0.5	0.5	0.5	0.4	0.4	0.4	0.4	0.3	0.3	0.2	0.2
	$\frac{1}{2}$	1.4	1.3	1.1	1.1	1.1	1.0	1.0	1.0	0.8	0.8	0.8	0.8	0.7	0.7	0.5	0.5
	$\frac{3}{4}$	2.4	2.1	1.9	1.9	1.9	1.6	1.6	1.6	1.3	1.3	1.3	1.3	1.1	1.1	0.9	0.9
$1\frac{1}{4}$		—	18	17	16	15	15	14	14	13	12	11	11	9.7	9.0	8.3	7.3
	$\frac{3}{8}$[a]	—	0.9	0.9	0.8	0.8	0.8	0.7	0.7	0.7	0.7	0.6	0.6	0.4	0.4	0.4	0.3
	$\frac{1}{2}$	—	1.8	1.8	1.6	1.5	1.5	1.4	1.4	1.3	1.2	1.1	1.1	0.9	0.9	0.8	0.7
	$\frac{3}{4}$	—	2.9	2.9	2.6	2.4	2.4	2.3	2.3	2.1	2.0	1.8	1.8	1.5	1.5	1.3	1.2
$1\frac{1}{2}$		—	—	25	24	23	22	21	20	19	18	17	16	15	13	12	11
	$\frac{3}{8}$[a]	—	—	1.0	1.0	1.0	0.9	0.9	0.8	0.8	0.8	0.7	0.7	0.7	0.6	0.5	0.4
	$\frac{1}{2}$	—	—	2.0	2.0	1.9	1.8	1.7	1.6	1.6	1.5	1.4	1.3	1.2	1.1	1.0	0.9
	$\frac{3}{4}$	—	—	2.7	2.7	2.6	2.5	2.4	2.2	2.1	2.0	1.9	1.8	1.6	1.5	1.4	1.3

Source: The Hydronics Institute.

To use this table: In the lower portion of this table the gal/min capacities of the top line in each section are for sizing the circuit main or trunk. The gal/min capacities below the top line in a section are for sizing the branch pipes.

1. *Circuit main size.* Enter left-hand column of table with the available head pressure. Read to the right to the figure closest to the total length of the circuit. Read down this column into the lower portion of the table to a gal/min in the top line of a section that is equal to or greater than that required for the circuit.
2. *Branch pipe sizes.* Stay in the same column and section that was used to determine the circuit main size and use the gal/min capacities immediately below the top line to size the branches.
3. *Trunk.* Enter upper left-hand column with available head pressure, read to the right to the figure closest to the length of the longest circuit. Read down this column to a gal/min in the top line of a section that is equal to or greater than that required for the trunk. On this line read to the left to the left-hand column for the trunk size. If multiple pumps are installed, size the trunk based on the lowest pump head, the length of the longest circuit, and the gal/min being carried by the trunk.

Note 1: If the heat distributing units are located below the circuit, multiply the gal/min flow rate shown in the table for branches by 0.85.

Note 2: Flow rates shown for branches are based on a standard branch circuit composed of 15 ft of horizontal pipe, one heat distributing unit having a resistance of 3 elbow equivalents, radiator valve and elbows totaling 12 elbow equivalents. If the branch piping contains an unusual number of elbows or more than 15 ft of horizontal piping, decrease the gal/min branch flow rate indicated in this table 1% for each additional elbow equivalent and for each foot of horizontal pipe in excess of 15 ft.

[a] $\frac{3}{8}$-in. copper tubing only, with $\frac{1}{2}$ in. × $\frac{3}{8}$ in. bushing.

APPENDIX C

MAXIMUM RECOMMENDED LOOP OR CIRCUIT CAPACITIES USING BASEBOARDS

Baseboard size (in.)	Total load (Btu/hr)	Velocity based on 20°F temperature drop
Cast iron baseboard	40,000	2.4
$\frac{3}{4}$		
Steel pipe baseboard		0ab1.9
$\frac{1}{2}$	18,000	
$\frac{3}{4}$	40,000	2.4
1	72,000	2.7
$1\frac{1}{4}$	160,000	3.5
$1\frac{1}{2}$	230,000	3.7
2	450,000	4.6
Copper tube baseboard		
$\frac{1}{2}$	15,000	2.1
$\frac{3}{4}$	35,000	2.3
1	75,000	3.0
$1\frac{1}{4}$	130,000	3.5

Source: Taco, Inc.

APPENDIX D

AVERAGE MONTHLY AND YEARLY DEGREE-DAYS FOR CITIES IN THE UNITED STATES AND CANADA[a,b,c] (BASE 65°F)

State	Station	Avg. Winter Temp[d]	July	Aug.	Sept.	Oct.	Nov.	Dec.	Jan.	Feb.	Mar.	Apr.	May	June	Yearly Total
Ala.	Birmingham A	54.2	0	0	6	93	363	555	592	462	363	108	9	0	2551
	Huntsville A	51.3	0	0	12	127	426	663	694	557	434	138	19	0	3070
	Mobile A	59.9	0	0	0	22	213	357	415	300	211	42	0	0	1560
	Montgomery A	55.4	0	0	0	68	330	527	543	417	316	90	0	0	2291
Alaska	Anchorage A	23.0	245	291	516	930	1284	1572	1631	1316	1293	879	592	315	10864
	Fairbanks A	6.7	171	332	642	1203	1833	2254	2359	1901	1739	1068	555	222	14279
	Juneau A	32.1	301	338	483	725	921	1135	1237	1070	1073	810	601	381	9075
	Nome A	13.1	481	496	693	1094	1455	1820	1879	1666	1770	1314	930	573	14171
Ariz.	Flagstaff A	35.6	46	68	201	558	867	1073	1169	991	911	651	437	180	7152
	Phoenix A	58.5	0	0	0	22	234	415	474	328	217	75	0	0	1765
	Tucson A	58.1	0	0	0	25	231	406	471	344	242	75	6	0	1800
	Winslow A	43.0	0	0	6	245	711	1008	1054	770	601	291	96	0	4782
	Yuma A	64.2	0	0	0	0	108	264	307	190	90	15	0	0	974
Ark.	Fort Smith A	50.3	0	0	12	127	450	704	781	596	456	144	22	0	3292
	Little Rock A	50.5	0	0	9	127	465	716	756	577	434	126	9	0	3219
	Texarkana A	54.2	0	0	0	78	345	561	626	468	350	105	0	0	2533
Calif.	Bakersfield A	55.4	0	0	0	37	282	502	546	364	267	105	19	0	2122
	Bishop A	46.0	0	0	48	260	576	797	874	680	555	306	143	36	4275
	Blue Canyon A	42.2	28	37	108	347	594	781	896	795	806	597	412	195	5596
	Burbank A	58.6	0	0	6	43	177	301	366	277	239	138	81	18	1646
	Eureka C	49.9	270	257	258	329	414	499	546	470	505	438	372	285	4643
	Fresno A	53.3	0	0	0	84	354	577	605	426	335	162	62	6	2611
	Long Beach A	57.8	0	0	9	47	171	316	397	311	264	171	93	24	1803
	Los Angeles A	57.4	28	28	42	78	180	291	372	302	288	219	158	81	2061
	Los Angeles C	60.3	0	0	6	31	132	229	310	230	202	123	68	18	1349
	Mt. Shasta C	41.2	25	34	123	406	696	902	983	784	738	525	347	159	5722
	Oakland A	53.5	53	50	45	127	309	481	527	400	353	255	180	90	2870
	Red Bluff A	53.8	0	0	0	53	318	555	605	428	341	168	47	0	2515
	Sacramento A	53.9	0	0	0	56	321	546	583	414	332	178	72	0	2502
	Sacramento C	54.4	0	0	0	62	312	533	561	392	310	173	76	0	2419
	Sandberg C	46.8	0	0	30	202	480	691	778	661	620	426	264	57	4209
	San Diego A	59.5	9	0	21	43	135	236	298	235	214	135	90	42	1458
	San Francisco A	53.4	81	78	60	143	306	462	508	395	363	279	214	126	3015
	San Francisco C	55.1	192	174	102	118	231	388	443	336	319	279	239	180	3001
	Santa Maria A	54.3	99	93	96	146	270	391	459	370	363	282	233	165	2967
Colo.	Alamosa A	29.7	65	99	279	639	1065	1420	1476	1162	1020	696	440	168	8529
	Colorado Springs A	37.3	9	25	132	456	825	1032	1128	938	893	582	319	84	6423
	Denver A	37.6	6	9	117	428	819	1035	1132	938	887	558	288	66	6283
	Denver C	40.8	0	0	90	366	714	905	1004	851	800	492	254	48	5524
	Grand Junction A	39.3	0	0	30	313	786	1113	1209	907	729	387	146	21	5641
	Pueblo A	40.4	0	0	54	326	750	986	1085	871	772	429	174	15	5462
Conn.	Bridgeport A	39.9	0	0	66	307	615	986	1079	966	853	510	208	27	5617
	Hartford A	37.3	0	12	117	394	714	1101	1190	1042	908	519	205	33	6235
	New Haven A	39.0	0	12	87	347	648	1011	1097	991	871	543	245	45	5897
Del.	Wilmington A	42.5	0	0	51	270	588	927	980	874	735	387	112	6	4930
D.C.	Washington A	45.7	0	0	33	217	519	834	871	762	626	288	74	0	4224
Fla.	Apalachicola C	61.2	0	0	0	16	153	319	347	260	180	33	0	0	1308
	Daytona Beach A	64.5	0	0	0	0	75	211	248	190	140	15	0	0	879
	Fort Myers A	68.6	0	0	0	0	24	109	146	101	62	0	0	0	442
	Jacksonville A	61.9	0	0	0	12	144	310	332	246	174	21	0	0	1239
	Key West A	73.1	0	0	0	0	0	28	40	31	9	0	0	0	108
	Lakeland C	66.7	0	0	0	0	57	164	195	146	99	0	0	0	661
	Miami A	71.1	0	0	0	0	0	65	74	56	19	0	0	0	214

[a]Data for United States cities from a publication of the United States Weather Bureau, *Monthly Normals of Temperature, Precipitation and Heating Degree Days*, 1962, are for the period 1931 to 1960 inclusive. These data also include information from the 1963 revisions to this publication, where available.
[b]Data for airport stations, A, and city stations, C, are both given where available.
[c]Data for Canadian cities were computed by the Climatology Division, Department of Transport from normal monthly mean temperatures, and the monthly values of heating degree days data were obtained using the National Research Council computer and a method devised by H. C. S. Thom of the United States Weather Bureau. The heating degree days are based on the period from 1931 to 1960.
[d]For period October to April, inclusive.

Source: ASHRAE, *Handbook of Fundamentals*, 1981.

APPENDIX D *(continued)*

State	Station	Avg. Winter Temp[d]	July	Aug.	Sept.	Oct.	Nov.	Dec.	Jan.	Feb.	Mar.	Apr.	May	June	Yearly Total
Fla. (Cont'd)	Miami Beach C	72.5	0	0	0	0	0	40	56	36	9	0	0	0	141
	Orlando A	65.7	0	0	0	0	72	198	220	165	105	6	0	0	766
	Pensacola A	60.4	0	0	0	19	195	353	400	277	183	36	0	0	1463
	Tallahassee A	60.1	0	0	0	28	198	360	375	286	202	36	0	0	1485
	Tampa A	66.4	0	0	0	0	60	171	202	148	102	0	0	0	683
	West Palm Beach A	68.4	0	0	0	0	6	65	87	64	31	0	0	0	253
Ga.	Athens A	51.8	0	0	12	115	405	632	642	529	431	141	22	0	2929
	Atlanta A	51.7	0	0	18	124	417	648	636	518	428	147	25	0	2961
	Augusta A	54.5	0	0	0	78	333	552	549	445	350	90	0	0	2397
	Columbus A	54.8	0	0	0	87	333	543	552	434	338	96	0	0	2383
	Macon A	56.2	0	0	0	71	297	502	505	403	295	63	0	0	2136
	Rome A	49.9	0	0	24	161	474	701	710	577	468	177	34	0	3326
	Savannah A	57.8	0	0	0	47	246	437	437	353	254	45	0	0	1819
	Thomasville C	60.0	0	0	0	25	198	366	394	305	208	33	0	0	1529
Hawaii	Lihue A	72.7	0	0	0	0	0	0	0	0	0	0	0	0	0
	Honolulu A	74.2	0	0	0	0	0	0	0	0	0	0	0	0	0
	Hilo A	71.9	0	0	0	0	0	0	0	0	0	0	0	0	0
Idaho	Boise A	39.7	0	0	132	415	792	1017	1113	854	722	438	245	81	5809
	Lewiston A	41.0	0	0	123	403	756	933	1063	815	694	426	239	90	5542
	Pocatello A	34.8	0	0	172	493	900	1166	1324	1058	905	555	319	141	7033
Ill.	Cairo C	47.9	0	0	36	164	513	791	856	680	539	195	47	0	3821
	Chicago (O'Hare) A	35.8	0	12	117	381	807	1166	1265	1086	939	534	260	72	6639
	Chicago (Midway) A	37.5	0	0	81	326	753	1113	1209	1044	890	480	211	48	6155
	Chicago C	38.9	0	0	66	279	705	1051	1150	1000	868	489	226	48	5882
	Moline A	36.4	0	9	99	335	774	1181	1314	1100	918	450	189	39	6408
	Peoria A	38.1	0	6	87	326	759	1113	1218	1025	849	426	183	33	6025
	Rockford A	34.8	6	9	114	400	837	1221	1333	1137	961	516	236	60	6830
	Springfield A	40.6	0	0	72	291	696	1023	1135	935	769	354	136	18	5429
Ind.	Evansville A	45.0	0	0	66	220	606	896	955	767	620	237	68	0	4435
	Fort Wayne A	37.3	0	9	105	378	783	1135	1178	1028	890	471	189	39	6205
	Indianapolis A	39.6	0	0	90	316	723	1051	1113	949	809	432	177	39	5699
	South Bend A	36.6	0	6	111	372	777	1125	1221	1070	933	525	239	60	6439
Iowa	Burlington A	37.6	0	0	93	322	768	1135	1259	1042	859	426	177	33	6114
	Des Moines A	35.5	0	6	96	363	828	1225	1370	1137	915	438	180	30	6588
	Dubuque A	32.7	12	31	156	450	906	1287	1420	1204	1026	546	260	78	7376
	Sioux City A	34.0	0	9	108	369	867	1240	1435	1198	989	483	214	39	6951
	Waterloo A	32.6	12	19	138	428	909	1296	1460	1221	1023	531	229	54	7320
Kans.	Concordia A	40.4	0	0	57	276	705	1023	1163	935	781	372	149	18	5479
	Dodge City A	42.5	0	0	33	251	666	939	1051	840	719	354	124	9	4986
	Goodland A	37.8	0	6	81	381	810	1073	1166	955	884	507	236	42	6141
	Topeka A	41.7	0	0	57	270	672	980	1122	893	722	330	124	12	5182
	Wichita A	44.2	0	0	33	229	618	905	1023	804	645	270	87	6	4620
Ky.	Covington A	41.4	0	0	75	291	669	983	1035	893	756	390	149	24	5265
	Lexington A	43.8	0	0	54	239	609	902	946	818	685	325	105	0	4683
	Louisville A	44.0	0	0	54	248	609	890	930	818	682	315	105	9	4660
La.	Alexandria A	57.5	0	0	0	56	273	431	471	361	260	69	0	0	1921
	Baton Rouge A	59.8	0	0	0	31	216	369	409	294	208	33	0	0	1560
	Lake Charles A	60.5	0	0	0	19	210	341	381	274	195	39	0	0	1459
	New Orleans A	61.0	0	0	0	19	192	322	363	258	192	39	0	0	1385
	New Orleans C	61.8	0	0	0	12	165	291	344	241	177	24	0	0	1254
	Shreveport A	56.2	0	0	0	47	297	477	552	426	304	81	0	0	2184
Me.	Caribou A	24.4	78	115	336	682	1044	1535	1690	1470	1308	858	468	183	9767
	Portland A	33.0	12	53	195	508	807	1215	1339	1182	1042	675	372	111	7511
Md.	Baltimore A	43.7	0	0	48	264	585	905	936	820	679	327	90	0	4654
	Baltimore C	46.2	0	0	27	189	486	806	859	762	629	288	65	0	4111
	Frederich A	42.0	0	0	66	307	624	955	995	876	741	384	127	12	5087
Mass.	Boston A	40.0	0	9	60	316	603	983	1088	972	846	513	208	36	5634
	Nantucket A	40.2	12	22	93	332	573	896	992	941	896	621	384	129	5891
	Pittsfield A	32.6	25	59	219	524	831	1231	1339	1196	1063	660	326	105	7578
	Worcester A	34.7	6	34	147	450	774	1172	1271	1123	998	612	304	78	6969

(continued)

APPENDIX D *(continued)*

State	Station	Avg. Winter Temp[d]	July	Aug.	Sept.	Oct.	Nov.	Dec.	Jan.	Feb.	Mar.	Apr.	May	June	Yearly Total
Mich.	Alpena A	29.7	68	105	273	580	912	1268	1404	1299	1218	777	446	156	8506
	Detroit (City) A	37.2	0	0	87	360	738	1088	1181	1058	936	522	220	42	6232
	Detroit (Wayne) A	37.1	0	0	96	353	738	1088	1194	1061	933	534	239	57	6293
	Detroit (Willow Run) A	37.2	0	0	90	357	750	1104	1190	1053	921	519	229	45	6258
	Escanaba C	29.6	59	87	243	539	924	1293	1445	1296	1203	777	456	159	8481
	Flint A	33.1	16	40	159	465	843	1212	1330	1198	1066	639	319	90	7377
	Grand Rapids A	34.9	9	28	135	434	804	1147	1259	1134	1011	579	279	75	6894
	Lansing A	34.8	6	22	138	431	813	1163	1262	1142	1011	579	273	69	6909
	Marquette C	30.2	59	81	240	527	936	1268	1411	1268	1187	771	468	177	8393
	Muskegon A	36.0	12	28	120	400	762	1088	1209	1100	995	594	310	78	6696
	Sault Ste. Marie A	27.7	96	105	279	580	951	1367	1525	1380	1277	810	477	201	9048
Minn.	Duluth A	23.4	71	109	330	632	1131	1581	1745	1518	1355	840	490	198	10000
	Minneapolis A	28.3	22	31	189	505	1014	1454	1631	1380	1166	621	288	81	8382
	Rochester A	28.8	25	34	186	474	1005	1438	1593	1366	1150	630	301	93	8295
Miss.	Jackson A	55.7	0	0	0	65	315	502	546	414	310	87	0	0	2239
	Meridian A	55.4	0	0	0	81	339	518	543	417	310	81	0	0	2289
	Vicksburg C	56.9	0	0	0	53	279	462	512	384	282	69	0	0	2041
Mo.	Columbia A	42.3	0	0	54	251	651	967	1076	874	716	324	121	12	5046
	Kansas City A	43.9	0	0	39	220	612	905	1032	818	682	294	109	0	4711
	St. Joseph A	40.3	0	6	60	285	708	1039	1172	949	769	348	133	15	5484
	St. Louis A	43.1	0	0	60	251	627	936	1026	848	704	312	121	15	4900
	St. Louis C	44.8	0	0	36	202	576	884	977	801	651	270	87	0	4484
	Springfield A	44.5	0	0	45	223	600	877	973	781	660	291	105	6	4900
Mont.	Billings A	34.5	6	15	186	487	897	1135	1296	1100	970	570	285	102	7049
	Glasgow A	26.4	31	47	270	608	1104	1466	1711	1439	1187	648	335	150	8996
	Great Falls A	32.8	28	53	258	543	921	1169	1349	1154	1063	642	384	186	7750
	Havre A	28.1	28	53	306	595	1065	1367	1584	1364	1181	657	338	162	8700
	Havre C	29.8	19	37	252	539	1014	1321	1528	1305	1116	612	304	135	8182
	Helena A	31.1	31	59	294	601	1002	1265	1438	1170	1042	651	381	195	8129
	Kalispell A	31.4	50	99	321	654	1020	1240	1401	1134	1029	639	397	207	8191
	Miles City A	31.2	6	6	174	502	972	1296	1504	1252	1057	579	276	99	7723
	Missoula A	31.5	34	74	303	651	1035	1287	1420	1120	970	621	391	219	8125
Neb.	Grand Island A	36.0	0	6	108	381	834	1172	1314	1089	908	462	211	45	6530
	Lincoln C	38.8	0	6	75	301	726	1066	1237	1016	834	402	171	30	5864
	Norfolk A	34.0	9	0	111	397	873	1234	1414	1179	983	498	233	48	6979
	North Platte A	35.5	0	6	123	440	885	1166	1271	1039	930	519	248	57	6684
	Omaha A	35.6	0	12	105	357	828	1175	1355	1126	939	465	208	42	6612
	Scottsbluff A	35.9	0	0	138	459	876	1128	1231	1008	921	552	285	75	6673
	Valentine A	32.6	9	12	165	493	942	1237	1395	1176	1045	579	288	84	7425
Nev.	Elko A	34.0	9	34	225	561	924	1197	1314	1036	911	621	409	192	7433
	Ely A	33.1	28	43	234	592	939	1184	1308	1075	977	672	456	225	7733
	Las Vegas A	53.5	0	0	0	78	387	617	688	487	335	111	6	0	2709
	Reno A	39.3	43	87	204	490	801	1026	1073	823	729	510	357	189	6332
	Winnemucca A	36.7	0	34	210	536	876	1091	1172	916	837	573	363	153	6761
N.H.	Concord A	33.0	6	50	177	505	822	1240	1358	1184	1032	636	298	75	7383
	Mt. Washington Obsv.	15.2	493	536	720	1057	1341	1742	1820	1663	1652	1260	930	603	13817
N.J.	Atlantic City A	43.2	0	0	39	251	549	880	936	848	741	420	133	15	4812
	Newark A	42.8	0	0	30	248	573	921	983	876	729	381	118	0	4589
	Trenton C	42.4	0	0	57	264	576	924	989	885	753	399	121	12	4980
N. M.	Albuquerque A	45.0	0	0	12	229	642	868	930	703	595	288	81	0	4348
	Clayton A	42.0	0	6	66	310	699	899	986	812	747	429	183	21	5158
	Raton A	38.1	9	28	126	431	825	1048	1116	904	834	543	301	63	6228
	Roswell A	47.5	0	0	18	202	573	806	840	641	481	201	31	0	3793
	Silver City A	48.0	0	0	6	183	525	729	791	605	518	261	87	0	3705
N.Y.	Albany A	34.6	0	19	138	440	777	1194	1311	1156	992	564	239	45	6875
	Albany C	37.2	0	9	102	375	699	1104	1218	1072	908	498	186	30	6201
	Binghamton A	33.9	22	65	201	471	810	1184	1277	1154	1045	645	313	99	7286
	Binghamton C	36.6	0	28	141	406	732	1107	1190	1081	949	543	229	45	6451
	Buffalo A	34.5	19	37	141	440	777	1156	1256	1145	1039	645	329	78	7062
	New York (Cent. Park) C	42.8	0	0	30	233	540	902	986	885	760	408	118	9	4871
	New York (La Guardia) A	43.1	0	0	27	223	528	887	973	879	750	414	124	6	4811

APPENDIX D *(continued)*

State	Station	Avg. Winter Temp[d]	July	Aug.	Sept.	Oct.	Nov.	Dec.	Jan.	Feb.	Mar.	Apr.	May	June	Yearly Total
	New York (Kennedy) .... A	41.4	0	0	36	248	564	933	1029	935	815	480	167	12	5219
	Rochester .... A	35.4	9	31	126	415	747	1125	1234	1123	1014	597	279	48	6748
	Schenectady C	35.4	0	22	123	422	756	1159	1283	1131	970	543	211	30	6650
	Syracuse .... A	35.2	6	28	132	415	744	1153	1271	1140	1004	570	248	45	6756
N. C.	Asheville C	46.7	0	0	48	245	555	775	784	683	592	273	87	0	4042
	Cape Hatteras	53.3	0	0	0	78	273	521	580	518	440	177	25	0	2612
	Charlotte .... A	50.4	0	0	6	124	438	691	691	582	481	156	22	0	3191
	Greensboro .... A	47.5	0	0	33	192	513	778	784	672	552	234	47	0	3805
	Raleigh .... A	49.4	0	0	21	164	450	716	725	616	487	180	34	0	3393
	Wilmington .... A	54.6	0	0	0	74	291	521	546	462	357	96	0	0	2347
	Winston-Salem .... A	48.4	0	0	21	171	483	747	753	652	524	207	37	0	3595
N. D.	Bismarck .... A	26.6	34	28	222	577	1083	1463	1708	1442	1203	645	329	117	8851
	Devils Lake C	22.4	40	53	273	642	1191	1634	1872	1579	1345	753	381	138	9901
	Fargo .... A	24.8	28	37	219	574	1107	1569	1789	1520	1262	690	332	99	9226
	Williston .... A	25.2	31	43	261	601	1122	1513	1758	1473	1262	681	357	141	9243
Ohio	Akron-Canton .... A	38.1	0	9	96	381	726	1070	1138	1016	871	489	202	39	6037
	Cincinnati C	45.1	0	0	39	208	558	862	915	790	642	294	96	6	4410
	Cleveland .... A	37.2	9	25	105	384	738	1088	1159	1047	918	552	260	66	6351
	Columbus .... A	39.7	0	6	84	347	714	1039	1088	949	809	426	171	27	5660
	Columbus C	41.5	0	0	57	285	651	977	1032	902	760	396	136	15	5211
	Dayton .... A	39.8	0	6	78	310	696	1045	1097	955	809	429	167	30	5622
	Mansfield .... A	36.9	9	22	114	397	768	1110	1169	1042	924	543	245	60	6403
	Sandusky C	39.1	0	6	66	313	684	1032	1107	991	868	495	198	36	5796
	Toledo .... A	36.4	0	16	117	406	792	1138	1200	1056	924	543	242	60	6494
	Youngstown .... A	36.8	6	19	120	412	771	1104	1169	1047	921	540	248	60	6417
Okla.	Oklahoma City .... A	48.3	0	0	15	164	498	766	868	664	527	189	34	0	3725
	Tulsa .... A	47.7	0	0	18	158	522	787	893	683	539	213	47	0	3860
Ore.	Astoria .... A	45.6	146	130	210	375	561	679	753	622	636	480	363	231	5186
	Burns C	35.9	12	37	210	515	867	1113	1246	988	856	570	366	177	6957
	Eugene .... A	45.6	34	34	129	366	585	719	803	627	589	426	279	135	4726
	Meacham .... A	34.2	84	124	288	580	918	1091	1209	1005	983	726	527	339	7874
	Medford .... A	43.2	0	0	78	372	678	871	918	697	642	432	242	78	5008
	Pendleton .... A	42.6	0	0	111	350	711	884	1017	773	617	396	205	63	5127
	Portland .... A	45.6	25	28	114	335	597	735	825	644	586	396	245	105	4635
	Portland C	47.4	12	16	75	267	534	679	769	594	536	351	198	78	4109
	Roseburg .... A	46.3	22	16	105	329	567	713	766	608	570	405	267	123	4491
	Salem .... A	45.4	37	31	111	338	594	729	822	647	611	417	273	144	4754
Pa.	Allentown .... A	38.9	0	0	90	353	693	1045	1116	1002	849	471	167	24	5810
	Erie .... A	36.8	0	25	102	391	714	1063	1169	1081	973	585	288	60	6451
	Harrisburg .... A	41.2	0	0	63	298	648	992	1045	907	766	396	124	12	5251
	Philadelphia .... A	41.8	0	0	60	297	620	965	1016	889	747	392	118	40	5144
	Philadelphia C	44.5	0	0	30	205	513	856	924	823	691	351	93	0	4486
	Pittsburgh .... A	38.4	0	9	105	375	726	1063	1119	1002	874	480	195	39	5987
	Pittsburgh C	42.2	0	0	60	291	615	930	983	885	763	390	124	12	5053
	Reading C	42.4	0	0	54	257	597	939	1001	885	735	372	105	0	4945
	Scranton .... A	37.2	0	19	132	434	762	1104	1156	1028	893	498	195	33	6254
	Williamsport .... A	38.5	0	9	111	375	717	1073	1122	1002	856	468	177	24	5934
R. I.	Block Island .... A	40.1	0	16	78	307	594	902	1020	955	877	612	344	99	5804
	Providence .... A	38.8	0	16	96	372	660	1023	1110	988	868	534	236	51	5954
S. C.	Charleston .... A	56.4	0	0	0	59	282	471	487	389	291	54	0	0	2033
	Charleston C	57.9	0	0	0	34	210	425	443	367	273	42	0	0	1794
	Columbia .... A	54.0	0	0	0	84	345	577	570	470	357	81	0	0	2484
	Florence .... A	54.5	0	0	0	78	315	552	552	459	347	84	0	0	2387
	Greenville-Spartenburg .... A	51.6	0	0	6	121	399	651	660	546	446	132	19	0	2980
S. D.	Huron .... A	28.8	9	12	165	508	1014	1432	1628	1355	1125	600	288	87	8223
	Rapid City .... A	33.4	22	12	165	481	897	1172	1333	1145	1051	615	326	126	7345
	Sioux Falls .... A	30.6	19	25	168	462	972	1361	1544	1285	1082	573	270	78	7839
Tenn.	Bristol .... A	46.2	0	0	51	236	573	828	828	700	598	261	68	0	4143
	Chattanooga .... A	50.3	0	0	18	143	468	698	722	577	453	150	25	0	3254
	Knoxville .... A	49.2	0	0	30	171	489	725	732	613	493	198	43	0	3494
	Memphis .... A	50.5	0	0	18	130	447	698	729	585	456	147	22	0	3232

(continued)

APPENDIX D *(continued)*

State or Prov.	Station	Avg. Winter Temp[d]	July	Aug.	Sept.	Oct.	Nov.	Dec.	Jan.	Feb.	Mar.	Apr.	May	June	Yearly Total
	Memphis C	51.6	0	0	12	102	396	648	710	568	434	129	16	0	3015
	Nashville A	48.9	0	0	30	158	495	732	778	644	512	189	40	0	3578
	Oak Ridge C	47.7	0	0	39	192	531	772	778	669	552	228	56	0	3817
Tex.	Abilene A	53.9	0	0	0	99	366	586	642	470	347	114	0	0	2624
	Amarillo A	47.0	0	0	18	205	570	797	877	664	546	252	56	0	3985
	Austin A	59.1	0	0	0	31	225	388	468	325	223	51	0	0	1711
	Brownsville A	67.7	0	0	0	0	66	149	205	106	74	0	0	0	600
	Corpus Christi A	64.6	0	0	0	0	120	220	291	174	109	0	0	0	914
	Dallas A	55.3	0	0	0	62	321	524	601	440	319	90	6	0	2363
	El Paso A	52.9	0	0	0	84	414	648	685	445	319	105	0	0	2700
	Fort Worth A	55.1	0	0	0	65	324	536	614	448	319	99	0	0	2405
	Galveston A	62.2	0	0	0	6	147	276	360	263	189	33	0	0	1274
	Galveston C	62.0	0	0	0	0	138	270	350	258	189	30	0	0	1235
	Houston A	61.0	0	0	0	6	183	307	384	288	192	36	0	0	1396
	Houston C	62.0	0	0	0	0	165	288	363	258	174	30	0	0	1278
	Laredo A	66.0	0	0	0	0	105	217	267	134	74	0	0	0	797
	Lubbock A	48.8	0	0	18	174	513	744	800	613	484	201	31	0	3578
	Midland A	53.8	0	0	0	87	381	592	651	468	322	90	0	0	2591
	Port Arthur A	60.5	0	0	0	22	207	329	384	274	192	39	0	0	1447
	San Angelo A	56.0	0	0	0	68	318	536	567	412	288	66	0	0	2255
	San Antonio A	60.1	0	0	0	31	204	363	428	286	195	39	0	0	1546
	Victoria A	62.7	0	0	0	6	150	270	344	230	152	21	0	0	1173
	Waco A	57.2	0	0	0	43	270	456	536	389	270	66	0	0	2030
	Wichita Falls A	53.0	0	0	0	99	381	632	698	518	378	120	6	0	2832
Utah	Milford A	36.5	0	0	99	443	867	1141	1252	988	822	519	279	87	6497
	Salt Lake City A	38.4	0	0	81	419	849	1082	1172	910	763	459	233	84	6052
	Wendover A	39.1	0	0	48	372	822	1091	1178	902	729	408	177	51	5778
Vt.	Burlington A	29.4	28	65	207	539	891	1349	1513	1333	1187	714	353	90	8269
Va.	Cape Henry C	50.0	0	0	0	112	360	645	694	633	536	246	53	0	3279
	Lynchburg A	46.0	0	0	51	223	540	822	849	731	605	267	78	0	4166
	Norfolk A	49.2	0	0	0	136	408	698	738	655	533	216	37	0	3421
	Richmond A	47.3	0	0	36	214	495	784	815	703	546	219	53	0	3865
	Roanoke A	46.1	0	0	51	229	549	825	834	722	614	261	65	0	4150
Wash.	Olympia A	44.2	68	71	198	422	636	753	834	675	645	450	307	177	5236
	Seattle-Tacoma A	44.2	56	62	162	391	633	750	828	678	657	474	295	159	5145
	Seattle C	46.9	50	47	129	329	543	657	738	599	577	396	242	117	4424
	Spokane A	36.5	9	25	168	493	879	1082	1231	980	834	531	288	135	6655
	Walla Walla C	43.8	0	0	87	310	681	843	986	745	589	342	177	45	4805
	Yakima A	39.1	0	12	144	450	828	1039	1163	868	713	435	220	69	5941
W. Va.	Charleston A	44.8	0	0	63	254	591	865	880	770	648	300	96	9	4476
	Elkins A	40.1	9	25	135	400	729	992	1008	896	791	444	198	48	5675
	Huntington A	45.0	0	0	63	257	585	856	880	764	636	294	99	12	4446
	Parkersburg C	43.5	0	0	60	264	606	905	942	826	691	339	115	6	4754
Wisc.	Green Bay A	30.3	28	50	174	484	924	1333	1494	1313	1141	654	335	99	8029
	La Crosse A	31.5	12	19	153	437	924	1339	1504	1277	1070	540	245	69	7589
	Madison A	30.9	25	40	174	474	930	1330	1473	1274	1113	618	310	102	7863
	Milwaukee A	32.6	43	47	174	471	876	1252	1376	1193	1054	642	372	135	7635
Wyo.	Casper A	33.4	6	16	192	524	942	1169	1290	1084	1020	657	381	129	7410
	Cheyenne A	34.2	28	37	219	543	909	1085	1212	1042	1026	702	428	150	7381
	Lander A	31.4	6	19	204	555	1020	1299	1417	1145	1017	654	381	153	7870
	Sheridan A	32.5	25	31	219	539	948	1200	1355	1154	1051	642	366	150	7680
Alta.	Banff C	—	220	295	498	797	1185	1485	1624	1364	1237	855	589	402	10551
	Calgary A	—	109	186	402	719	1110	1389	1575	1379	1268	798	477	291	9703
	Edmonton A	—	74	180	411	738	1215	1603	1810	1520	1330	765	400	222	10268
	Lethbridge A	—	56	112	318	611	1011	1277	1497	1291	1159	696	403	213	8644
B. C.	Kamloops A	—	22	40	189	546	894	1138	1314	1057	818	462	217	102	6799
	Prince George* A	—	236	251	444	747	1110	1420	1612	1319	1122	747	468	279	9755
	Prince Rupert C	—	273	248	339	539	708	868	936	808	812	648	493	357	7029
	Vancouver* A	—	81	87	219	456	657	787	862	723	676	501	310	156	5515
	Victoria* A	—	136	140	225	462	663	775	840	718	691	504	341	204	5699
	Victoria C	—	172	184	243	426	607	723	805	668	660	487	354	250	5579

APPENDIX D *(continued)*

State or Prov.	Station		Avg. Winter Temp[d]	July	Aug.	Sept	Oct.	Nov.	Dec.	Jan.	Feb.	Mar.	Apr.	May	June	Yearly Total
Man.	Brandon*	A	—	47	90	357	747	1290	1792	2034	1737	1476	837	431	198	11036
	Churchill	A	—	360	375	681	1082	1620	2248	2558	2277	2130	1569	1153	675	16728
	The Pas	C	—	59	127	429	831	1440	1981	2232	1853	1624	969	508	228	12281
	Winnipeg	A	—	38	71	322	683	1251	1757	2008	1719	1465	813	405	147	10679
N. B.	Fredericton*	A	—	78	68	234	592	915	1392	1541	1379	1172	753	406	141	8671
	Moncton	C	—	62	105	276	611	891	1342	1482	1336	1194	789	468	171	8727
	St. John	C	—	109	102	246	527	807	1194	1370	1229	1097	756	490	249	8219
Nfld.	Argentia	A	—	260	167	294	564	750	1001	1159	1085	1091	879	707	483	8440
	Corner Brook	C	—	102	133	324	642	873	1194	1358	1283	1212	885	639	333	8978
	Gander	A	—	121	152	330	670	909	1231	1370	1266	1243	939	657	366	9254
	Goose*	A	—	130	205	444	843	1227	1745	1947	1689	1494	1074	741	348	11887
	St. John's*	A	—	186	180	342	651	831	1113	1262	1170	1187	927	710	432	8991
N. W. T.	Aklavik	C	—	273	459	807	1414	2064	2530	2632	2336	2282	1674	1063	483	18017
	Fort Norman	C	—	164	341	666	1234	1959	2474	2592	2209	2058	1386	732	294	16109
	Resolution Island	C	—	843	831	900	1113	1311	1724	2021	1850	1817	1488	1181	942	16021
N. S.	Halifax	C	—	58	51	180	457	710	1074	1213	1122	1030	742	487	237	7361
	Sydney	A	—	62	71	219	518	765	1113	1262	1206	1150	840	567	276	8049
	Yarmouth	A	—	102	115	225	471	696	1029	1156	1065	1004	726	493	258	7340
Ont.	Cochrane	C	—	96	180	405	760	1233	1776	1978	1701	1528	963	570	222	11412
	Fort William	A	—	90	133	366	694	1140	1597	1792	1557	1380	876	543	237	10405
	Kapuskasing	C	—	74	171	405	756	1245	1807	2037	1735	1562	978	580	222	11572
	Kitchener	C	—	16	59	177	505	855	1234	1342	1226	1101	663	322	66	7566
	London	A	—	12	43	159	477	837	1206	1305	1198	1066	648	332	66	7349
	North Bay	C	—	37	90	267	608	990	1507	1680	1463	1277	780	400	120	9219
	Ottawa	C	—	25	81	222	567	936	1469	1624	1441	1231	708	341	90	8735
	Toronto	C	—	7	18	151	439	760	1111	1233	1119	1013	616	298	62	6827
P.E.I.	Charlottetown	C	—	40	53	198	518	804	1215	1380	1274	1169	813	496	204	8164
	Summerside	C	—	47	84	216	546	840	1246	1438	1291	1206	841	518	216	8488
Que.	Arvida	C	—	102	136	327	682	1074	1659	1879	1619	1407	891	521	231	10528
	Montreal*	A	—	9	43	165	521	882	1392	1566	1381	1175	684	316	69	8203
	Montreal	C	—	16	28	165	496	864	1355	1510	1328	1138	657	288	54	7899
	Quebec*	A	—	56	84	273	636	996	1516	1665	1477	1296	819	428	126	9372
	Quebec	C	—	40	68	243	592	972	1473	1612	1418	1228	780	400	111	8937
Sasks	Prince Albert	A	—	81	136	414	797	1368	1872	2108	1763	1559	867	446	219	11630
	Regina	A	—	78	93	360	741	1284	1711	1965	1687	1473	804	409	201	10806
	Saskatoon	C	—	56	87	372	750	1302	1758	2006	1689	1463	798	403	186	10870
Y. T.	Dawson	C	—	164	326	645	1197	1875	2415	2561	2150	1838	1068	570	258	15067
	Mayo Landing	C	—	208	366	648	1135	1794	2325	2427	1992	1665	1020	580	294	14454

*The data for these normals were from the full ten-year period 1951–1960, adjusted to the standard normal period 1931–1960.

APPENDIX E

VOLUMETRIC CAPACITIES OF STEEL PIPE AND COPPER TUBE (GAL/FT)

Pipe size	$\frac{1}{2}$	$\frac{3}{4}$	1	$1\frac{1}{4}$	$1\frac{1}{2}$	2	$2\frac{1}{2}$
Copper tube	0.012	0.025	0.043	0.065	0.092	0.161	0.250
Steel pipe	0.016	0.028	0.045	0.078	0.105	0.172	0.250
Pipe size	**3**	**4**	**5**	**6**	**8**	**10**	**12**
Copper tube	0.357	0.625	1.00	1.40	2.43	3.78	5.40
Steel pipe	0.385	0.667	1.00	1.50	2.63	4.20	5.90

Source: Burnham Corp.

To use this table: (1) Enter the left-hand column at the desired type of pipe; (2) move to the right to the desired pipe size; (3) read the resulting gallons per foot of pipe.

Index

A

B

C

D

E

F

G

H

I

L

M

N

O

P

R

S

T

U

V

W

Y

Z